ENTERING RESEARCH

A Curriculum to Support Undergraduate and Graduate Research Trainees

Second Edition

Janet L. Branchaw
University of Wisconsin–Madison, Department of Kinesiology and Wisconsin Institute for Science Education and Community Engagement

Amanda R. Butz
University of Wisconsin–Madison, Wisconsin Institute for Science Education and Community Engagement

Amber R. Smith
University of Wisconsin–Madison, Wisconsin Institute for Science Education and Community Engagement

Part of the
Navigating Research and Mentoring Series

w.h.freeman
Macmillan Learning
New York

Vice President, STEM:	Daryl Fox
Program Director:	Andrew Dunaway
Developmental Editor:	Andrew Newton
Editorial Assistant:	Justin Jones
Marketing Manager:	Nancy Bradshaw
Marketing Assistant:	Madeleine Inskeep
Director of Digital Production:	Keri deManigold
Director, Content Management Enhancement:	Tracey Kuehn
Senior Managing Editor:	Lisa Kinne
Project Manager:	Sumathy Kumaran, Lumina Datamatics, Inc.
Director of Design, Content Management:	Diana Blume
Design Services Manager:	Natasha Wolfe
Art Manager:	Matthew McAdams
Cover Design Manager:	John Callahan
Permissions Manager:	Jennifer MacMillan
Senior Workflow Manager:	Susan Wein
Production Supervisor:	Lawrence Guerra
Composition:	Lumina Datamatics, Inc.
Printing and Binding:	LSC Communications
Cover Images:	haveseen/Getty Images and sirup/Getty Images
Icons:	Modified from -VICTOR-/Getty Images

Library of Congress Preassigned Control Number: 2018967570

ISBN-13: 978-1-319-26368-3
ISBN-10: 1-319-26368-2

Printed in the United States of America

1 2 3 4 5 6 23 22 21 20 19

W. H. Freeman and Company
One New York Plaza
Suite 4600
New York, NY 10004-1562
www.macmillanlearning.com

Contents

Foreword

The excitement of making discoveries, solving problems, deciphering data, and further understanding the natural world—this is what attracted many of us into careers in science. When we started down this path, we came to understand what research is and how to do it—some with thoughtful guidance from more experienced scientists and some without it. Almost a decade ago, *Entering Research* broke new ground by codifying this guidance, making it available as a curriculum that could be implemented in pieces or as a complete course for all students getting started in research to benefit. In the years since the first edition was published, there has been substantial growth in our understanding of student learning and professional development as well as how to promote equity and inclusion and broaden participation in the sciences. The new edition of *Entering Research* transforms this current body of knowledge into practical guidance for supporting diverse students in successfully navigating their first research experience.

The guide follows the full trajectory of gaining research experience, from finding a mentor to formulating a project to transitioning to advanced forms of research, such as graduate work. The guide is designed "backward"—meaning that it starts with learning objectives related to the nature and practice of science and follows with ready-to-use activities and assessment tools for promoting and gauging student growth. At its core remains the bread and butter of learning to do research, including how to read and make sense of scientific literature, how to think critically and deeply about a project, and how to connect immediate work with the broader research going on within a research group and across a scientific community. *Entering Research* also offers guidance on developing skills critical to being a scientist, such as writing reports and papers, presenting scientific work to other scientists and the general public, and engaging responsibly and ethically in the research process. The updates to *Entering Research* make additional elements of research transparent, including how research is funded, how papers get published, and how to respond constructively to feedback.

These foundational materials are now supplemented with materials that go further in fostering students' psychosocial development. Although scientific thinking skills are vital to success as a researcher, factors such as confidence in one's ability to do research, tolerance for ambiguity, and perseverance in the face of failure may be equally if not more influential to students' educational and career trajectories. *Entering Research* takes this into account by providing materials for fostering students' research self-efficacy and their abilities to cope with the setbacks, navigate experiences of stereotype threat and imposter syndrome, and build personal and professional networks that foster success in science.

Entering Research has been expanded to help facilitators serve a broader group of students, with curricular plans tailored for novice and intermediate undergraduates as well as early graduate students. Facilitator guidance has also been expanded to address sticky issues we all face as researchers, such as making decisions about authorship and managing interpersonal conflicts. All of the instructional materials are coupled with sensible advice for prompting discussion and for responding when discussion goes off course or doesn't go at all.

Using the systematic, evidence-based support provided here will help students to make the most of their research experiences. *Entering Research* will also enable the scientific community to cultivate all available talent to forge new scientific frontiers and address the most pressing societal challenges of our time.

Erin L. Dolan
Professor, Biochemistry and Molecular Biology
Georgia Athletic Association Professor of Innovative Science Education
University of Georgia

Preface

Effective research training programs are key to our efforts to diversify the next generation of scholars. In the traditional apprentice-style research training model, the responsibility for guiding research trainee development is often left to an individual mentor (e.g., an undergraduate or graduate thesis advisor). This can inadvertently lead to the selection and persistence of trainees from a narrow range of backgrounds, namely, those whose experiences and values align most closely with their mentor's. By contrast, we propose that providing opportunities for the entire disciplinary community to contribute to a scholar's training, specifically stakeholders who have diverse experiences and areas of expertise, will lead to the recruitment and persistence of more diverse cohorts of research trainees.

Research training programs can use the *Entering Research* curriculum to build diverse networks of mentors by inviting peers, scientists, instructors, and other disciplinary experts to facilitate the curricular activities. Through facilitating, they provide guidance and advice to cohorts of trainees as they navigate their research experiences. The curricular activities can also be used by individual research mentors, especially those who are still developing the knowledge and skills they need to effectively mentor trainees from diverse backgrounds. We believe the entire research community is needed to train a more diverse generation of scholars and have designed the *Entering Research* curriculum to be used by a broad array of stakeholders working with undergraduate and graduate research trainees.

We have learned much about research training, trainee development, and how people use the *Entering Research* curriculum since the first edition of this book was published. Beyond incorporating our learning about trainee development and modifying the format to maximize flexible use of the second edition, we invited research training program colleagues from across the country who have experience working with graduate trainees and trainees from diverse backgrounds to contribute to this edition. The scope has broadened to include graduate materials and materials that address important issues not explicitly addressed in the first edition such as researcher identity, research self-efficacy, and equity and inclusion awareness and skills. We have thoroughly enjoyed working with and learning from our colleagues and anticipate that the materials in this volume will both inspire and provide research training program directors and mentors with the resources they need to support trainees from a variety of backgrounds.

We will continue to study the impact of mentored research experiences and the *Entering Research* curriculum on the development of undergraduate and graduate research trainees and invite those who are interested in this research to contact us. We are eager to collaborate.

Janet Branchaw
Amanda Butz
Amber Smith

Acknowledgments

We thank our team of colleagues from across the country who helped to revise the original *Entering Research* materials, generate new materials, and serve as reviewers during development. Their expertise and diverse perspectives have contributed significantly to this edition.

- Lori Adams, University of Iowa
- Gustavo Arrizabalaga, Indiana University School of Medicine
- Cheri Barta, University of Wisconsin–Madison
- Jay Bhatt, University of Texas, El Paso
- Ali Bramson, University of Arizona
- Molly Carnes, University of Wisconsin–Madison
- Gabriela Chavira, California State University, Northridge
- Kate Eskine, Wheaton College
- Evelyn Frazier, Florida Atlantic University
- Jennifer Gleason, University of Kansas
- Jessica Harrell, University of North Carolina–Chapel Hill
- Anna Kaatz, University of Wisconsin–Madison
- Shireen Keyl, Utah State University
- Charnell Long, University of Wisconsin–Madison
- Beronda Montgomery, Michigan State University
- Anna O'Connell, University of North Carolina–Chapel Hill
- Christine Pfund, University of Wisconsin–Madison
- Christine Pribbenow, University of Wisconsin–Madison
- Amy Prunuske, Medical College of Wisconsin–Central Campus
- Carrie Saetermoe, California State University, Northridge
- Rochelle Smith, Washington University in St. Louis
- Anna Sokac, Baylor College of Medicine
- John Svaren, University of Wisconsin–Madison
- David Wassarman, University of Wisconsin–Madison

We thank the University of Wisconsin–Madison professional staff members and student assistants and researchers, who helped us keep everything and everyone organized and on track for publication.

- Monica Montano, University Services Program Associate, *Entering Research* Team
- Amanda Sagen, Project Assistant, *Entering Research* Team
- Kimberly Spencer, Assistant Researcher, National Research Mentoring Network
- Cathy Thornburg, Program Coordinator, National Research Mentoring Network
- Hao Yuan, Student Research Assistant, *Entering Research* Team

We thank the many individuals who participated in pilot testing studies and facilitator training workshops. They provided incredibly valuable feedback on the usability of the activities and participated in collecting data for validation of our learning assessment tool.

- Lori Adams, University of Iowa
- Diana Azurdia, University of California, Los Angeles
- Cheri Barta, University of Wisconsin–Madison
- Robert Bell, Michigan State University
- Bruce Birren, Broad Institute
- Maureen Noonen Bischof, University of Wisconsin–Madison
- Elizabeth Borda, Texas A&M University–Galveston Campus
- Chris Brace, University of Wisconsin–Madison
- Ali Bramson, University of Arizona
- Britt Carlson, Parkland College
- Liza Chang, University of Wisconsin–Madison
- Laurie Connor, Baylor College of Medicine
- Tiera Coston, Xavier University of Louisiana
- Jen Eklund, Institute for Systems Biology
- Kyla Esguerra, University of Wisconsin–Milwaukee
- Matt Evans, University of Wisconsin–Eau Claire
- Tanya Faltens, Purdue University
- Evelyn Frazier, Florida Atlantic University
- Peter Gergen, Stony Brook University
- Lindsey Hamilton, University of Colorado, Denver
- Karen Harholm, University of Wisconsin–Eau Claire
- Jessica Harrell, University of North Carolina, Chapel Hill
- Lisa Harrison-Bernard, Louisiana State University School of Medicine
- Jazmyn Haywood, University of Rochester
- Audra Hernandez, University of Wisconsin–Madison
- Paul Hernandez, West Virginia University
- Tara Hobson-Prater, Indiana University School of Medicine
- Tania Houjeiry, Clemson University
- Tim Huffaker, Cornell University
- Torey Jacques, San Francisco State University
- Diana José-Edwards, Washington University in St. Louis
- Anna Kaatz, University of Wisconsin–Madison
- Ann Kimble-Hill, Indiana University School of Medicine
- Pamela Kling, University of Wisconsin–Madison
- Heather Lavender, Louisiana State University
- Paul Laybourn, Colorado State University
- Nathan Lysne, University of Arizona
- Heather Matthews, Colorado State University
- Melissa McDaniels, Michigan State University
- Michele Nishiguchi, New Mexico State University
- Tino Nyawelo, University of Utah
- Carrie Oberland Owens, Rice University
- Anna O'Connell, University of North Carolina–Chapel Hill
- Leslie O'Neill, Georgia Institute of Technology
- Ruthanne Paradise, University of Massachusetts–Amherst
- Fred Pereira, Baylor College of Medicine
- Brian Popp, West Virginia University

- Shauna Price, Field Museum of Natural History
- Kimberly Quedado, West Virginia University
- Giselle Sandi, Rush University
- Nathan Schroeder, University of Illinois at Urbana–Champaign
- Jessica Schuld, University of Wisconsin–Milwaukee
- Linda Sealy, Vanderbilt University School of Medicine
- Sophia Seifert, University of Pennsylvania
- Megan Shannahan, University of Utah
- Morgan Shields, University of Wisconsin–Madison
- Karen Singer-Freeman, Purchase College–State University of New York
- Rochelle Smith, Washington University in St. Louis
- Anna Sokac, Baylor College of Medicine
- Rachel Steward, University of South Carolina; Rocky Mountain Biological Lab
- Vera Tsenkova, University of Wisconsin–Madison
- Ka Vang, University of Wisconsin–Eau Claire
- Ken Voglesonger, Northeast Illinois University
- Letha Woods, Meharry Medical School
- Nina Wright, Texas State University
- Darlene Yanez, University of Texas at Austin
- Wade Znosko, Longwood University

Finally, and most importantly, we thank our families for their tireless support and patience as we worked to complete this project.

This book would not have been possible without the generous support of our funders:

- National Research Mentoring Network (NRMN, NIH award U54 GM119023)
- The Wisconsin Institute for Science Education and Community Engagement (WISCIENCE), University of Wisconsin–Madison
- The development of *Undergraduate Thesis, 1, 2, and 3* was supported by the National Institute of General Medical Sciences of the National Institutes of Health under linked Award Numbers RL5GM118969, TL4GM118971, and UL1GM118970.

NOTE: This work is solely the responsibility of the authors and does not necessarily represent the official view of the National Institutes of Health or the University of Wisconsin–Madison.

CHAPTER 1

The *Entering Research* Framework

This book contains active learning materials to support undergraduate and graduate research trainees across various stages of individual and career development as they do research. The activities are designed to be mixed and matched to create customized curricula for a training program or to create professional development seminars and workshops to support trainees across training programs. Ideally, the activities are implemented with small groups of 12–15 trainees; however, they can be adapted for use with larger groups, or by research mentors for use with individual trainees.

Importantly, the *Entering Research* curriculum is designed to complement the one-on-one mentorship that research trainees receive from their primary research mentors. The activities help trainees understand and contextualize their research through formal presentations (e.g., research project proposals, poster presentations) and through informal conversations with their peers. When the curricular materials are implemented as a formal course or seminar series, they also support the creation of a learning community, where trainees can discuss challenges that arise in their research groups and where they can celebrate their research achievements.

The original version of the *Entering Research* curriculum guided novice undergraduate research trainees through the first two semesters of a research experience. This edition has been expanded to include materials for graduate student trainees and to address a broader range of topics important to trainee development. Evidence of the effectiveness of the activities has been collected through extensive pilot testing with practitioners across the country. The materials are organized in a conceptual framework of seven areas of development drawn from research on factors that contribute to research trainee development. This organization offers maximal flexibility for creating custom curricula of varying intensities and lengths. The framework also aligns the assessment tools with the activities. The seven areas of trainee development are summarized below and listed with their specific learning objectives.

Research Comprehension and Communication Skills—Activities in this area help trainees develop a deep understanding of their research project by requiring them to write about and present their research to others. Trainees are challenged to contextualize their research in the discipline, as well as explain their research content. Through writing and presenting, trainees interact with colleagues, in particular their mentor, and have the opportunity to establish and cultivate professional relationships and their interpersonal and research communication skills.

Learning Objectives:

- Develop Effective Interpersonal Communication Skills
- Develop Disciplinary Knowledge
- Develop Research Communication Skills
- Develop Logical/Critical Thinking Skills
- Develop an Understanding of the Research Environment

Practical Research Skills—Activities in this area provide scaffolded support to trainees as they progress through their research project. Trainees are required to explain the methods they are using to conduct research, the rationale for selecting those methods, the nature and value of the data they are collecting, and their approach to analyzing the data.

Learning Objectives:

- Develop Ability to Design a Research Project
- Develop Ability to Conduct a Research Project

Research Ethics—Activities in this area raise awareness and provide opportunities to explore how to recognize and respond to unethical behavior in the research environment.

Learning Objective:

- Develop Responsible and Ethical Research Practices

Researcher Identity—Activities in this area help trainees to develop their identity as a researcher in the discipline and provide opportunities to explore how their emerging identity as a researcher can integrate with other identities that are important to them (e.g., cultural, ethnic, and socioeconomic identities).

Learning Objective:

- Develop Identity as a Researcher

Researcher Confidence and Independence—Activities in this area encourage trainees to take ownership of their research projects and to become progressively more confident and independent in conducting research, analyzing and interpreting data, and making decisions about the next steps in their projects.

Learning Objectives:

- Develop Confidence as a Researcher
- Develop Independence as a Researcher

Equity and Inclusion Awareness and Skills—Activities in this area provide opportunities to explore the benefits and challenges of individual differences in the research environment. Various perspectives are explored and trainees learn strategies to engage cultural capital to support their progress, successfully navigate challenges, and contribute as members of diverse research teams.

Learning Objectives:

- Develop Skills to Deal with Personal Differences in the Research Environment
- Advance Equity and Inclusion in the Research Environment

Professional and Career Development Skills—Activities in this area raise trainee awareness of a variety of research and research-related careers, allowing them to explore which careers may be right for them. Trainees also consider the skills, knowledge, and types of experiences they will need to prepare for various career paths.

Learning Objectives:

- Explore and Pursue a Research Career
- Develop Confidence in Pursuing a Research Career

CHAPTER 2

Active Learning: Facilitating, Not Teaching

The *Entering Research* active learning activities are designed to be facilitated, rather than taught. Facilitators act as a guide as trainees reflect on and navigate their individual research experiences. Facilitators also structure discussions that build communities where trainees learn from one another's experiences. Ideally, the facilitator asks, rather than answers, questions that lead to deeper exploration of topics and learning.

Facilitator Roles

Importantly, the role of a facilitator is not to teach, but to guide.

As a facilitator, you should:

- **Make it safe.** Take time to tell the group members that the workshops are a safe place to be honest about their experiences, ideas, and feelings. Everyone's story is worth hearing. For it to feel safe, the content of the sessions must be treated respectfully and kept in confidence.
- **Keep it constructive and positive.** Remind trainees to keep things constructive and positive. Ask the group how they want to deal with negativity and pointless venting. Remind them that the workshop is about working together to learn and about supporting one another as they navigate the challenges inherent in doing research.
- **Make the discussion functional.** At the start of each session, explain the goals of the session to the group. Try to keep the group on task without rushing them. If the conversation begins to move beyond the main topic, the facilitator must decide whether the new topic is important enough to redirect the discussion or to set aside time in the future to address the topic and bring the discussion back to the main theme.
- **Give all participants a voice.** In a group, there are likely to be issues that can play out in ways that allow certain members of the group to dominate while others remain silent. At the start of the conversation, point out that a diversity of perspectives is an essential part of the process. Remind trainees to respect one another and that it is important that everyone's voice is heard. You may want to consider using a community-building activity like "Constructive and Destructive Behaviors" to encourage trainees to be aware of how their behaviors can impact the dynamics of the group.

Facilitating Groups

Each group takes on its own feel and personality based on the trainees and your approach as a facilitator. It helps if you are able to release your expectations for how a workshop or session should go, and instead focus on core aspects of the process. Your role as facilitator is to be intentional and explicit, while remaining flexible and not overly prescriptive. To a large extent it is up to the trainees to take ownership of their own learning. Individual ownership, self-reflection, and shared discovery will promote the deepest learning.

As challenges and normal group dynamics surface, the group will look to you to fix problems. Part of your role is to help others see that they are also responsible for fixing problems. You can help them realize this by adhering to the following core ideas of group dynamics:

- Respectful interactions (listening, nonjudging, nondominating, genuine questioning, etc.) are essential.
- You need to keep moving ahead, but there is no need to push the schedule if the trainees need time to reflect or slow down. If you slow down or skip something, you can anticipate that trainees may feel they are behind or missing out, so reassure them this is normal and that there will be time to revisit topics if needed.
- If you try something and it doesn't go well, don't abandon it right away. Step back and think about what went wrong, talk to the group, learn from it, and try it again. It often takes a time or two to get the group warmed up to something new.
- Discomfort and silence are OK, but with a clearly stated context and purpose. Silence may seem like a waste of time during the workshops, but it gives people a chance to think, digest, and reflect. Allow for a few silent breaks before, during, and at the end of each workshop.
- Make it easy, rewarding, and fun for trainees to participate. You may consider building in multiple modes of participation through individual reflection, small-group work, and large-group discussion.

Handling Facilitation Challenges

How can I build trust and community in the group?

- In the first meeting, establish ground rules for how your group will work together. You can offer a basic list to get them started. For example:
  - Everyone participates
  - No dominating the discussion
  - Do not interrupt
  - Respect others' views and opinions
  - What is said in the group, stays in the group
  - Do not have side conversations
  - Listen attentively and carefully
  - Criticize ideas, not individuals; criticize constructively
  - Use "Yes, and..." statements when responding to others' comments
- If possible, sit at a table or put chairs in a circle, so all participants can see one another. Sit with the participants, instead of standing in front of the room.

- Use index cards with participants' names to randomly form new mini-groups for discussion each week. Over time, all participants will get to know one another. The index cards can also be used to identify "volunteers" to do things like taking notes for the group or writing important points from the discussion on the board.
- Rotate responsibility among the group members for facilitating simple check-in or close-out activities for each session.

What do I do when no one talks?

- Have participants write an idea or answer to a question on a piece of paper and toss it in the middle of the table. Each participant then draws a piece of paper from the center of the table (excluding their own) and reads it out loud. All ideas are read out loud before any open discussion begins.
- Have participants discuss a topic in pairs for 3 to 5 minutes before opening the discussion to the larger group.

What do I do when one person is dominating the conversation?

- Use a talking stone to guide the discussion. Participants may only talk when holding the stone. Each person in the group is given a chance to speak before anyone else can have a second turn with the stone. Participants may pass if they choose not to talk. Importantly, each person holding the stone should share their own ideas and respond to someone else's ideas. Generally, once everyone has a chance to speak, the group can move into open discussion without the stone.
- Use the Constructive/Destructive Group Behaviors activity. Each participant chooses their most constructive and most destructive group behavior from the list and writes the two behaviors on the back of their name tag or table tent. Then participants share their choices with the larger group and explain why they chose those behaviors. This exercise also helps provide the group with a vocabulary so they may name these behaviors later in themselves and others. It provides a lighthearted and nonthreatening way to help each other stay on track.

What do I do when participants direct all their questions and comments to me (the facilitator) instead of their fellow participants?

- Each time a group member talks to you, move your eye contact to someone else in the group to help the speaker direct their attention elsewhere.
- Ask the participants for help in resolving one of your research challenges. For example, ask them for advice on how to deal with a lab member who is not keeping their work area clean. This helps the group members stop looking to you for the right answers and redirects the problem-solving and discussion focus to the entire group.

What do I do when a certain person never talks?

- Have a different participant initiate each discussion to ensure that different people have the chance to speak.

- Assign participants different roles in a scenario or case study and ask them to adopt a certain perspective for the discussion. For example, some participants could consider the perspective of the trainee, while others could consider the perspective of the mentor.
- Try smaller group discussions (two to three participants per group), since some participants may feel more comfortable talking in smaller groups or without certain group members present.

What do I do if a participant makes a statement or raises a question that makes other participants visibly uncomfortable, upset, etc.?

- Acknowledge that the statement or question is intriguing and invite the participant to elaborate on what they meant or are looking for in response. For example, "That is very interesting. Can you tell us a little more about why you feel or think that way?"
- Invite, but don't insist, that others respond to the statement or question.
- If the statement or question is inappropriate, not just intriguing, invite the participant to meet with you individually to discuss why they brought it up.

What should I do if a participant shows signs of emotional distress?

- Acknowledge that the participant is upset with a statement, such as "I understand that may be upsetting, frustrating, etc.," or "That is a very difficult situation, emotion, etc., for you to struggle with."
- Invite other participants to share how they are feeling.
- Offer that it is OK to step away from the group for a while if it would help.
- Invite the participant to meet with you individually to discuss the issue in more depth.

What do I do when the group gets off topic?

- Have everyone write about the ideas they want to share on a given topic for 3 minutes. This short writing time will help participants collect their ideas and decide what thoughts they would most like to share with the group so they can focus on that point.
- Ask someone to take notes and recap the discussion at the halfway and end points of the session to keep the conversation focused.
- If the topic is important but there is not enough time to discuss it in the current session, encourage the group to establish a "parking lot" where ideas can be listed and revisited at a future time.

How do I prepare for facilitating conversations or subjects that are difficult for me?

- Identify resources about the subject to review before the session so that you feel more confident in your knowledge and ability to handle questions that arise.
- At the beginning of the session, remind all participants of the ground rules about respect.
- Remember, as the facilitator you are not expected to be the expert in the room. Acknowledging that you are learning about this topic along with the participants can help invite everyone to participate in the conversation and look to one another for support and information.

How do I effectively facilitate with a diverse group?

- Treat everyone as an individual. Do not ask individuals to speak on behalf of a particular group.
- Be aware that differences are sometimes hidden. Although a group may appear homogeneous, there are many different dimensions to diversity and identity that are not visually apparent.
- Understand the role that stereotype threat (i.e., the fear that individuals may have about confirming a stereotype about their race, gender, or other aspect of their identity) may play in group dynamics.

References and Resources

Branchaw, J., Pfund, C., and Rediske, R. (2010). *Entering Research: A Facilitator's Manual: Workshops for Students Beginning Research in Science.* New York, NY: W.H. Freeman & Company.

Center for the Integration of Research, Teaching, and Learning. *Creating a Collaborative Learning Environment Guidebook.* http://www.cirtl.net/files/Guidebook_CreatingACollaborativeLearningEnvironment.pdf.

Center for Research on Teaching and Learning, University of Michigan. "Responding to Difficult Moments." http://www.crlt.umich.edu/multicultural-teaching/difficult-moments.

Pfund, C., Branchaw, J., and Handelsman, J. (2014). *Entering Mentoring.* New York, NY: W.H. Freeman & Company.

CHAPTER 3

Building a Trainee Curriculum Using Backward Design

The *Entering Research* activities offer a rich palette from which to create tailored active learning experiences for undergraduate and graduate research trainees. By identifying the areas of trainee development and learning objectives that address their research trainees' needs, research program directors and mentors can use backward design to choose activities and modules that will address their larger goals (Wiggins et al., 1998). Backward design is a process by which the outcomes or learning objectives are identified first, then assessment tools to measure the outcomes and activities to support the learning are selected. Using this process, custom *Entering Research* curricula for workshops, research training programs, or standalone courses taken in parallel with research credits can be created. Alternatively, *Entering Research* activities may be incorporated into existing curricula to broaden and deepen the training offered.

Each activity in the *Entering Research* curriculum has a full set of facilitator (instructor) notes and trainee materials. All activities were designed for a small cohort (i.e., 12–15 trainees), but can be scaled up for use with larger groups if additional time is allowed and multiple facilitators are available. To aid in sorting and identifying *Entering Research* activities of interest, each activity is categorized by the area(s) of trainee development and learning objectives in the conceptual framework (see chapter 1). Each activity includes: specific learning objectives, trainee stage (undergraduate and/or graduate), trainee level of experience (novice, intermediate or advanced), resources, and assessment tools. In addition, each activity includes "Inclusion Considerations" that are designed to help facilitators promote an inclusive environment by providing advice for making sure all trainees feel welcome to participate in the activity.

Custom Curricula

To build a custom curriculum for your course or program, use the backward design process outlined below to align your learning objectives, activities, and assessments. The same design process is used for one-time workshops and multiple session seminars.

Step 1—Identify what you want your trainees to learn and the constraints and opportunities afforded by your program or course implementation.

1. Given your trainees' career stage (undergraduate or graduate), previous research experience (novice, intermediate, or advanced), and the goals of your program or course, what do you want trainees to know or be able to do by the end of their research experience? What are your learning objectives for trainees?
2. What are the constraints and opportunities of your program or course implementation? Are you building a curriculum from scratch, or do you have an existing curriculum into which you would like to integrate *Entering Research* activities? How much total time and over how many meetings will trainees be available to engage in learning activities? Will trainees volunteer or earn credit for their participation?
3. Which *Entering Research* areas of trainee development and associated learning objectives (chapter 1) align with your trainee goals and are feasible within your program or course implementation?

Step 2—Select the materials you will use to support your trainees to achieve the learning objectives.

1. Prioritize your learning objectives. Which areas of trainee development and learning objectives identified in step 1 are the most important for your trainees?
2. To what extent will trainees be simultaneously doing research while participating in your program or course? Are there expectations associated with the research experience (e.g., participating in weekly research group meetings, writing research proposals or papers, giving presentations)? How can you design your curriculum to help trainees meet these expectations and get the most out of their research experience?
3. If you are incorporating *Entering Research* activities into an existing curriculum, what current activities will you retain? Which, if any, will you remove or replace?
4. Which *Entering Research* activities will best support your trainees' achievement of the learning objectives?

Step 3—Select the instruments you will use to assess whether your trainees have achieved the learning objectives.

1. Which of the formative assessment strategies and tools recommended in the activities selected in step 2 will you use to track trainee learning and, if necessary, make adjustments to your implementation along the way?
2. Which of the summative assessment strategies and tools recommended in the activities selected in step 2 will you use to determine whether trainees have achieved the learning objectives at the end of the research experience? Will you use the *Entering Research* learning assessment tool?
3. If you are incorporating *Entering Research* activities into an existing curriculum, what current assessment strategies and tools will you retain? Which, if any, will you remove or replace?
4. Which *Entering Research* assessment strategies and tools (chapter 4) will you use to assess trainee learning?

Using a table like the one below will allow you to easily align the learning objectives, activities, and assessments in your curriculum. The power of backward design is in the alignment, which ensures that everything in the learning experience supports achievement of the learning objectives.

Learning Objectives	Activities (Curricular & Research)	Assessments

To be sure that your trainees' curricular and research learning experiences are integrated, list the research project activities the trainees will do as part of their mentored research experience as well as the *Entering Research* curricular activities they will do in your program or course in the middle column. This will allow you to identify synergies and to schedule curricular activities when they will best complement your trainees' research project activities. In short, the curricular activities should be scheduled to provide "just in time" support as your trainees navigate their research experience.

Complete Curricula

Ready-to-use complete curricula developed using the backward design process are available at the end of this chapter for four types of implementation:

1. 10-week summer program for undergraduate student researchers
2. 1-semester (15 weeks) seminar for novice undergraduate student researchers
3. 1-semester (15 weeks) seminar for intermediate undergraduate student researchers
4. 1-semester (15 weeks) first-year seminar for graduate student researchers

The activities and assessments referred to in the complete curricula tables can be found in chapters 4 and 5. The complete curricula can be implemented as designed or adapted for specific groups of trainees.

Reference

Wiggins, G. P., McTighe, J., Kiernan, L. J., Frost, F., and Association for Supervision and Curriculum Development. (1998). *Understanding by design.* Alexandria, VA: Association for Supervision and Curriculum Development.

COMPLETE CURRICULUM #1

10-Week Summer Research Program for Undergraduate Students

Career Stage: Undergraduate Trainees

Trainees' Prior Level of Research Experience: Intermediate

Implementation Description: In this 10-week summer research program students are engaging in full-time research (~40 hrs./wk.). Professional development sessions are offered for 90 minutes each week.

Meta Learning Objectives and Areas of Trainee Development: The learning objectives addressed in this curriculum are listed below. The percentage (%) following each learning objective indicates the proportion of sessions that address that learning objective.

Research Comprehension and Communication Skills

- Develop Effective Interpersonal Communication Skills (60%)
- Develop Disciplinary Knowledge (50%)
- Develop Research Communication Skills (60%)
- Develop Logical/Critical Thinking Skills (50%)
- Develop an Understanding of the Research Environment (40%)

Practical Research Skills

- Develop Ability to Conduct a Research Project (20%)

Research Ethics

- Develop Responsible and Ethical Research Practices (30%)

Researcher Identity

- Develop Identity as a Researcher (40%)

Researcher Confidence and Independence

- Develop Confidence as a Researcher (40%)
- Develop Independence as a Researcher (50%)

Equity and Inclusion Awareness and Skills

- Develop Skills to Deal with Personal Differences in the Research Environment (20%)
- Advance Equity and Inclusion in the Research Environment (10%)

Professional and Career Development Skills

- Explore and Pursue a Research Career (40%)
- Develop Confidence in Pursuing a Research Career (20%)

Week 1: Program Orientation and Meeting Your Mentor

Research Comprehension and Communication Skills

- Develop Effective Interpersonal Communication Skills
- Develop an Understanding of the Research Environment

Researcher Identity

- Develop Identity as a Researcher

Activities (In-Session)	Assignments Due	Assessment Tools
• Introductions • Research Experience Reflections 1: Entering Research? • Prioritizing Research Mentor Roles		

Week 2: Defining Your Research Project

Research Comprehension and Communication Skills

- Develop Effective Interpersonal Communication Skills
- Develop Disciplinary Knowledge
- Develop Research Communication Skills
- Develop Logical/Critical Thinking Skills
- Develop an Understanding of the Research Environment

Practical Research Skills

- Develop Ability to Conduct a Research Project

Research Ethics

- Develop Responsible and Ethical Research Practices

Researcher Identity

- Develop Identity as a Researcher

Research Confidence and Independence

- Develop Independence as a Researcher

Professional and Career Development Skills

- Explore and Pursue a Research Career

Activities (In-Session)	Assignments Due	Assessment Tools
• Research Group Diagram • Elevator Sentences • Aligning Mentor and Trainee Expectations • Communicating Research Findings 1: Poster Presentations	• Mentor-Trainee Expectations Agreement • Research Group Diagram • Elevator Sentences • Safety Training Checklist • Mentor-Trainee Check in #1	• Elevator Sentences: Peer Review Assessment Rubric

Week 3: Documenting Your Research and Persistence in Research

Research Comprehension and Communication Skills

- Develop Effective Interpersonal Communication Skills
- Develop Disciplinary Knowledge
- Develop Research Communication Skills

Practical Research Skills

- Develop Ability to Conduct a Research Project

Research Ethics

- Develop Responsible and Ethical Research Practices

Researcher Confidence and Independence

- Develop Independence as a Researcher
- Develop Confidence as a Researcher

Professional and Career Development Skills

- Develop Confidence in Pursuing a Research Career

Activities (In-Session)	Assignments Due	Assessment Tools
• Research Documentation Process • Case Study: Keeping the Data • Case Study: Overwhelmed • Case Study: Frustrated	• Draft of research proposal poster to reviewers • Research Documentation Process	• Research Writing 6: Peer Review of Research Proposal Assessment Rubric

Week 4: Presenting Your Research Project

Research Comprehension and Communication Skills

- Develop Research Communication Skills
- Develop Logical/Critical Thinking Skills

Researcher Confidence and Independence

- Develop Independence as a Researcher

Activities (In-Session)	Assignments Due	Assessment Tools
• Sharing poster feedback from peers • Research Articles 3: Practical Reading Strategies • Research Articles 2: Guided Reading	• Practice research proposal poster presentations • Watch Poster Presentation Tips (from Communicating Research Findings 1) • Read assigned paper and complete guided reading guide	• Communicating Research Findings 3: Poster Presentation Assessment Rubric • Communicating Research Findings 3: Poster Presentation Peer Review
Friday of Week 4: Proposal Poster Session		

Week 5: Ethics in Research

Research Comprehension and Communication Skills
- Develop Effective Interpersonal Communication Skills
- Develop an Understanding of the Research Environment

Research Ethics
- Develop Responsible and Ethical Research Practices

Researcher Identity
- Develop Identity as a Researcher

Activities (In-Session)	Assignments Due	Assessment Tools
• Debrief Research Proposal poster session • Ethics Case Study Discussion with Mentor • Truth and Consequences Article Discussion • Case Study: Credit Where Credit Is Due	• Discussion of ethics case study with mentor • Read "Truth and Consequences" • Mentor-Trainee Check in #2	• Assessment of mentor-trainee relationship quality/Reflection essay

Week 6: Visiting Peer Research Groups

Research Comprehension and Communication Skills
- Develop Effective Interpersonal Communication Skills
- Develop an Understanding of the Research Environment

Researcher Identity
- Develop Identity as a Researcher

Equity and Inclusion Awareness and Skills
- Develop Skills to Deal with Personal Differences in the Research Environment

Professional and Career Development Skills
- Explore and Pursue a Research Career

Activities (In-Session)	Assignments Due	Assessment Tools
• Letters of Recommendation • Visiting Peer Research Groups	• Letter of Recommendation reflection and discussion with mentor • Write short reflection essay on relationship with mentor	

Week 7: Graduate School 101/Equity and Inclusion in STEM

Researcher Confidence and Independence

- Develop Confidence as a Researcher

Equity and Inclusion Awareness and Skills

- Develop Skills to Deal with Personal Differences in the Research Environment
- Advance Equity and Inclusion in the Research Environment

Professional and Career Development Skills

- Develop Confidence in Pursuing a Research Career

Activities (In-Session)	Assignments Due	Assessment Tools
• Stereotype Threat • Graduate school panel	• Read Stereotype Threat summary	

Week 8: Writing Your Final Paper

Research Comprehension and Communication Skills

- Develop Disciplinary Knowledge
- Develop Research Communication Skills
- Develop Logical/Critical Thinking Skills

Activities (In-Session)	Assignments Due	Assessment Tools
• Peer review research paper draft #1 with faculty and students • Discussion of research paper feedback	• Research Paper Draft: title, affiliations, introduction, methods, and references • Mentor-Trainee Check in #3	• Research Writing 7: Research Paper Assessment Rubric

Week 9: Preparing Your Final Presentation/Exploring Research Careers

Research Comprehension and Communication Skills

- Develop Disciplinary Knowledge
- Develop Research Communication Skills
- Develop Logical/Critical Thinking Skills

Research Confidence and Independence

- Develop Independence as a Researcher
- Develop Confidence as a Researcher

Professional and Career Development Skills

- Explore and Pursue a Research Career

Activities (In-Session)	Assignments Due	Assessment Tools
• Feedback on research paper draft #2 • Career Panel Discussion	• Research Paper: Complete draft including results, figures, discussion • Questions for career panel • Draft PowerPoint presentation	• Research Writing 7: Research Paper Assessment Rubric

Week 10: Final Symposium

Research Comprehension and Communication Skills

- Develop Effective Interpersonal Communication Skills
- Develop Disciplinary Knowledge
- Develop Research Communication Skills
- Develop Logical/Critical Thinking Skills

Research Confidence and Independence

- Develop Independence as a Researcher
- Develop Confidence as a Researcher

Professional and Career Development Skills

- Explore and Pursue a Research Career

Activities (In-Session)	Assignments Due	Assessment Tools
• Final Presentation practice talks	• Research Writing 7: Research Paper • Program evaluation • Mentor–Trainee Check in #4	• Research Writing 7: Research Paper Assessment Rubric • Communicating Research Findings 3: Oral Presentation Peer Review • Assessment of mentor–trainee relationship quality/Reflection essay

Comprehensive Program Evaluation and *Entering Research* Learning Assessment (ERLA) Surveys	
• Trainee Survey • Mentor Survey	• Trainee self-assessment of learning gains • Mentor assessment of trainee learning gains

COMPLETE CURRICULUM #2

15-Week Seminar for Novice Undergraduate Students

Career Stage: Undergraduate Trainees

Trainees' Prior Level of Research Experience: Novice

Implementation Description: This 1 credit seminar course, which meets for 50 minutes each week, is designed for undergraduate students with 0–1 semesters of prior research experience. The students are expected to have found a research mentor prior to the beginning of the semester. In addition to registering for this 1 credit seminar, students simultaneously enroll in 1–3 credits of independent study with their research mentor, equivalent to 4–12 hours of research per week.

Meta Learning Objectives and Areas of Trainee Development: The learning objectives addressed in this curriculum are listed below. The percentage (%) following each learning objective indicates the proportion of sessions that address that learning objective.

Research Comprehension and Communication Skills

- Develop Effective Interpersonal Communication Skills (47%)
- Develop Disciplinary Knowledge (27%)
- Develop Research Communication Skills (60%)
- Develop Logical/Critical Thinking Skills (20%)
- Develop an Understanding of the Research Environment (33%)

Practical Research Skills

- Develop Ability to Design a Research Project (6%)
- Develop Ability to Conduct a Research Project (6%)

Research Ethics

- Develop Responsible and Ethical Research Practices (13%)

Researcher Identity

- Develop Identity as a Researcher (33%)

Researcher Confidence and Independence

- Develop Confidence as a Researcher (20%)
- Develop Independence as a Researcher (27%)

Equity and Inclusion Awareness and Skills

- Develop Skills to Deal with Personal Differences in the Research Environment (6%)

Professional and Career Development Skills

- Explore and Pursue a Research Career (13%)
- Develop Confidence in Pursuing a Research Career (6%)

Week 1: Introduction to *Entering Research* and Research Experience Expectations

Research Comprehension and Communication Skills

- Develop an Understanding of the Research Environment

Researcher Identity

- Develop Identity as a Researcher

Activities (In-Session)	Assignments Due	Assessment Tools
• Introductions • Course Overview • Research Experience Reflections 1: Entering Research?		

Week 2: Navigating the Mentoring Relationship

Research Comprehension and Communication Skills

- Develop Effective Interpersonal Communication Skills
- Develop an Understanding of the Research Environment

Researcher Identity

- Develop Identity as a Researcher

Activities (In-Session)	Assignments Due	Assessment Tools
• Prioritizing Research Mentor Roles • Three Mentors	• Safety Training Checklist • Reminder: meet with your mentor to create an expectations agreement	

Week 3: Aligning Mentor and Trainee Expectations

Research Comprehension and Communication Skills

- Develop Effective Interpersonal Communication Skills
- Develop Research Communication Skills
- Develop Logical/Critical Thinking Skills

Researcher Identity

- Develop Identity as a Researcher

Researcher Confidence and Independence

- Develop Independence as a Researcher

Activities (In-Session)	Assignments Due	Assessment Tools
• Aligning Mentor and Trainee Expectations • Elevator Sentences	• Mentor–Trainee Expectations Agreement • Elevator Sentence (1st draft)	• Elevator Sentences: Peer Review Assessment Rubric

Week 4: Research Group Focus

<table>
<tr><td colspan="3">Research Comprehension and Communication Skills
• Develop Effective Interpersonal Communication Skills
• Develop Disciplinary Knowledge
• Develop Research Communication Skills

Practical Research Skills
• Develop Ability to Design a Research Project</td></tr>
<tr><td>Activities (In-Session)</td><td>Assignments Due</td><td>Assessment Tools</td></tr>
<tr><td>• Your Research Group's Focus</td><td>• Research Group Diagram</td><td>• Your Research Group's Focus: Assessment Rubric</td></tr>
</table>

Week 5: Documenting Your Research and Persistence in Research

<table>
<tr><td colspan="3">Research Comprehension and Communication Skills
• Develop Effective Interpersonal Communication Skills
• Develop Disciplinary Knowledge
• Develop Research Communication Skills
• Develop an Understanding of the Research Environment

Practical Research Skills
• Develop Ability to Conduct a Research Project

Research Ethics
• Develop Responsible and Ethical Research Practices

Researcher Confidence and Independence
• Develop Independence as a Researcher
• Develop Confidence as a Researcher

Professional and Career Development Skills
• Develop Confidence in Pursuing a Research Career</td></tr>
<tr><td>Activities (In-Session)</td><td>Assignments Due</td><td>Assessment Tools</td></tr>
<tr><td>• Research Documentation Process
• Elevator Sentence practice
• Case Study: Overwhelmed
• Case Study: Frustrated</td><td>• Research Documentation Process
• Revise Elevator Sentences</td><td>• Post-Activity Mini-Reflection and Assessment Rubric</td></tr>
</table>

Week 6: Reading Research Literature

Research Comprehension and Communication Skills

- Develop Research Communication Skills
- Develop Logical/Critical Thinking Skills

Researcher Confidence and Independence

- Develop Independence as a Researcher

Activities (In-Session)	Assignments Due	Assessment Tools
• Research Articles 3: Practical Reading Strategies • Elevator Sentence practice	• Research Articles 2: Guided Reading • Read article assigned in class • Research Experience Reflections 2: Reflection Exercise	• Research Experience Reflections 2: Assessment Rubric

Week 7: Research Self-Efficacy

Research Comprehension and Communication Skills

- Develop Effective Interpersonal Communication Skills

Researcher Confidence and Independence

- Developing Confidence as a Researcher

Activities (In-Session)	Assignments Due	Assessment Tools
• Fostering Your Own Research Self-Efficacy • The Power of Social Persuasion	• Read Self-Efficacy article	• Post-Activity Mini-Reflection and Assessment Rubric

Week 8: Research Posters

Research Comprehension and Communication Skills

- Develop Disciplinary Knowledge
- Develop Research Communication Skills
- Develop Logical/Critical Thinking Skills

Activities (In-Session)	Assignments Due	Assessment Tools
• Communicating Research Findings 1: Poster Presentations	• Communicating Research Findings 1: Poster Hunt worksheet • Mentor Interview about Making Research Posters	

Week 9: Communicating Research

Research Comprehension and Communication Skills
- Develop Research Communication Skills

Activities (In-Session)	Assignments Due	Assessment Tools
• Communicating Research to the General Public • Research Outline check-in	• Draft of Introduction for Research Poster	

Week 10: Research Ethics

Research Comprehension and Communication Skills
- Develop Effective Interpersonal Communication Skills
- Develop an Understanding of the Research Environment

Research Ethics
- Develop Responsible and Ethical Research Practices

Activities (In-Session)	Assignments Due	Assessment Tools
• Ethics Case: Discussion with Mentor • Case Study: Credit Where Credit Is Due • Case Study: Keeping the Data	• Draft of Methods for Research Poster • Discussion of case study with mentor	• Post-Activity Mini-Reflection and Assessment Rubric

Week 11: Visiting Peer Research Groups

Research Comprehension and Communication Skills
- Develop Effective Interpersonal Communication Skills
- Develop an Understanding of the Research Environment

Equity and Inclusion Awareness and Skills
- Develop Skills to Deal with Personal Differences in the Research Environment

Activities (In-Session)	Assignments Due	Assessment Tools
• Visiting Peer Research Groups	• Draft of Results for Research Poster	• Assessment of mentor–trainee relationship quality/Reflection essay

Week 12: Developing a Professional Development Plan

Researcher Identity
- Develop Identity as a Researcher

Researcher Confidence and Independence
- Develop Independence as a Researcher

Professional and Career Development Skills
- Explore and Pursue a Research Career

Activities (In-Session)	Assignments Due	Assessment Tools
• Professional Development Plans	• Professional Development Plans: Individual Development Plan worksheet	

Week 13: Poster Peer Review and Presentations

Research Comprehension and Communication Skills
- Develop Research Communication Skills

Professional and Career Development Skills
- Explore and Pursue a Research Career

Activities (In-Session)	Assignments Due	Assessment Tools
• Communicating Research Findings 3: Poster peer review • Elevator Sentence practice	• Complete draft of poster • Revised elevator sentence to be used during poster presentation	• Communicating Research Findings 3: Poster Presentation Assessment Rubric and Peer Review forms

Week 14: Science and Society

Research Comprehension and Communication Skills
- Develop Research Communication Skills

Activities (In-Session)	Assignments Due	Assessment Tools
• Science and Society	• Read articles on science communication • Revise and print final poster	

Week 15: Poster Session

<table>
<tr><td colspan="3">Research Comprehension and Communication Skills
• Develop Disciplinary Knowledge
• Develop Research Communication Skills
Researcher Identity
• Develop Identity as a Researcher
Researcher Confidence and Independence
• Develop Confidence as a Researcher</td></tr>
<tr><td>Activities (In-Session)</td><td>Assignments Due</td><td>Assessment Tools</td></tr>
<tr><td>• Present poster
• Listen and ask questions of peers</td><td></td><td>• Communicating Research Findings 3: Poster Presentation Assessment Rubric and Peer Review forms</td></tr>
<tr><td colspan="2">Comprehensive Seminar Evaluation and Entering Research Learning Assessment (ERLA) Surveys
• Trainee Survey
• Mentor Survey</td><td>• Trainee self-assessment of learning gains
• Mentor assessment of trainee learning gains</td></tr>
</table>

COMPLETE CURRICULUM #3

15-Week Seminar for Intermediate Undergraduate Students

Career Stage: Undergraduate Trainees

Trainees' Prior Level of Research Experience: Intermediate

Implementation Description: This 1-credit seminar course, which meets for 50 minutes each week, was designed for undergraduate students with 2+ semesters of prior research experience. The students are expected to have a research mentor prior to the beginning of the semester. Students simultaneously enroll in 1–3 credits of independent study equivalent to 4–12 hours of research per week.

Meta Learning Objectives and Areas of Trainee Development: The learning objectives addressed in this curriculum are listed below. The percentage (%) following each learning objective indicates the proportion of sessions that address that learning objective.

Research Comprehension and Communication Skills

- Develop Effective Interpersonal Communication Skills (47%)
- Develop Disciplinary Knowledge (27%)
- Develop Research Communication Skills (33%)
- Develop Logical/Critical Thinking Skills (13%)
- Develop an Understanding of the Research Environment (27%)

Practical Research Skills

- Develop Ability to Design a Research Project (6%)

Research Ethics

- Develop Responsible and Ethical Research Practices (6%)

Researcher Identity

- Develop Identity as a Researcher (40%)

Researcher Confidence and Independence

- Develop Independence as a Researcher (27%)
- Develop Confidence as a Researcher (13%)

Equity and Inclusion Awareness and Skills

- Develop Skills to Deal with Personal Differences in the Research Environment (13%)
- Advance Equity and Inclusion in the Research Environment (20%)

Professional and Career Development Skills

- Explore and Pursue a Research Career (47%)
- Develop Confidence in Pursuing a Research Career (6%)

Week 1: Introduction to *Entering Research* and Research Experience Expectations

Research Comprehension and Communication Skills
- Develop an Understanding of the Research Environment

Researcher Identity
- Develop Identity as a Researcher

Activities (In-Session)	Assignments Due	Assessment Tools
• Introductions • Course Overview • Research Experience Reflections 1: Entering Research?		

Week 2: Research Group Focus

Research Comprehension and Communication Skills
- Develop Effective Interpersonal Communication Skills
- Develop Disciplinary Knowledge
- Develop Research Communication Skills
- Develop an Understanding of the Research Environment

Practical Research Skills
- Develop Ability to Design a Research Project

Professional and Career Development Skills
- Explore and Pursue a Research Career

Activities (In-Session)	Assignments Due	Assessment Tools
• Your Research Group's Focus	• Research Group Diagram	• Your Research Group's Focus: Assessment Rubric

Week 3: Aligning Mentor and Trainee Expectations

Research Comprehension and Communication Skills
- Develop Effective Interpersonal Communication Skills

Researcher Identity
- Develop Identity as a Researcher

Researcher Confidence and Independence
- Develop Independence as a Researcher

Activities (In-Session)	Assignments Due	Assessment Tools
• Aligning Mentor and Trainee Expectations • Elevator Sentences	• Mentor-Trainee Expectations Agreement • Elevator Sentence (1st draft)	• Elevator Sentences: Peer Review Assessment Rubric

Week 4: Efficient Research Article Reading

Research Comprehension and Communication Skills

- Develop Disciplinary Knowledge

Professional and Career Development Skills

- Explore and Pursue a Research Career

Activities (In-Session)	Assignments Due	Assessment Tools
• Article Organization, Comprehension, and Recall	• Article Organization, Comprehension and Recall: Tools for Handling Your Papers	

Week 5: Research Ethics

Research Comprehension and Communication Skills

- Develop an Understanding of the Research Environment

Research Ethics

- Develop Responsible and Ethical Research Practices

Researcher Identity

- Develop Identity as a Researcher

Activities (In-Session)	Assignments Due	Assessment Tools
• Truth and Consequences Article	• Read "Truth or Consequences"	• Post-Activity Mini-Reflection and Assessment Rubric

Week 6: Diversity in STEM

Equity and Inclusion Awareness and Skills

- Develop Skills to Deal with Personal Differences in the Research Environment
- Advance Equity and Inclusion in the Research Environment

Activities (In-Session)	Assignments Due	Assessment Tools
• Diversity in STEM	• Diversity in STEM: read "Dr. Quiñones' Story" • Diversity in STEM: watch "What's up with chicks in science?"	

Week 7: Why Diversity Matters

Equity and Inclusion Awareness and Skills

- Advance Equity and Inclusion in the Research Environment

Activities (In-Session)	Assignments Due	Assessment Tools
• Why Diversity Matters in STEM	• Why Diversity Matters in STEM: read "Supreme Court Hits a Nerve with Comments on Diversity" • Why Diversity Matters in STEM: read "An Open Letter to SCOTUS from Professional Physicists"	• Post-Activity Mini-Reflection and Assessment Rubric

Week 8: Coping Efficacy

Researcher Confidence and Independence

- Develop Confidence as a Researcher

Equity and Inclusion Awareness and Skills

- Advance Equity and Inclusion in the Research Environment

Activities (In-Session)	Assignments Due	Assessment Tools
• Coping Efficacy		

Week 9: Mini-Grant Proposal and Mentor Check-In

Research Comprehension and Communication Skills

- Develop Effective Interpersonal Communication Skills
- Develop Disciplinary Knowledge
- Develop Research Communication Skills
- Develop Logical/Critical Thinking Skills

Researcher Confidence and Independence

- Develop Independence as a Researcher
- Develop Confidence as a Researcher

Equity and Inclusion Awareness and Skills

- Develop Skills to Deal with Personal Differences in the Research Environment

Professional and Career Development Skills

- Develop Confidence in Pursuing a Research Career

Activities (In-Session)	Assignments Due	Assessment Tools
• Mini-Grant Proposal • Reflecting on Your Mentoring Relationship	• Mentoring relationship reflection and meeting with your mentor	

Week 10: Research Group Funding

Research Comprehension and Communication Skills

- Develop an Understanding of the Research Environment

Researcher Confidence and Independence

- Develop Independence as a Researcher

Professional and Career Development Skills

- Explore and Pursue a Research Career

Activities (In-Session)	Assignments Due	Assessment Tools
• Research Group Funding	• Research Group Funding Worksheet	

Week 11: Research Careers: The Informal Interview

Research Comprehension and Communication Skills

- Develop Effective Interpersonal Communication Skills
- Develop Research Communication Skills

Professional and Career Development Skills

- Explore and Pursue a Research Career

Activities (In-Session)	Assignments Due	Assessment Tools
• Research Careers: The Informational Interview	• Research Career Interview Write-Up	

Week 12: Letter of Recommendation

Research Comprehension and Communication Skills

- Develop Effective Interpersonal Communication Skills

Researcher Identity

- Develop Identity as a Researcher

Professional and Career Development Skills

- Explore and Pursue a Research Career

Activities (In-Session)	Assignments Due	Assessment Tools
• Letter of Recommendation	• Letter of Recommendation Worksheet	

Week 13: Peer Review of Mini-Grant Proposal

<table>
<tr><td colspan="3">Research Comprehension and Communication Skills
• Develop Effective Interpersonal Communication Skills
• Develop Disciplinary Knowledge
• Develop Research Communication Skills
• Develop Logical/Critical Thinking Skills</td></tr>
<tr><td>Activities (In-Session)</td><td>Assignments Due</td><td>Assessment Tools</td></tr>
<tr><td>• Mini-Grant Proposal
• Personal Statement: overview</td><td>• Draft mini-grant proposal</td><td>• Mini-Grant Proposal: Assessment Rubric
• Mini-Grant Proposal: Peer Review form</td></tr>
</table>

Week 14: Personal Statements

<table>
<tr><td colspan="3">Researcher Identity
• Develop Identity as a Researcher
Professional and Career Development Skills
• Explore and Pursue a Research Career</td></tr>
<tr><td>Activities (In-Session)</td><td>Assignments Due</td><td>Assessment Tools</td></tr>
<tr><td>• Personal Statement: peer review</td><td>• Personal Statement draft</td><td></td></tr>
</table>

Week 15: Networking

<table>
<tr><td colspan="3">Research Comprehension and Communication Skills
• Develop Effective Interpersonal Communication Skills
• Develop Research Communication Skills
Researcher Identity
• Develop Identity as a Researcher
Researcher Confidence and Independence
• Develop Independence as a Researcher
Professional and Career Development Skills
• Explore and Pursue a Research Career</td></tr>
<tr><td>Activities (In-Session)</td><td>Assignments Due</td><td>Assessment Tools</td></tr>
<tr><td>• Networking 2: What Should Your Network Look Like?</td><td>• Mini-Grant Proposal</td><td>• Mini-Grant Proposal: Assessment Rubric</td></tr>
<tr><td colspan="2">Comprehensive Seminar Evaluation and Entering Research Learning Assessment (ERLA) Surveys
• Trainee Survey
• Mentor Survey</td><td>• Trainee self-assessment of learning gains
• Mentor assessment of trainee learning gains</td></tr>
</table>

COMPLETE CURRICULUM #4

15-Week Seminar for Novice Graduate Students

Career Stage: Graduate Student

Trainees' Prior Level of Research Experience: Novice

Implementation Description: This seminar course was designed to be taken by first semester graduate students while they engage in research group rotations. However, the activities are also beneficial for direct admit students.

Meta Learning Objectives and Areas of Trainee Development: The learning objectives addressed in this curriculum are listed below. The percentage (%) following each learning objective indicates the proportion of sessions that address that learning objective.

Research Comprehension and Communication Skills

- Develop Effective Interpersonal Communication Skills (27%)
- Develop Disciplinary Knowledge (27%)
- Develop Research Communication Skills (20%)
- Develop Logical/Critical Thinking Skills (13%)
- Develop an Understanding of the Research Environment (20%)

Practical Research Skills

- Develop Ability to Design a Research Project (6%)
- Develop Ability to Conduct a Research Project (6%)

Research Ethics

- Develop Responsible and Ethical Research Practices (13%)

Researcher Identity

- Develop Identity as a Researcher (40%)

Researcher Confidence and Independence

- Develop Independence as a Researcher (20%)
- Develop Confidence as a Researcher (13%)

Equity and Inclusion Awareness and Skills

- Develop Skills to Deal with Personal Differences in the Research Environment (13%)
- Advance Equity and Inclusion in the Research Environment (20%)

Professional and Career Development Skills

- Explore and Pursue a Research Career (40%)
- Develop Confidence in Pursuing a Research Career (6%)

Week 1: Course Introduction and Research Expectations

Research Comprehension and Communication Skills

- Develop an Understanding of the Research Environment

Researcher Identity

- Develop Identity as a Researcher

Activities (In-Session)	Assignments Due	Assessment Tools
• Introductions • Course overview • Research Experience Reflections 1: Entering Research?		

Week 2: Finding Potential Research Rotation Groups and Mentors

Research Comprehension and Communication Skills

- Develop Effective Interpersonal Communication Skills
- Develop an Understanding of the Research Environment

Researcher Identity

- Develop Identity as a Researcher

Professional and Career Development Skills

- Explore and Pursue a Research Career

Activities (In-Session)	Assignments Due	Assessment Tools
• Finding Potential Research Rotation Groups and Mentors • Research Rotation Evaluation: discussion	• Identifying research rotation groups • Contacting potential rotation mentors	• Post-Activity Mini-Reflection and Assessment Rubric

Week 3: Aligning Expectations

Research Comprehension and Communication Skills

- Develop Effective Interpersonal Communication Skills

Researcher Identity

- Develop Identity as a Researcher

Researcher Confidence and Independence

- Develop Independence as a Researcher

Activities (In-Session)	Assignments Due	Assessment Tools
• Aligning Mentor and Trainee Expectations	• Mentor-Trainee Expectations Agreement	

Week 4: Searching Online Databases

<table>
<tr><td colspan="3">Research Comprehension and Communication Skills
• Developing Logical/Critical Thinking Skills
Practical Research Skills
• Develop Ability to Design a Research Project</td></tr>
<tr><td>Activities (In-Session)</td><td>Assignments Due</td><td>Assessment Tools</td></tr>
<tr><td>• Searching Online Databases</td><td>• Generate a list of keywords for your research area</td><td></td></tr>
</table>

Week 5: Importance of Reading in Graduate School

<table>
<tr><td colspan="3">Research Comprehension and Communication Skills
• Develop Disciplinary Knowledge
Researcher Confidence and Independence
• Develop Independence as a Researcher
Professional and Career Development Skills
• Explore and Pursue a Research Career</td></tr>
<tr><td>Activities (In-Session)</td><td>Assignments Due</td><td>Assessment Tools</td></tr>
<tr><td>• Importance of Reading in Graduate School</td><td>• Read Parker, R. (2012). Skill development in graduate education. Molecular Cell, 46: 377–381.
• Importance of Reading in Graduate School: My Lab's Core Journal List</td><td></td></tr>
</table>

Week 6: Article Organization, Comprehension, and Critique

<table>
<tr><td colspan="3">Research Comprehension and Communication Skills
• Develop Disciplinary Knowledge
Professional and Career Development Skills
• Explore and Pursue a Research Career</td></tr>
<tr><td>Activities (In-Session)</td><td>Assignments Due</td><td>Assessment Tools</td></tr>
<tr><td>• Article Organization, Comprehension, and Recall</td><td>• Article Organization, Comprehension and Recall: Tools for Handling Your Papers</td><td></td></tr>
</table>

Week 7: Writing in Science

Research Comprehension and Communication Skills

- Develop Disciplinary Knowledge
- Develop Research Communication Skills

Professional and Career Development Skills

- Explore and Pursue a Research Career

Activities (In-Session)	Assignments Due	Assessment Tools
• Research Writing 4: Research Literature Review and Publishing Process • Tips for Technical Writers	• Research Writing 4: Journal Jargon handout • Write one paragraph summarizing a task or experiment you have done in the lab	• Tips for Technical Writers: Assessment Rubric

Week 8: Data Rigor and Reproducibility

Research Comprehension and Communication Skills

- Develop Disciplinary Knowledge
- Develop Research Communication Skills

Practical Research Skills

- Develop Ability to Conduct a Research Project

Research Ethics

- Develop Responsible and Ethical Research Practices

Activities (In-Session)	Assignments Due	Assessment Tools
• Research Documentation Process • Research Documentation: Can You Decipher This?	• Research Documentation Process • Bring in copies of your lab notebook	

Week 9: Research Ethics

Research Comprehension and Communication Skills

- Develop Logical/Critical Thinking Skills
- Develop an Understanding of the Research Environment

Research Ethics

- Develop Responsible and Ethical Research Practices

Researcher Identity

- Develop Identity as a Researcher

Activities (In-Session)	Assignments Due	Assessment Tools
• Truth and Consequences Article • Case Study: Selection of Data	• Read Couzin, J. (September 2006). Truth and Consequences. *Science, 313*: 1222–1226. doi:10.1126/science.313.5791.1222	• Post-Activity Mini-Reflection and Rubric

Week 10: The Power of Social Persuasion

<table>
<tr><td colspan="3">Research Comprehension and Communication Skills
• Develop Effective Interpersonal Communication Skills
Researcher Confidence and Independence
• Develop Confidence as a Researcher
Professional and Career Development Skills
• Develop Confidence in Pursuing a Research Career</td></tr>
<tr><td>Activities (In-Session)</td><td>Assignments Due</td><td>Assessment Tools</td></tr>
<tr><td>• Messages Sent and Received
• The Power of Social Persuasion</td><td></td><td></td></tr>
</table>

Week 11: Coping Efficacy

<table>
<tr><td colspan="3">Researcher Confidence and Independence
• Develop Confidence as a Researcher
Equity and Inclusion Awareness and Skills
• Advance Equity and Inclusion in the Research Environment</td></tr>
<tr><td>Activities (In-Session)</td><td>Assignments Due</td><td>Assessment Tools</td></tr>
<tr><td>• Coping Efficacy</td><td></td><td></td></tr>
</table>

Week 12: Diversity in STEM

<table>
<tr><td colspan="3">Equity and Inclusion Awareness and Skills
• Develop Skills to Deal with Personal Differences in the Research Environment
• Advance Equity and Inclusion in the Research Environment</td></tr>
<tr><td>Activities (In-Session)</td><td>Assignments Due</td><td>Assessment Tools</td></tr>
<tr><td>• Diversity in STEM</td><td>• Diversity in STEM: read Dr. Quiñones's story
• Diversity in STEM: watch Dr. deGrasse Tyson's response: "What's up with chicks and science?"</td><td></td></tr>
</table>

Week 13: Challenges Facing Diverse Teams

<table>
<tr><td colspan="3">Equity and Inclusion Awareness and Skills
• Develop Skills to Deal with Personal Differences in the Research Environment
• Advance Equity and Inclusion in the Research Environment</td></tr>
<tr><td>Activities (In-Session)</td><td>Assignments Due</td><td>Assessment Tools</td></tr>
<tr><td>• Challenges Facing Diverse Teams</td><td></td><td></td></tr>
</table>

Week 14: Networking

Research Comprehension and Communication Skills

- Develop Effective Interpersonal Communication Skills
- Develop Research Communication Skills

Researcher Identity

- Develop Identity as a Researcher

Researcher Confidence and Independence

- Develop Independence as a Researcher

Professional and Career Development Skills

- Explore and Pursue a Research Career

Activities (In-Session)	Assignments Due	Assessment Tools
• Networking 2: What Should Your Network Look Like? • Developing a Curriculum Vitae: introduction		

Week 15: Developing a Curriculum Vitae

Researcher Identity

- Develop Identity as a Researcher

Professional and Career Development Skills

- Explore and Pursue a Research Career

Activities (In-Session)	Assignments Due	Assessment Tools
• Developing a Curriculum Vitae	• Draft Curriculum Vitae	
Comprehensive Seminar Evaluation and *Entering Research* Learning Assessment (ERLA) Surveys • Trainee Survey • Mentor Survey		• Trainee self-assessment of learning gains • Mentor assessment of trainee learning gains

CHAPTER 4

Strategies and Tools for Assessing Trainee Learning and Evaluating Implementations

Defining trainee learning objectives and selecting activities are important when building a custom curriculum. Equally important is determining how to assess whether your trainees have achieved the learning objectives and to evaluate whether your workshop, program, or course implementation has achieved its goals. The data gathered through assessment and evaluation provide evidence of the effectiveness of a curriculum and its implementation and can guide revisions of the curriculum or implementation when needed. Assessment and evaluation data are frequently required for institutional or funding agency reports, so designing an effective assessment and evaluation plan can be essential to securing continued support for your program or course.

The terms *assessment* and *evaluation* are often used interchangeably. In this chapter, we provide specific definitions for assessment and evaluation in the context of the *Entering Research* curriculum and we provide guidance on how to develop an effective learning assessment and implementation evaluation plan. Several tools are presented, including a set of validated post-implementation assessment survey instruments for research trainees and their mentors based on the *Entering Research* conceptual framework outlined in chapter 1.

Designing an Assessment and Evaluation Plan

Assessment is the process of determining how and at what level trainees comprehend and can apply the knowledge and skills they gained through their learning experiences, including their participation in the *Entering Research* activities and their participation in research. The *Entering Research* assessment tools measure the extent to which trainees have achieved the research learning experience objectives.

Evaluation is the process of determining whether and to what extent an implementation (workshop, program or course) has met its stated goals. The *Entering Research* evaluation tools measure the quality of the implementation by asking participants to rate their satisfaction. These tools also collect participant demographic information to document who has participated and to measure whether different populations report different levels of satisfaction.

Assessment and evaluation data should always be collected. However, the depth and breadth of data collection will be determined by how the data will be used later. For example, data could be used to document trainee development, to guide course or program improvement, or to report outcomes to an institution or funding agency. Regardless, assessment and evaluation data should align with trainee learning objectives and implementation goals, respectively (see the alignment table in chapter 3). Consider the following questions to determine which assessment and evaluation strategies and tools to use with your implementation.

What type of data are you interested in collecting?

Are you most interested in evaluating the quality of the implementation, trainee learning, or both? How do you intend to use the data? Will the data be used for research purposes as well as assessment and evaluation? If used for research, there may be certain methods that should be used to collect data and Institutional Review Board approval may be required before data collection begins.

When determining the type of data you wish to collect, consider the amount of time you can dedicate to assessment and evaluation. For example, facilitators implementing *Entering Research* activities as part of a course, seminar, or workshop series may be able to incorporate multiple assessment rubrics over time into their evaluation and assessment plan to track trainee development. Conversely, facilitators implementing stand-alone workshops may be more interested in brief evaluation surveys and assessments that provide a snapshot of the quality of the implementation and trainee learning.

What do you need to be able to show with the data you collect? What type of data are your stakeholders or funders requiring you to report?

Do you need to provide evidence that you are meeting your grant or program aims? Do you want to demonstrate student learning or growth in a particular area? If your stakeholders or the organization funding your program are requesting specific information (e.g., demographics, enrollment data, evidence of program effectiveness), be sure that the information you are collecting and the assessment and evaluation questions you are asking speak directly to those requests.

Which areas of trainee development are you addressing? What assignments and activities are you using with your trainees?

The length of your implementation and the number of areas of trainee development that are addressed in the implementation are important considerations when developing an assessment and evaluation plan. A stand-alone workshop using one or two *Entering Research* activities will require a different assessment and evaluation plan than a one- or two-semester course in which students regularly engage with *Entering Research* content, while simultaneously participating in a mentored research experience.

Choosing Assessment and Evaluation Tools

Once a plan has been outlined for your implementation, begin to identify the specific assessment and evaluation tools you will use. Three different types of assessment and evaluation tools are available in this book:

1. **Evaluation survey to assess effectiveness of *Entering Research* implementation.** This survey provides example questions that assess satisfaction and request feedback from trainees in your course, workshop, or seminar. These questions were derived from established measures used by the Mentorship Training Core of the National Research Mentoring Network (NRMN; NIH U54GM119023).
2. **Trainee and research mentor surveys to assess trainee learning gains.** The *Entering Research* Leaning Assessment (ERLA) was developed as a tool to assess trainee learning gains. The ERLA aligns with the seven areas of trainee development included in the *Entering Research* conceptual framework (see chapter 1) and was developed and validated with undergraduate and graduate trainees and their mentors (Butz & Branchaw, in preparation). The areas of trainee development scales in the ERLA may be used independently if only certain areas of trainee development need to be assessed. Trainees' self-assessment scores of their own learning gains may be compared to the research mentor's assessment scores of their trainee's learning gains. The ERLA includes newly developed items as well as adapted items from existing validated measures (e.g., Mentor Competency Assessment [MCA], Fleming et al., 2013; URSSA, Weston, & Laursen, 2015; Estrada et al., 2011).

What is validity?

"The degree to which evidence and theory support the interpretation of test scores for proposed uses of tests" (American Educational Research Association et al., 2014, p. 11).

3. **Assessment Rubrics.** Rubrics may be used as formative or summative assessments to objectively measure trainee learning across all areas of trainee development. They have been incorporated into activities throughout the curriculum and may be used for peer feedback and/or facilitator assessment of trainee learning gains on individual *Entering Research* assignments, reflections, and presentations.
   a. **Rubrics** available for specific activities are listed in the rubric index below.
   b. A **general post-activity mini-reflection and companion rubric** is available to use with any *Entering Research* activity to assess the extent to which trainees can connect and apply what they learned from the activity and group discussions to their own research experiences.
   c. **Customizable reflection activities and associated rubrics** can be used to assess trainee growth at any time during the research experience (see Research Experience Reflection activities 1, 2, and 3 in chapter 5). Facilitators can select from a series of reflection questions based on the seven areas of trainee development at different time points throughout the research experience to assess growth in areas addressed by their curriculum. Importantly, facilitators can use these reflection questions and rubrics to assess psychosocial areas of development that can be difficult to fully assess with standard survey instruments.

Rubric Index

Activity Title	Peer Review	Assessment
3-Minute Research Story	✔	✔
Article Organization, Comprehension, and Recall	✔	✔
Communicating Research Findings 3: Developing Your Presentation	✔	✔
Developing a Curriculum Vitae	✔	✔
Elevator Sentences	✔	
General Public Abstract	✔	
Mini-Grant Proposal	✔	✔
Personal Statement	✔	
Research Experience Reflections 2: Reflection Exercise		✔
Research Writing 1: Background Information and Hypothesis/Research Question	✔	
Research Writing 6: Research Proposal	✔	✔
Research Writing 7: Research Paper	✔	✔
Tips for Technical Writers	✔	✔
Your Research Group's Focus		✔

References

American Educational Research Association, American Psychological Association, National Council on Measurement in Education, & Joint Committee on Standards for Educational and Psychological Testing (U.S.) (2014). Standards for educational and psychological testing. Washington, DC: AERA.

Estrada, M., Woodcock, A., Hernandez, P. R., & Schultz, P. W. (2011). Toward a model of social influence that explains minority student integration into the scientific community. *Journal of Educational Psychology, 103,* 206–222. doi:10.1037/a0020743

Fleming, M., House, S., Shewakramani, V., Yu, L., Garbutt, J., McGee, R., Kroenke, K., Abedin, Z., & Rubrio, D. (2013). The mentoring competency assessment: Validation of a new instrument to evaluation skills of research mentors. *Academic Medicine, 88*(7): 1002–1008. doi:10.1097/ACM.0b013e318295e298

Weston, T. J., & Laursen, S. L. (2015). The undergraduate research student self-assessment (URSSA): Validation for use in program evaluation. *CBE-Life Sciences Education, 14*: 1–10. doi:10.1187/cbe.14-11-0206

POST-ACTIVITY MINI-REFLECTION

Instructions: How can you apply what was learned and discussed as part of this activity to your research and/or personal experience? How does the topic(s) addressed relate to your own professional development as a researcher?

POST-ACTIVITY MINI-REFLECTION

Assessment Rubric

5	Excellent	Trainee clearly articulates and analyzes connections between topic(s) addressed in the activity and their research and/or personal experiences.
4	Very Good	Trainee articulates connections between topic(s) addressed in the activity and their research and/or personal experiences with limited analysis.
3	Good	Trainee articulates connections between topic(s) addressed in the activity and their research and/or personal experiences, but fails to analyze the situation.
2	Fair	Trainee articulates limited connections between topic(s) addressed in the activity and their research and/or personal experiences.
1	Poor	Trainee articulates no connections between topic(s) addressed in their activity and their research and/or personal experiences.

ENTERING RESEARCH IMPLEMENTATION EVALUATION SURVEY

NOTE: *Replace [implementation] with course, seminar, or workshop.*

1. Overall how effective were the facilitators in guiding discussion during this [*implementation*]?

very ineffective	ineffective	neither ineffective nor effective	effective	very effective

2. How likely are you to recommend participation in this [*implementation*]?

very unlikely	unlikely	undecided	likely	very likely

3. Overall, was your participation in this [*implementation*] a valuable use of your time? [Yes/No]
4. What aspects of this [*implementation*] did you find most useful?
5. What aspects of this [*implementation*] could be improved?
6. Considering your current research experience and your participation in this [*implementation*], how would you rate each of the following?

Your experience in this [*implementation*]	poor	fair	neutral	good	excellent
Your research experience while participating in this [*implementation*]	poor	fair	neutral	good	excellent
Your overall experience as a research trainee while participating in this [*implementation*]	poor	fair	neutral	good	excellent

NOTE: *Additional Questions to evaluate implementations at the activity or area of trainee development level:*

7. How effective was [*activity name*] in helping you gain knowledge and/or improve your ability to do research?

very ineffective	ineffective	neither ineffective nor effective	effective	very effective

8. How much did your experience in this [*implementation*] contribute to your skills in each of the following areas?

Understanding Research	not at all	a little	a moderate amount	a lot	a great deal
Communicating About Research	not at all	a little	a moderate amount	a lot	a great deal
Conducting Research	not at all	a little	a moderate amount	a lot	a great deal
Ethics in Research	not at all	a little	a moderate amount	a lot	a great deal
Research Confidence	not at all	a little	a moderate amount	a lot	a great deal
Research Independence	not at all	a little	a moderate amount	a lot	a great deal
Equity and Inclusion Awareness	not at all	a little	a moderate amount	a lot	a great deal
Professional Development	not at all	a little	a moderate amount	a lot	a great deal

ENTERING RESEARCH LEARNING ASSESSMENT (ERLA) SURVEYS

The *Entering Research* Learning Assessment (ERLA) surveys were validated with undergraduate and graduate research trainees and their mentors (Butz & Branchaw, in preparation). The seven scales, each of which assesses an area of trainee development, may be used independently or together as a comprehensive assessment of trainee learning gains. The degree of alignment between trainee and mentor assessment items can provide information about the quality of the mentoring relationship. This instrument is also available from the authors. For more information on the psychometric properties of the scale, item ordering and scoring, see Butz and Branchaw (in preparation) or contact the authors.

Entering Research Learning Assessment—Trainee

How much did you gain in your ability to do the following over the course of your research experience?

	no gain	a little gain	moderate gain	good gain	great gain
1. Understand the theory and concepts guiding your research project.	O	O	O	O	O
2. Practice regular and open communication with your research mentor.	O	O	O	O	O
3. Think of yourself as a scientist/researcher.	O	O	O	O	O
4. Identify forms of unethical practices or research misconduct.	O	O	O	O	O
5. Determine the next steps in your research project.	O	O	O	O	O
6. Analyze data.	O	O	O	O	O
7. Identify the biases and prejudices that you have about others.	O	O	O	O	O
8. Set research career goals.	O	O	O	O	O
9. Ask questions to clarify your understanding of your research project.	O	O	O	O	O
10. Design a research project.	O	O	O	O	O
11. Fit in with the research culture of your discipline.	O	O	O	O	O
12. Formulate a research question/hypothesis.	O	O	O	O	O
13. Demonstrate understanding and comprehension regarding your research project.	O	O	O	O	O
14. Tailor your research communications for different audiences (e.g., general public, disciplinary conference, etc.).	O	O	O	O	O
15. Do experiments.	O	O	O	O	O
16. Confidence in staying motivated and committed to your research project when things do not go as planned.	O	O	O	O	O
17. Determine the appropriate experimental approach to investigate your research question.	O	O	O	O	O
18. Keep detailed research records (e.g., a lab/field notebook).	O	O	O	O	O

(continued)

How much did you gain in your ability to do the following over the course of your research experience? (continued)

	no gain	a little gain	moderate gain	good gain	great gain
19. Communicate the relevance of your research to others.	O	O	O	O	O
20. Understand the consequences of unethical practices or research misconduct.	O	O	O	O	O
21. Use the tools, materials, and equipment needed to conduct research.	O	O	O	O	O
22. Work independently on your research project.	O	O	O	O	O
23. Confidence in pursuing a career in research.	O	O	O	O	O
24. Use logic and evidence to interpret data.	O	O	O	O	O
25. Communicate the context, methods, and results of your research.	O	O	O	O	O
26. Advocate for others who may be marginalized or excluded from the research environment.	O	O	O	O	O
27. Develop a plan to pursue a research career (determine the next step in your training).	O	O	O	O	O
28. Work in the research environment comfortably.	O	O	O	O	O
29. Understand that the process of discovery is iterative and never ending.	O	O	O	O	O
30. Collect data.	O	O	O	O	O
31. Understand the safety precautions relating to your research.	O	O	O	O	O
32. Fit in with the culture of your research group.	O	O	O	O	O
33. Make a case for your research question based on the literature.	O	O	O	O	O
34. Practice regular and open communication with your research team members.	O	O	O	O	O
35. Identify the biases and prejudices that others may have about you.	O	O	O	O	O
36. Confidence in conducting research.	O	O	O	O	O
37. Work effectively with the subject of study (e.g., mathematical models, mice, plants, rock formations).	O	O	O	O	O
38. Call yourself a researcher when talking to others.	O	O	O	O	O
39. Explore possible research career pathways.	O	O	O	O	O
40. Behave like a researcher in your discipline.	O	O	O	O	O
41. Align your research experience goals and expectations with your research mentors.	O	O	O	O	O
42. Take action to address unethical practices or research misconduct.	O	O	O	O	O
43. Feel like you belong in research.	O	O	O	O	O
44. Use logic and evidence to build arguments and draw conclusions from data.	O	O	O	O	O

(continued)

How much did you gain in your ability to do the following over the course of your research experience? (continued)

	no gain	a little gain	moderate gain	good gain	great gain
45. Accept and use criticism of your research to improve your research.	O	O	O	O	O
46. Understand the impact of biases on your interactions with others in a research environment.	O	O	O	O	O
47. Conduct a research project.	O	O	O	O	O
48. Confidence in coping with challenges when they arise in your research project.	O	O	O	O	O
49. Determine an analysis plan/statistical methods to analyze your data.	O	O	O	O	O
50. Investigate problems when they arise in your research (e.g., troubleshoot).	O	O	O	O	O
51. Understand how others might experience research differently based on their identity (e.g., race, socioeconomic status, first-generation status, etc.).	O	O	O	O	O
52. Confidence in completing your research training.	O	O	O	O	O
53. Make detailed observations.	O	O	O	O	O

ERLA—Trainee Scoring: The item numbers corresponding to each area of trainee development are listed below. Area of trainee development sub-scores can be calculated by summing the score for each item and dividing by the total number of items. Individual items should be scored as follows: no gain (1); a little gain (2); moderate gain (3); good gain (4); great gain (5).

Research Comprehension and Communication Skills (15 items): 1, 2, 6, 9, 13, 14, 19, 24, 25, 28, 29, 34, 41, 44, 45.
Practical Research Skills (13 items): 10, 12, 15, 17, 18, 21, 30, 31, 33, 37, 47, 49, 53.
Research Ethics (3 items): 4, 20, 42.
Researcher Identity (6 items): 3, 11, 32, 38, 40, 43.
Researcher Confidence and Independence (7 items): 5, 16, 22, 36, 48, 50, 52.
Equity and Inclusion Awareness and Skills (5 items): 7, 26, 35, 46, 51.
Professional and Career Development Skills (4 items): 8, 23, 27, 39.

Entering Research Learning Assessment—Mentor

How much did your [trainee/student researcher] gain in their ability to do the following over the course of their research experience? If you did not observe your [trainee/student researcher] engaged in a particular skill, please select "did not observe."

	no gain	a little gain	moderate gain	good gain	great gain	did not observe
1. Understand the theory and concepts guiding their research project.	O	O	O	O	O	O
2. Practice regular and open communication with you.	O	O	O	O	O	O
3. Identify forms of unethical practices or research misconduct.	O	O	O	O	O	O
4. Determine the next steps in their research project.	O	O	O	O	O	O
5. Analyze data.	O	O	O	O	O	O
6. Set research career goals.	O	O	O	O	O	O
7. Ask questions to clarify their understanding of their research project.	O	O	O	O	O	O
8. Design a research project.	O	O	O	O	O	O
9. Fit in with the research culture of your discipline.	O	O	O	O	O	O
10. Formulate a research question/hypothesis.	O	O	O	O	O	O
11. Demonstrate understanding and comprehension regarding their research project.	O	O	O	O	O	O
12. Tailor their research communications for different audiences (e.g., general public, disciplinary conference, etc.).	O	O	O	O	O	O
13. Confidence in staying motivated and committed to their research project when things do not go as planned.	O	O	O	O	O	O
14. Determine the appropriate experimental approach to investigate their research question.	O	O	O	O	O	O
15. Keep detailed research records (e.g., a lab/field notebook).	O	O	O	O	O	O
16. Communicate the relevance of their research to others.	O	O	O	O	O	O
17. Demonstrate understanding of the consequences of unethical practices or research misconduct.	O	O	O	O	O	O
18. Use the tools, materials, and equipment needed to conduct research.	O	O	O	O	O	O
19. Work independently on their research project.	O	O	O	O	O	O
20. Confidence in pursuing a career in research.	O	O	O	O	O	O
21. Use logic and evidence to interpret data.	O	O	O	O	O	O
22. Communicate the context, methods, and results of their research.	O	O	O	O	O	O
23. Advocate for others who may be marginalized or excluded from the research environment.	O	O	O	O	O	O
24. Develop a plan to pursue a research career (determine the next step in their training).	O	O	O	O	O	O

(continued)

How much did your [trainee/student researcher] gain in their ability to do the following over the course of their research experience? If you did not observe your [trainee/student researcher] engaged in a particular skill, please select "did not observe." (continued)

	no gain	a little gain	moderate gain	good gain	great gain	did not observe
25. Work in the research environment comfortably.	O	O	O	O	O	O
26. Demonstrate understanding that the process of discovery is iterative and never ending.	O	O	O	O	O	O
27. Collect data.	O	O	O	O	O	O
28. Demonstrate understanding of the safety precautions relating to their research.	O	O	O	O	O	O
29. Fit in with the culture of your research group.	O	O	O	O	O	O
30. Make a case for their research question based on literature.	O	O	O	O	O	O
31. Practice regular and open communication with your research team members.	O	O	O	O	O	O
32. Confidence in conducting research.	O	O	O	O	O	O
33. Work effectively with the subject of study (e.g., chemicals, mathematical models, mice, plants, rock formations).	O	O	O	O	O	O
34. Demonstrate understanding of possible research career pathways.	O	O	O	O	O	O
35. Behave like a researcher in your discipline.	O	O	O	O	O	O
36. Align their research experience goals and expectations with your goals and expectations.	O	O	O	O	O	O
37. Take action to address unethical practices or research misconduct.	O	O	O	O	O	O
38. Act like they belong in research.	O	O	O	O	O	O
39. Use logic and evidence to build arguments and draw conclusions from data.	O	O	O	O	O	O
40. Accept and use criticism of their research to improve their research.	O	O	O	O	O	O
41. Conduct a research project.	O	O	O	O	O	O
42. Confidence in coping with challenges when they arise in their research project.	O	O	O	O	O	O
43. Determine an analysis plan/statistical methods to analyze their data.	O	O	O	O	O	O
44. Investigate problems when they arise in their research (e.g., troubleshoot).	O	O	O	O	O	O
45. Demonstrate understanding of how others might experience research differently based on their identity (e.g., race, socioeconomic status, first-generation status, etc.).	O	O	O	O	O	O
46. Confidence in completing their research training.	O	O	O	O	O	O
47. Make detailed observations.	O	O	O	O	O	O

ERLA—Mentor Scoring: The item numbers corresponding to each area of trainee development are listed below. Area of trainee development sub-scores can be calculated by summing the score for each item and dividing by the total number of items. Individual items should be scored as follows: no gain (1); a little gain (2); moderate gain (3); good gain (4); great gain (5); did not observe (1).

Research Comprehension and Communication Skills (15 items): 1, 2, 5, 7, 11, 12, 16, 21, 22, 25, 26, 31, 36, 39, 40.
Practical Research Skills (12 items): 8, 10, 14, 15, 18, 27, 28, 30, 33, 41, 43, 47.
Research Ethics (3 items): 3, 17, 37.
Researcher Identity (4 items): 9, 29, 35, 38.
Researcher Confidence and Independence (7 items): 4, 13, 19, 32, 42, 44, 46.
Equity and Inclusion Awareness and Skills (2 items): 23, 45.
Professional and Career Development Skills (4 items): 6, 20, 24, 34.

CHAPTER 5

The *Entering Research* Activities

To easily identify *Entering Research* activities to use with trainees at different career stages and in different types of implementation, each activity is listed in the table below with specific learning objectives, intended audience [undergraduate (U) or graduate (G)], time needed to complete, and area(s) of trainee development addressed. Three introductory activities recommended for use at the beginning of implementations that span multiple weeks are presented first.

SUMMARY OF ENTERING RESEARCH ACTIVITIES

Page	Introductory Activities			
63	**Introductory Activities** • Learn about other members in the group and begin to build a learning community. • Reflect on group dynamics and ways to make the group functional.	U, G	varies	Research Comprehension and Communication Skills
65	**Constructive and Destructive Group Behaviors** • Identify their own constructive and destructive behaviors. • Discuss ways to promote an inclusive and engaging group by holding each other accountable for destructive behaviors.	U, G	35"	Equity and Inclusion Awareness and Skills Research Comprehension and Communication Skills
68	**Setting the Stage for Inclusive Discussions** • Understand how to effectively discuss social issues in the context of STEM. • Contribute to establishing a welcoming space for all members of the group to fully participate.	U, G	50"	Equity and Inclusion Awareness and Skills Research Comprehension and Communication Skills

Page	Entering Research Activities			
71	**Addressing Conflict** • Learn how to manage conflict in their research experiences.	U, G	1'5"	Research Comprehension and Communication Skills Equity and Inclusion Awareness and Skills
76	**Aligning Mentor and Trainee Expectations** • Understand that their expectations of their research mentoring relationship may be different from their mentors. • Establish an open line of communication with their mentor in order to address any differences and align their expectations.	U, G	1'10"	Research Comprehension and Communication Skills Researcher Identity Research Confidence and Independence
96	**Article Organization, Comprehension, and Recall** • Learn about electronic tools for organizing papers, citing papers, and taking notes on papers. • Learn to use a framing funnel as a strategy to actively read papers. • Learn to use guided questions to understand and evaluate scholarly papers.	U, G	1'	Research Comprehension and Communication Skills Professional and Career Development Skills
107	**Case Study: Authorship** • Identify the authorship guidelines for a discipline/career stage. • Practice strategies for discussing authorship with mentors and members of the research group.	U, G	20"	Research Ethics Research Comprehension and Communication Skills
111	**Case Study: Awkward Mentor** • Discuss challenges that may arise in mentoring relationships. • Develop strategies to address mentoring relationship challenges.	U, G	25"	Equity and Inclusion Awareness and Skills Research Comprehension and Communication Skills
114	**Barriers to Effective Communication** • Learn strategies for improving communication (in person, at a distance, across multiple mentors, and within proper personal boundaries).	U, G	35"	Research Comprehension and Communication Skills Equity and Inclusion Awareness and Skills
118	**Bias Literacy: *Fair Play* Video Game** • Learn about the different types of bias individuals encounter in research and in everyday life. • Develop awareness of the potential for bias to impact all individuals' judgment and behavior, even if it is unintentional. • Identify specific strategies to counteract bias.	U, G	2'	Equity and Inclusion Awareness and Skills

129	**Challenges Facing Diverse Teams** • Learn about three common challenges that individuals working in diverse teams may face. • Identify practical strategies for recognizing and mitigating challenges to create more inclusive research groups.	U, G	1'	Equity and Inclusion Awareness and Skills
137	**Communicating Research Findings 1: Poster Presentations** • Identify the characteristics of effective research posters. • Learn how to give a poster presentation of research findings. • Recognize that research presentations are opportunities for discussing and networking with colleagues about research, not only for reporting research.	U, G	1'	Research Comprehension and Communication Skills
143	**Communicating Research Findings 2: Oral Presentations** • Identify the characteristics of effective oral presentations. • Learn how to give an oral presentation of research findings. • Recognize that research presentations are opportunities for discussing and networking with colleagues about research, not only for reporting research.	U, G	1'	Research Comprehension and Communication
147	**Communicating Research Findings 3: Developing Your Presentation** • Prepare and confidently present a research poster or talk summarizing research findings. • Recognize that scientific presentations are opportunities for discussion and networking with colleagues about research, not only for reporting research.	U, G	varies	Research Comprehension and Communication Skills Researcher Confidence and Independence
166	**Communicating Research to the General Public** • Understand the importance of communicating research in a way that is accessible. • Define the "general public" and consider researchers' responsibility to communicate with this audience. • Compare and contrast ways in which researchers communicate with each other and with nonresearcher audiences.	U, G	55"	Research Comprehension and Communication Skills

(*Continued*)

Page	Entering Research Activities			
170	**Coping Efficacy** • Be able to distinguish between emotion- and problem-focused coping strategies. • Identify when it is appropriate to apply emotion- and problem-focused coping strategies.	U, G	55"	Research Confidence and Independence Equity and Inclusion Awareness and Skills
178	**Counter-Storytelling** • Recognize that "double consciousness" can generate stress and issues of identity convergence from nonmajority people. • Imagine scenarios of equity and mutual respect in research and other professional interactions by reframing and reenvisioning responses to microaggressions, stereotype threat, and messages of inferiority. • Become more aware of and challenge negative stereotypes. • Reframe racial/ethnic narratives to become more empowered.	U, G	S1: 1' S2: 1'	Equity and Inclusion Awareness and Skills Researcher Confidence and Independence Professional and Career Development Skills
188	**Case Study: Credit Where Credit is Due** • Discuss and practice ethical research decision making.	U, G	25"	Research Ethics Research Comprehension and Communication Skills
191	**Developing a Curriculum Vitae** • Learn the differences between a resume and curriculum vitae (CV). • Draft a CV.	U, G	S1: 1' S2: 1'	Researcher Identity Professional and Career Development Skills
199	**Discussion of the Nature of Science** • Explain and debate the role of science and the relationship between research and knowledge.	U, G	20–30"	Research Comprehension and Communication Skills
202	**Discussion with Experienced Undergraduate Researchers** • Refine and expand a definition of research. • Increase confidence about how to identify a research mentor.	U	30"–1'	Professional and Career Development Skills Researcher Identity Research Comprehension and Communication Skills
205	**Diversity in STEM** • Be more aware of disparities between majority and historically underrepresented groups in STEM careers. • Consider different reasons for disparities in STEM career and degree attainment. • Articulate strategies to reduce these disparities in STEM career and degree attainment. • Learn about the prejudice encountered by researchers belonging to historically underrepresented groups.	U, G	S1: 20" S2: 35"	Equity and Inclusion Awareness and Skills

210	**Elevator Sentences** • Engage general and expert audiences in conversations about research.	U, G	45"	Research Comprehension and Communication Skills Researcher Identity
214	**Establishing Your Ideal Thesis Committee** • Identify the primary and secondary roles of a graduate thesis committee. • Explore the roles that can be played by different members of a thesis committee. • Begin identifying members of a thesis committee.	G	1'	Professional and Career Development Skills Research Comprehension and Communication Skills Researcher Confidence and Independence
220	**Ethics Case: Discussion with Mentor** • Discuss with a mentor ethical issues associated with working in a research group. • Learn strategies to deal with ethical situations associated with working in a research group. • Learn about consequences of unethical behavior.	U	20"	Research Ethics Research Comprehension and Communication Skills
224	**Finding a Research Mentor** • Learn how to identify and contact potential research mentors.	U	Pre-work	Professional and Career Development Skills Research Comprehension and Communication Skills
230	**Finding Potential Research Rotation Groups and Mentors** • Identify potential research groups for rotation. • Learn about strategies for contacting potential research mentors. • Learn about important considerations when choosing research groups for rotation.	G	1'35"	Professional and Career Development Skills Research Comprehension and Communication Skills Researcher Identity
237	**Fostering Your Own Research Self-Efficacy** • Define self-efficacy and its sources. • Clearly articulate steps to complete a research-related task. • Articulate how the four sources of self-efficacy can be used to build confidence to complete research-related tasks.	U, G	55"	Researcher Confidence and Independence
243	**Case Study: Frustrated** • Practice strategies for communicating with mentors and research group members.	U, G	20"	Research Comprehension and Communication Skills Researcher Confidence and Independence
247	**Funding Your Research** • Identify funding sources available to support research.	U, G	35"	Professional and Career Development Skills Research Comprehension and Communication Skills

(*Continued*)

Page	Entering Research Activities			
250	**General Public Abstract** • Identify the attributes of a good public abstract. • Learn to identify and eliminate or define research jargon. • Give constructive feedback to peers on general public abstracts.	U, G	S1: 1'10" S2: 1'10"	Research Comprehension and Communication Skills Research Ethics
259	**Importance of Reading in Graduate School** • Learn how research literature can play an important role in research success. • Develop strategies to read and comprehend appropriate and relevant research literature.	G	45"	Research Comprehension and Communication Skills Professional and Career Development Skills Researcher Confidence and Independence
265	**Interviewing for Graduate School** • Learn what to expect when interviewing for graduate school. • Articulate skills and areas of interest in research. • Become empowered to succeed in graduate school interviews.	U	1'	Professional and Career Development Skills Research Comprehension and Communication Skills
270	**Case Study: Keeping the Data** • Explain why it is important to accurately document research. • Identify key elements in research documentation. • Understand the ethical implications of documenting research.	U, G	15"	Research Ethics
273	**Letter of Recommendation** • Identify the characteristics and attributes desired in a letter of recommendation. • Learn how to request a letter of recommendation from a mentor.	U,G	25–45"	Professional and Career Development Skills Research Comprehension and Communication Skills Researcher Identity
281	**Mentor Biography** • Begin to establish a positive relationship with a research mentor by getting to know them as a researcher and a person. • Learn about the diversity of experiences in research careers by comparing a mentor's experiences to their own.	U, G	S1: 10–15" S2: 10–15"	Researcher Identity Research Comprehension and Communication Skills Equity and Inclusion Awareness and Skills Professional and Career Development Skills
285	**Mentor Interview About Making Research Posters** • Learn how to create a research poster. • Learn disciplinary norms for graph or image construction. • Set deadline(s) for poster completion.	U, G	10"	Research Comprehension and Communication Skills

288	**Messages Sent and Received** • Identify the intent behind statements and questions. • Practice effective ways to communicate with their research mentor.	U, G	25"	Researcher Confidence and Independence Professional and Career Development Skills Research Comprehension and Communication Skills
292	**Mini-Case Studies: Sticky Situations** • Develop strategies to deal with difficult situations that may arise during the course of their research experience.	U, G	35"	Research Comprehension and Communication Skills Equity and Inclusion Awareness and Skills Research Ethics Professional and Career Development Skills
297	**Mini-Grant Proposal** • Learn how to find and apply for grants. • Develop a logical progression of ideas. • Develop disciplinary written communication skills.	U, G	S1: 30" S2: 1'	Research Comprehension and Communication Skills Research Confidence and Independence
306	**My Mentoring and Support Network** • Define current support network. • Explore how to establish professional relationships. • Discuss how trainees can engage a mentoring network to advance their research career. • Articulate the role(s) that each mentor in a network plays and develop strategies to fill any missing roles.	U, G	45"	Professional and Career Development Skills Research Comprehension and Communication Skills
311	**Networking 1: Introduction to Networking** • Learn the positive career impact of building local, regional, and national networks. • Begin to learn strategies for effective networking.	U, G	1'10"	Professional and Career Development Skills Research Comprehension and Communication Skills Researcher Confidence and Independence Researcher Identity
315	**Networking 2: What Should Your Network Look Like?** • Think strategically about the individuals and types of individuals who should be in a professional network.	U, G	25"	Professional and Career Development Skills Research Comprehension and Communication Skills Researcher Confidence and Independence Researcher Identity

(Continued)

Page	Entering Research Activities			
319	**Networking 3: Your Brand** • Learn the positive career impact of building local, regional, and national networks. • Identify what their "brand" is and the network that is required to promote their future success.	U, G	45"	Researcher Identity Research Comprehension and Communication Skills Professional and Career Development Skills Researcher Confidence and Independence
324	**Networking 4: Planning for Networking Opportunities and Engaging in Purposeful Interactions** • Learn the positive career impact of building local, regional, and national networks. • Discuss and employ strategies for effective networking.	U, G	S1: 40" S2: 40" S3: 55"	Researcher Confidence and Independence Professional and Career Development Skills Research Comprehension and Communication Skills Researcher Identity
329	**Case Study: Overwhelmed** • Practice strategies to ask for help from mentor(s). • Identify additional resources for help.	U, G	20"	Researcher Confidence and Independence Researcher Comprehension and Communication Skills Professional and Career Development Skills
333	**Personal Statement** • Articulate career goals and reflect on research experiences. • Draft a personal statement. • Review peers' personal statements.	U, G	S1: 15" S2: 45"	Researcher Identity Professional and Career Development Skills
338	**Prioritizing Research Mentor Roles** • Identify the different roles that research mentors can play and prioritize those roles based on needs. • Establish and align expectations with a mentor(s).	U, G	35"	Professional and Career Development Skills Research Comprehension and Communication Skills Researcher Identity
342	**Privilege and White Fragility** • Learn about privileges often not afforded to members of groups historically underrepresented in STEM. • Learn about the myth of meritocracy. • Learn about the concept of "White Fragility" (DiAngelo, 2011).	U, G	1'25"	Equity and Inclusion Awareness and Skills
347	**Professional Development Plans** • Develop a plan to guide career and professional development.	U, G	S1: 35" S2: 25"	Professional and Career Development Skills Researcher Confidence and Independence Researcher Identity

356	**Reflecting on Your Mentoring Relationship** • Revisit the goals and expectations established with a research mentor. • Identify and address any issues that have arisen in the mentor–trainee relationship.	U, G	30"	Professional and Career Development Skills Research Comprehension and Communication Skills Equity and Inclusion Awareness and Skills Researcher Confidence and Independence
359	**Research Articles 1: Introduction** • Learn the basic structure of a research article.	U, G	1'	Research Comprehension and Communication Skills Researcher Confidence and Independence
362	**Research Articles 2: Guided Reading** • Learn strategies to effectively and efficiently read research articles in the discipline.	U, G	1'	Research Comprehension and Communication Skills Researcher Confidence and Independence
367	**Research Articles 3: Practical Reading Strategies** • Develop critical reading skills.	U, G	1'	Research Comprehension and Communication Skills Researcher Confidence and Independence
372	**Research Careers: The Informational Interview** • Explore possible research careers and consider how the skills learned by doing research may be transferable to other types of careers.	U, G	45"	Professional and Career Development Skills Research Comprehension and Communication Skills
376	**Research Documentation: Can You Decipher This?** • Develop an appreciation for the need to keep detailed notes. • Recognize the ethical implications of poor note taking.	U, G	15"	Practical Research Skills Research Ethics Research Comprehension and Communication Skills
379	**Research Documentation Process** • Explain why it is important to accurately document research. • Identify key elements in research documentation. • Identify commonalities and differences in documentation associated with different research fields. • Understand the ethical implications of documenting research.	U, G	1'	Practical Research Skills Research Ethics Research Comprehension and Communication Skills
383	**Research Experience Reflections 1: Entering Research?** • Explore realistic expectations for working in a research group. • Self-evaluate readiness for research.	U, G	45"	Researcher Identity Research Comprehension and Communication Skills

(*Continued*)

Page	Entering Research Activities			
387	**Research Experience Reflections 2: Reflection Exercise** • Critically analyze and articulate learning and growth through guided written reflections.	U, G	1'	Determined by reflection prompts used
394	**Research Experience Reflections 3: Research Experience Exit Interview** • Reflect on what was learned and the goals achieved during a research experience.	U, G	1'	Research Identity Researcher Confidence and Independence Professional and Career Development Skills
397	**Research Group Diagram** • Meet all of the people in a research group and learn about their areas of responsibility and research projects. • Articulate how research group personnel differ in education, responsibilities, and contribution to the research team.	U, G	50"	Research Comprehension and Communication Skills Professional and Career Development Skills
404	**Research Group Funding** • Explore how research groups are funded. • Identify where research funds come from. • Identify who is responsible for securing funding. • Identify what funding is available to undergraduate or graduate research trainees. • Gain an appreciation for the amount of funding necessary to sustain research.	U, G	35"	Professional and Career Development Skills Research Comprehension and Communication Skills Research Confidence and Independence
409	**Research Rotation Evaluation** • Learn about the factors that are important to consider when selecting a thesis research mentor and group. • Reflect on and compare research group rotation experiences.	G	S1: 10" S2: 30"	Professional and Career Development Skills Research Comprehension and Communication Skills
413	**Research Writing 1: Background Information and Hypothesis or Research Question** • Identify and summarize the key background information needed to explain and justify a research project. • Articulate the knowledge gap that the research will address. • Formulate specific hypotheses or research questions to address the gap.	U, G	1'25"	Research Comprehension and Communication Skills
421	**Research Writing 2: Research Project Outline and Abstract** • Write an outline summarizing a research project. • Write an abstract about a research project.	U, G	S1: 1' S2: 1'	Practical Research Skills Research Comprehension and Communication Skills

432	**Research Writing 3: Project Design** • Design activities or experiments to test a hypothesis or investigate a research question.	U, G	1'25"	Practical Research Skills Research Comprehension and Communication Skills
436	**Research Writing 4: Research Literature Review and Publishing Process** • Learn how peer-reviewed research papers are published.	U, G	30"	Research Comprehension and Communication Skills Professional and Career Development Skills
440	**Research Writing 5: The Peer-Review Process** • Learn about the peer-review process in STEM. • Categorize reviewer comments and use a framework to review peers' research.	U, G	1'	Research Ethics Research Comprehension and Communication Skills Professional and Career Development Skills
448	**Research Writing 6: Research Proposal** • Develop a logical progression of ideas. • Develop research writing skills.	U, G	S1: 35" S2: 50"	Research Comprehension and Communication Skills
456	**Research Writing 7: Research Paper** • Identify the characteristics of effective research papers. • Practice technical writing skills to communicate research. • Recognize that research writing provides opportunities for reporting research findings.	U, G	1'	Research Comprehension and Communication Skills
461	**Case Study: Responding to Feedback** • Understand the role of constructive feedback. • Communicate effectively across diverse backgrounds and cultures.	U, G	20"	Research Comprehension and Communication Skills Equity and Inclusion Awareness and Skills Professional and Career Development Skills
464	**Safety Training Checklist** • Become familiar with the measures necessary to safely engage in research in a discipline and participate in formal safety training as required.	U, G	5"	Research Ethics Practical Research Skills
471	**Science and Society** • Learn how science can be perceived differently by the scientific community and the general public. • Become aware of how science and society interact. • Consider the social implications of research. • Recognize the responsibility to communicate research to the general public. • Develop strategies to translate research to the general public.	U, G	35"	Research Comprehension and Communication Skills

(Continued)

Page	Entering Research Activities			
472	**Science Literacy Test** • Articulate what it means to be scientifically literate. • Reflect on individual levels of science literacy.	U, G	1'	Research Comprehension and Communication Skills
477	**Science or Pseudoscience?** • Evaluate scientific and pseudoscientific claims.	U, G	30"	Research Comprehension and Communication Skills
480	**Searching Online Databases** • Learn how to use online resources to search for scholarly articles.	U, G	1'	Practical Research Skills Research Comprehension and Communication Skills
484	**Case Study: Selection of Data** • Discuss and practice ethical research decision making.	U, G	25"	Research Ethics Research Comprehension and Communication Skills
487	**Steps to Researcher Independence** • Become familiar with the characteristics of researcher independence at various training stages.	U, G	15–20"	Researcher Confidence and Independence Professional and Career Development Skills
492	**Stereotype Threat** • Define *stereotype threat* and identify the different types of stereotype threat that may be present in a research experience. • Identify ways to mitigate stereotype threat in the research experience.	U, G	45"	Equity and Inclusion Awareness and Skills Researcher Confidence and Independence Professional and Career Development Skills
499	**Summer Undergraduate Research Programs** • Explore undergraduate research programs available to STEM students. • Evaluate programs of interest. • Compile information needed for applications.	U	55"	Professional and Career Development Skills
504	**The Next Step in Your Career** • Identify factors that are important when considering the next steps in professional development/career.	U, G	25–30"	Professional and Career Development Skills Researcher Identity
508	**The Power of Social Persuasion** • Assess the influence that mentors have on confidence in abilities. • Devise strategies to cope with and respond to feedback that negatively influences trainee confidence.	U, G	50"	Researcher Confidence and Independence Research Comprehension and Communication Skills
513	**Case Study: The Sharing of Research Materials** • Discuss and practice ethical research decision making.	U, G	25"	Research Ethics Research Comprehension and Communication Skills

517	**Three Mentors** • Identify different mentoring styles. • Identify preferred mentoring styles.	U, G	35"	Professional and Career Development Skills Research Comprehension and Communication Skills Researcher Identity
525	**Three-Minute Research Story** • Practice communicating research to a general audience in three minutes or less.	U, G	25"	Research Comprehension and Communication Skills Researcher Identity
530	**Tips for Technical Writers** • Learn strategies for improving technical writing. • Learn strategies and criterion for giving feedback on technical writing.	U, G	40"	Research Comprehension and Communication Skills
536	**Truth and Consequences Article** • Explore academic misconduct and the impact it can have on mentors and their trainees.	U, G	20"	Research Ethics Research Comprehension and Communication Skills Researcher Identity
540	**Undergraduate Thesis 1: Components of an Undergraduate Research Thesis** • Learn about components of an undergraduate research thesis. • Analyze research theses of previous or graduating students.	U	1'	Research Comprehension & Communication Skills
543	**Undergraduate Thesis 2: Thesis Writing Discussion Panel** • Learn about the thesis writing process with research mentors and graduated or soon to graduate trainees who have successfully written an undergraduate research thesis. • Identify writing strategies to guide the thesis writing process.	U	1'–1'30"	Research Comprehension and Communication Skills
546	**Undergraduate Thesis 3: Developing a Thesis Writing Plan** • Develop a thesis writing plan.	U	1'	Research Comprehension and Communication Skills
552	**Universalism in STEM: Case Study and Analysis** • Consider the social, political, and historical influences behind actions. • Examine how historical narratives can help us understand contemporary issues in STEM. • Explore how history can help the STEM community create a more inclusive environment.	U, G	1'20"	Equity and Inclusion Awareness and Skills Research Comprehension and Communication Skills

(*Continued*)

Page	Entering Research Activities			
560	**Visiting Peer Research Groups** • Develop an appreciation for differences in culture among research groups. • Identify valuable research group attributes.	U	20"	Research Comprehension and Communication Skills Equity and Inclusion Awareness and Skills
563	**What Happens to Research Results?** • Identify ways research findings are communicated to the public. • Reflect on how research findings affect society.	U, G	30"	Research Comprehension and Communication Skills Research Ethics
566	**Case Study: "Whatever you do, don't join our lab."** • Reflect on factors to consider when selecting a thesis mentor and research group. • Discuss strategies to use when faced with conflicting advice.	G	25"	Professional and Career Development Skills Research Comprehension and Communication Skills
569	**Why Diversity Matters in STEM Research** • Learn why including individuals from diverse backgrounds in the STEM community is important. • Be able to communicate to others the importance of diversity in STEM research.	U, G	1'	Equity and Inclusion Awareness and Skills
572	**Your Research Group's Focus** • Learn about the research and methods used by the research group. • Become familiar with the researchers in the group. • Create a graphical abstract that represents the research of the group.	U, G	1' 20"	Research Comprehension and Communication Skills Practical Research Skills

Note: PDF versions of all implementation guides and trainee materials are available through the Center for the Improvement of Mentored Experiences in Research (CIMER; http://cimerproject.org)

INTRODUCTORY ACTIVITIES

Learning Objectives

Trainees will:

- Learn about other members in the group and begin to build a learning community.
- Reflect on group dynamics and ways to make the group functional.

Trainee Level

undergraduate or graduate trainees
novice, intermediate, or advanced trainees

Areas of Trainee Development

- Research Comprehension and Communication Skills
  - Develop effective interpersonal communication skills.

Activity Components and Estimated Time for Completion:

- In Session Time: varies

Total time: varies

When to Use This Activity

These activities should be used at the beginning of a course or workshop series to help participants get to know one another and to begin to establish a community of practice.

Inclusion Considerations

Emphasize that listening carefully and respecting one another are key to open and inclusive discussions. Trainees should keep the important aspects of their peers' identities in mind when contributing to discussions and recognize that different perspectives can lead to different ideas about effective communication practices.

Contributed by J. Branchaw with information from Pfund, Branchaw, and Handelsman. (2015). *Entering Mentoring: A Seminar to Train a New Generation of Scientists* (2nd ed.). New York: W.H. Freeman & Co.

Implementation Guide

Facilitators may choose to implement one or more of the activities presented here. All activities were adapted from *Entering Mentoring: A Seminar to Train a New Generation of Scientists.*

1. **Visual Explorer**
   Spread out thirty or more pictures that broadly depict phenomena related to learning and research situations. Participants choose a visual representation in response to a question or statement, such as "Choose a picture that best represents research." Each participant explains their choice of picture. In addition to using Visual Explorer, pictures can also be obtained as a packet of postcards, pages from a magazine, or printed images from websites, or participants can be asked to find an image on their own and bring it in. (Adapted from Paulus, C. J., Horth, D. M., and Drath, W. H. (1999). *Visual Explorer: A Tool for Making Shared Sense of Complexity.* Center for Creative Leadership Press, http://www.ccl.org/leadership/index.aspx.)

2. **Significant Mentor**
   Have participants think of a mentor they have had who influenced their own practices. This could be a positive or negative example. Have each person briefly share what they learned.

3. **Who Are You?**
   Participants add fun information about themselves to the four corners of their name tags. Some examples include:
   Hometown; Favorite food; Favorite TV show; Hobby; Favorite kind of music; Number of people in their family (How each person defines family can be very interesting!)

4. **Interviews**
   Participants interview the person next to them, then introduce one another to the larger group.

5. **Truth or Lie?**
   Everyone tells two truths and one lie, then the group guesses the lie for each person.

6. **Memorable Moments**
   Each person shares something they're excited about with regard to their upcoming or current research experience and their motivation for participating in the *Entering Research* training.

7. **Letter Names**
   Each person says their name and shares characteristics that start with the first letter of their name.

8. **The Candy Game**
   Pass around a dish of small candies and tell participants to take as many as they would like. Ask them to introduce themselves by sharing as many characteristics about themselves as is equal to the number of candies they took from the dish.

Contributed by J. Branchaw with information from Pfund, Branchaw, and Handelsman. (2015). *Entering Mentoring: A Seminar to Train a New Generation of Scientists* (2nd ed.). New York: W.H. Freeman & Co.

CONSTRUCTIVE AND DESTRUCTIVE GROUP BEHAVIORS

Learning Objectives

Trainees will:

- Identify their own constructive and destructive behaviors.
- Discuss ways to promote an inclusive and engaging group by holding each other accountable for destructive behaviors.

Trainee Level

undergraduate or graduate trainees
novice, intermediate, or advanced trainees

Areas of Trainee Development

- Equity and Inclusion Awareness and Skills
  - Develop skills to deal with personal differences in the research environment.
- Research Comprehension and Communication Skills
  - Develop effective interpersonal communication skills.

Activity Components and Estimated Time for Completion

- Trainee Pre-Assignment Time: 15 minutes
- In Session Time: 20 minutes

Total time: 35 minutes

When to Use This Activity

This activity can be used with undergraduate and graduate trainees at any career stage and should be implemented at the start of a course, seminar, or workshop series.

Inclusion Considerations

Emphasize that listening carefully and respecting one another are key to open and inclusive discussions. Trainees should keep the important aspects of their peers' identities in mind when contributing to discussions and recognize that different backgrounds can contribute to different ideas about what good communication looks like. As a facilitator, challenge trainees to consider whether their interpretations of their peers' contributions are based on assumptions they may have made, encourage them to respect differences, even if they don't completely understand them, and seek to understand what others are saying.

Information from Brune. (1993). Facilitation skills for quality improvement. *Quality Enhancement Strategies.* 1008 Fish Hatchery Road. Madison WI 53715.

Implementation Guide

Workshop Session (20 minutes)

- Distribute the constructive and destructive behaviors handout included in the trainee materials. Ask trainees to review the handout and select one constructive and one destructive behavior that best describes them.
- Have participants share their constructive and destructive group behaviors. This can be done in many ways:
  - If you have name tents, they can write their constructive and destructive behaviors on the back side to remind themselves during discussions.
  - Participants can discuss their constructive and destructive behaviors in pairs or small groups. Then each participant can briefly introduce themselves to the larger group with their constructive and destructive behaviors.
  - The facilitator can tally the participants' constructive and destructive behaviors on a white/chalk board to give everyone a chance to see group strengths as well as identify any challenges the group might face.
- Once behaviors have been named, ask the group to identify group agreements/ground rules for the workshop. You may wish to ask participants to think about how constructive/destructive behaviors should inform any group agreements.

Information from Brune. (1993). Facilitation skills for quality improvement. *Quality Enhancement Strategies.* 1008 Fish Hatchery Road. Madison WI 53715.

CONSTRUCTIVE AND DESTRUCTIVE GROUP BEHAVIORS

Learning Objectives

Trainees will:

- Identify their own constructive and destructive behaviors.
- Discuss ways to promote an inclusive and engaging group by holding each other accountable for destructive behaviors.

Instructions: Choose one constructive and one destructive behavior that best describes you.

Constructive Group Behaviors

- *Cooperating:* Is interested in the views and perspectives of other group members and willing to adapt for the good of the group.
- *Clarifying:* Makes issues clear for the group by listening, summarizing, and focusing discussions.
- *Inspiring:* Enlivens the group, encourages participation and progress.
- *Harmonizing:* Encourages group cohesion and collaboration. For example, uses humor as relief after a particularly difficult discussion.
- *Risk Taking:* Is willing to risk possible personal loss or embarrassment for success of the overall group or project.
- *Process Checking:* Questions the group on process issues such as agenda, time frames, discussion topics, decision methods, use of information, etc.

Destructive Group Behaviors

- *Dominating:* Uses most of the meeting time to express personal views and opinions. Tries to take control by use of power, time, etc.
- *Rushing:* Encourages the group to move on before task is complete. Gets tired of listening to others and working with the group.
- *Withdrawing:* Removes self from discussions or decision making. Refuses to participate.
- *Discounting:* Disregards or minimizes group or individual ideas or suggestions. Severe discounting behavior includes insults, which are often in the form of jokes.
- *Digressing:* Rambles, tells stories, and takes group away from primary purpose.
- *Blocking:* Impedes group progress by obstructing all ideas and suggestions. "That will never work because . . ."

Information from Brune. (1993). Facilitation skills for quality improvement. *Quality Enhancement Strategies.* 1008 Fish Hatchery Road. Madison WI 53715.

SETTING THE STAGE FOR INCLUSIVE DISCUSSIONS

Learning Objectives

Trainees will:

- Understand how to effectively discuss social issues in the context of STEM.
- Contribute to establishing a welcoming space for all members of the group to fully participate.

Trainee Level

undergraduate or graduate trainees
novice, intermediate, or advanced trainees

Areas of Trainee Development

- Equity and Inclusion Awareness and Skills
  - Develop skills to deal with personal differences in the research environment.
- Research Comprehension and Communication Skills
  - Develop effective interpersonal communication skills.

Activity Components and Estimated Time for Completion:

- In Session Time: 50 minutes

Total time: 50 minutes

When to Use This Activity

This activity can be used at any time with trainees at any career stage and with any level of prior research experience. However, it is most effective when used as an introductory activity to help build rapport amongst trainees.

Additional Readings/Resources for Facilitators

1. National Multicultural Institute. "Diversity Terms." (2003). Available at: https://our.ptsem.edu/UploadedFiles/Multicultural/MCRDiversityTerms.pdf
2. The Office of Multicultural Affairs. University of Massachusetts–Lowell. "Diversity and Social Justice a glossary of working definitions." Available at: https://www.uml.edu/docs/Glossary_tcm18-55041.pdf

Inclusion Considerations

Emphasize that listening carefully and respecting one another are key to open and inclusive discussions. Trainees should keep the important aspects of their peers' identities in mind when contributing to discussions and recognize that different perspectives can lead to different ideas about effective communication practices. As a facilitator, challenge trainees to examine whether or not internalized assumptions are influencing how they interpret their peers' contributions and encourage them to respect differences. Make sure you push them to seek understanding of different ways of knowing, interpreting, and explaining the world humans have constructed.

Contributed by C. R. C. Long. (2018). *Setting the Stage for Inclusive Discussions.*

Implementation Guide

Workshop Session (50 minutes)

► **Introduction: Opening Assignment** (5 minutes)

- Have trainees free-write about important aspects of their identity that they would like share with the group. To help generate ideas, display the following questions:
  - What are your gendered preferred pronouns?
  - Do you identify with any particular social groups?
  - What aspects of your identity are most important to you?

► **Small-Group Discussion** (20 minutes)

- Divide everyone into groups randomly and distribute the small-group discussion handout (see trainee materials). Give groups enough time to come up with terms and best practices for discussion of social issues in the context of STEM.
- Provide groups with additional resources about diversity discourse and talking about information. Please use the resources listed under Additional Readings/Resources for Facilitators.

► **Large-Group Discussion** (20 minutes)

- Give each group the opportunity to share with the larger group the terms they feel are most appropriate to use. In the larger group finalize which terms will be used in the course and write them on poster paper.
- Invite the group to identify other ground rules for future workshop sessions. Come up with discussion norms or rules. A few examples include:

> - Invite all voices into the discussion and give everyone the opportunity to speak.
> - Respect confidentiality and do not share names or stories that are shared within the group.
> - Practice active listening and check your understanding of one another (e.g., "I think I heard you say . . . ").
> - Be present (limit distractions from computers, phones, etc.).

► **Wrap-up** (5 minutes)

- Summarize the terms and ground rules that were generated during the session. If you will be meeting with trainees on a regular basis, record these rules in a common document that all participants can access. Trainees can sign the rules to collectively endorse and agree upon them.

SETTING THE STAGE FOR INCLUSIVE DISCUSSIONS

Learning Objectives

Trainees will:

- Understand how to effectively discuss social issues in the context of STEM.
- Contribute to establishing a welcoming space for all members of the group to fully participate.

Small-Group Discussion Handout

Task 1: Personal Identities

When talking about identity and culture in groups, it is best to allow participants to self-identify. In your group, allow each person time to introduce themselves and share the social identities (gender, race, religion, nationality, ethnicity, etc.) with which they identify. Write down some of those identities here:

Task 2: Discussing Social Issues

Identify one person in the group to read the following out loud.

Although it can be challenging to create a space in which everyone feels comfortable sharing their beliefs, establishing discussion group norms can facilitate constructive, productive discussions. One important aspect is deciding on the words the group will use when discussing the identities of particular groups of people. For example, which term is most appropriate: minoritized, nondominant, underrepresented, non-White, people of color, minority, or marginalized? Should the group use the racial category African American or Black? These are very complicated and actively debated questions. Therefore, it is best to allow the discussion participants who identify with a particular social identity group to determine which term to use. If someone from a particular social group is not present, it is best to use the most contemporarily accepted terminology.

For example: The term *Latinx* is a gender neutral term often used to describe a group of people who are from Latin America. Using the term *Hispanic* could be perceived as problematic by some people because it privileges Spanish-speaking colonizers. Additionally, the term is not representative of the diverse heritage of Spanish speakers from the region. However, these sentiments are not universal and some people prefer to be called *Hispanic*.

Create a list of terms that your group would like to use during discussion of social issues (gender, race, religion, nationality, ethnicity, etc.). Feel free to use additional resources.

Contributed by C. R. C. Long. (2018). *Setting the Stage for Inclusive Discussions.*

ADDRESSING CONFLICT

Learning Objectives

Trainees will:
- ► Learn how to manage conflict in their research experiences.

Trainee Level

undergraduate or graduate trainees
novice or intermediate trainees

Areas of Trainee Development

- ► Research Comprehension and Communication Skills
  - Develop effective interpersonal communication skills.
- ► Equity and Inclusion Awareness and Skills
  - Develop skills to deal with personal differences in the research environment.

Activity Components and Estimated Time for Completion

- ► In Session Time: 1 hour, 5 minutes

Total time: 1 hour, 5 minutes

When to Use This Activity

This activity can be used after a trainee has been matched with a research mentor or has been working with a research mentor for any length of time.

Inclusion Considerations

Discuss how cross-cultural communications can play a role in how conflict is perceived and handled by individuals from different backgrounds. Trainees may respond to these case studies differently depending upon their background and previous experiences. For example, they may be quick to blame Manuel or Obgonda for the conflict or respond in other ways that may include implicit or explicit biases. Encourage them to take a "yes/and" approach when responding to others' comments about the case studies, rather than a "yes/but" approach.

As a facilitator, use the differences in trainee responses as an opportunity to guide trainees to reflect upon their assumptions, and invite other perspectives into the conversation. Guide them to find what is common, seek to understand what others are saying, and do not to rely on stereotypes when interpreting situations.

Contributed by E. Frazier. (2018). *Addressing Conflict.*

Implementation Guide

Workshop Session (1 hour, 5 minutes)

- **Introduction** (5 minutes)
  - The purpose of this activity is to help trainees learn how to manage conflict in their research experiences. Have trainees read and discuss two case studies about conflicts that can arise in the course of a research experience.
- **Case Study #1: Manuel's Dilemma**
  - Have trainees read the case study silently to themselves or have one or two volunteers read the case study to the group. (5 minutes)
  - **Discussion Questions** (20 minutes)
    - Did Manuel do the right thing?
    - What could Manuel have done to resolve this conflict earlier on in his research experience?
    - [Question for graduate trainees]: What could Janet have done to resolve this conflict?
    - How do you address inappropriate or unprofessional behavior in a research mentoring relationship when the relationship is with someone with higher rank than you in the hierarchy of your research group?
    - What are the resources available for you to address these situations when they arise? (i.e., talk to other research mentors, undergraduate research coordinators, or graduate students)
  - Emphasize to trainees that they can politely ask for a meeting with their mentor at any time.
- **Case Study #2: Cross-Cultural Challenge**
  - Have trainees read the case study silently to themselves or have one or two volunteers read the case study to the group. (5 minutes)
  - **Discussion Questions** (20 minutes)
    - Did Dr. McCliff or Obgonda do anything wrong?
    - Is there any indication of unprofessional behavior?
    - What should Obgonda do next?
    - Is it OK for a trainee to start a conversation with their mentor about unaligned expectations? Why or why not?
- **Wrap-up** (10 minutes)
  - Summarize key points raised in the discussion. Additional questions that can be raised around both case studies are provided below.
    - If you decide to leave or change research groups, how would you tell your research mentor?
      - Would you schedule a meeting, write an email or text message, or not say anything and just stop going to lab?
      - What should you say? This was a horrible experience. Thank the faculty member for the opportunity and move on. Explain in great detail why you have decided to leave the lab.
      - Should you be concerned about the kind of relationship you would have with that research mentor in the future?
  - If you find another research group to join, would you mention your previous research experience? What would you say to your new research mentor when they ask you why you left the previous research group?
    - Emphasize to students that it is important to maintain professional conduct, as much as possible, during conflicts. Remind them that the research world is a very small and interconnected group, and

that is it important to keep good professional relationships with most people because you may run into the same people again in your future career, such as in grant committees, hiring committees, at conferences, etc.

- For graduate trainees: When changing research groups, the stakes for graduate trainees are much higher than for undergraduate trainees. It can directly affect their career goals. Considerations when leaving/changing research groups as a graduate student include:
  - Which new research group will I join?
  - Is this new research group in direct competition with my previous research group? If so, what would be the implications of my choice of new research for my career?
  - Should I change fields of research to avoid conflicts between the research groups?
  - These are topics that you might consider discussing with other research mentors (faculty members of your thesis committee might be a great resource) in your field of research to help you evaluate the impact of changing research groups on your career.

Contributed by E. Frazier. (2018). *Addressing Conflict.*

ADDRESSING CONFLICT

Learning Objectives

Trainees will:

- Learn how to manage conflict in their research experiences.

Case Study 1: Manuel's Dilemma

Manuel is an undergraduate student in Dr. Smith's research group and is interested in the genetics of fruit flies and molecular biology. Janet, a graduate student in the research group, has been assigned as Manuel's bench mentor and is supposed to teach him the DNA isolation and amplification techniques required for his project. Though Manuel has made several appointments with Janet, she has repeatedly canceled or failed to show up. Two months go by and Manuel finally shares his frustration with Janet, who says that she is very busy and does not have time to teach him because she needs to finish her own work in order to graduate at the end of the semester. Manuel then asks that she address the issue with Dr. Smith so that he can be assigned to another graduate student, but Janet refuses saying that she does not want to upset Dr. Smith.

Another month goes by and Dr. Smith is getting annoyed that Manuel has not started his project. Dr. Smith asks to speak with him in private. He summons the courage to talk to her about his issue with Janet. She is disappointed that he did not come to her earlier, but he didn't want to get Janet in trouble. Since none of the other research group members knows how to do DNA isolation and amplification, Dr. Smith suggests that Manuel change his project to one that uses neurophysiological techniques. Another graduate student would be able to help him. Manuel is frustrated because he is really interested in genetics and molecular biology, not neurophysiology, and wonders if he should just try to find another research group.

Questions for Discussion

1. Did Manuel do anything wrong?

2. Did Janet do anything wrong?

3. Did Janet behave in a professional manner and communicate with Manuel efficiently?

4. Did Dr. Smith do anything wrong?

5. What should Manuel do?

6. What should Janet do?

Contributed by E. Frazier. (2018). *Addressing Conflict.*

Case Study 2: Cross-Cultural Challenge

Obgonda is an international student working in Dr. McCliff's laboratory. They agree on a research project and Dr. McCliff tells Obgonda to take two weeks to read scientific papers and then return to discuss the methodology that will be used. Two weeks later they meet and decide on the methodology, but Dr. McCliff tells Obgonda that she has to write grants to fund the study because she is not funded for that type of research.

Obgonda finds a funding source and starts drafting a grant proposal. She sends the proposal to Dr. McCliff, who writes her an email saying "This is not ready for me to read. Go find someone to proofread your papers before you send them to me." Obgonda is saddened, but pays a student majoring in English to proofread her grant proposal.

Two weeks later Obgonda sends Dr. McCliff the revised grant proposal and Dr. McCliff replies, "You have not read enough on this topic. You're missing major relevant scientific papers, and this is NOT scientific writing. You indicated during our interview that you had done this kind of work before, but if you're going to be successful in research in this country, you'll need to do better."

Obgonda feels terrible and does not know what to do to make Dr. McCliff happy. She was raised to show utmost respect for her elders and does not know how to address this situation without showing disrespect.

Questions for Discussion

1. Do you think that Obgonda did anything wrong?

2. Do you think that Dr. McCliff did anything wrong?

3. What should Obgonda do?

4. If you were Obgonda, would you change research groups?

5. If a student does choose to change research groups, what are some of the issues that the student should take into consideration?

Contributed by E. Frazier. (2018). *Addressing Conflict.*

ALIGNING MENTOR AND TRAINEE EXPECTATIONS

Learning Objectives

Trainees will:

- Understand that their expectations of their research mentoring relationship may be different from their mentors.
- Establish an open line of communication with their mentor in order to address any differences and align their expectations.

Trainee Level

undergraduate or graduate trainees
novice, intermediate, or advanced trainees

Areas of Trainee Development

- Research Comprehension and Communication Skills
  - Develop effective interpersonal communication skills.
- Researcher Identity
  - Develop an identity as a researcher.
- Research Confidence and Independence
  - Develop independence as a researcher.

Activity Components and Estimated Time for Completion

One Session Implementation

- Trainee Pre-Assignment Time: 1–2 hours
- In Session Time: 1 hour, 10 minutes

Total time: 2 hours, 10 minutes–2 hours, 30 minutes

Two Session Implementation

- Session 1 Time: 1 hour, 5 minutes
- Trainee Pre-Assignment Session 2 Time: 30 minutes
- Session 2 Time: 1 hour, 10 minutes

Total time: 2 hours, 45 minutes

When to Use This Activity

This activity is most beneficial when trainees first join a research team, regardless of their career stage (e.g., during lab rotations for graduate students). However, it can be used at any stage of a mentor–trainee relationship to address issues arising from misaligned expectations. An example mentor–trainee agreement can be found in the Trainee Materials. Other activities that may be used with this activity include:

- Professional Development Plans

Inclusion Considerations

Individuals from backgrounds historically underrepresented in STEM might perceive typical lab or mentoring dynamics such as occasional limited mentor access and a hands-off style of mentoring as related to their identity. Different mentor and mentee identities may create barriers to connecting and therefore to the development of effective mentor–mentee relationships. Help trainees

Contributed by E. Frazier, C. Pfund, and A. R. Butz with information from Branchaw, Pfund, and Rediske. (2010). *Entering Research: A Facilitator's Manual.* New York: W.H. Freeman & Co.

realize that even with different backgrounds and identities they can develop a positive and trusting relationship through open and honest communication.

Trainees may also be unsure of how to handle the power and authority dynamics they perceive in the relationship and therefore resist asking for exactly what they need from a faculty member who holds authority over them. Reassure them that their mentors are there to support their training and development and encourage them to ask for what they need.

Contributed by E. Frazier, C. Pfund, and A. R. Butz with information from Branchaw, Pfund, and Rediske. (2010). *Entering Research: A Facilitator's Manual.* New York: W.H. Freeman & Co.

Implementation Guide

This activity may be implemented in either one or two sessions.

One-Session Implementation

Trainee Pre-Assignment (1–2 hours)

- Have trainees write answers to the questions listed in the "Goals and Expectations" exercise.
- Have trainees meet with their mentor to discuss their answers to the questions. Send the research mentor the expectations worksheet and allow the mentor to review the questions before the meeting.
    - NOTE: If their mentor is not the leader of the research group (e.g., the Professor), then the leader should be invited to join the meeting. If it is not possible for all to meet together, then separate meetings can be scheduled.

Workshop Session (1 hour, 10 minutes)

- **In-Session Activity: Small-Group Discussion** (30 minutes)
    - In groups of four or five, have trainees share the expectation agreements they developed with their mentors.
- **In-Session Activity: Large-Group Discussion** (30 minutes)
    - Each small group shares the breadth of responses from their members from one of the questions. Example responses are provided in the box below.
- **Wrap-up** (10 minutes)
    - Summarize the main ideas generated from the large-group discussion.
    - Encourage trainees to revisit their agreements with their mentor regularly throughout the course of their relationship. Invite trainees to suggest how frequently this should occur (e.g., every 6 months, annually, at milestones during training experience—after a preliminary exam, after a conference presentation, after the publication of a paper, etc.)

Example Trainee Expectations of Mentor	Example Mentor Expectations of Trainee
*I expect my **mentor** to:* • meet with me at least every few weeks. • be open to my questions and to take time to think about them carefully. • be patient with me because I am new to research. • initially be directive but eventually let me design and do experiments on my own. • challenge and encourage me. • teach me basic research techniques/procedures and safety protocols. • help me define a project that is doable, yet relevant, and that keeps me busy. • help me understand the basic scientific concepts and the study design underlying my project.	*I expect my **trainee** to:* • be present and punctual when we have scheduled meeting times. • work hard and give their best effort. • manage their time efficiently and effectively when doing research. • keep up with coursework, but to let me know if they need a break from research to focus on courses. • make every effort on their own to understand the research our group does, but to ask questions when they do not understand. • listen carefully, take notes, and follow instructions when being taught new techniques. • follow all disciplinary ethics and safety procedures.

Contributed by E. Frazier, C. Pfund, and A. R. Butz with information from Branchaw, Pfund, and Rediske. (2010). *Entering Research: A Facilitator's Manual.* New York: W.H. Freeman & Co.

Example Trainee Expectations of Mentor	Example Mentor Expectations of Trainee
*I expect my **mentor** to:*	*I expect my **trainee** to:*
• understand when I need to take time away from research to focus on my coursework and allow me to take it. • help me network with other researchers in the group and/or discipline. • be willing to discuss possible career goals and/or future jobs that will utilize the skills being learned during this research opportunity.	• gradually gain independence, but to regularly communicate with me about what they are doing. • be able to analyze their experimental data, generate logical conclusions based on that analysis, and propose future experiments, with assistance. • work cooperatively, collaboratively, and respectfully with other members of the research team. • be attentive, creative, and contribute at research group meetings.

Contributed by E. Frazier, C. Pfund, and A. R. Butz with information from Branchaw, Pfund, and Rediske. (2010). *Entering Research: A Facilitator's Manual.* New York: W.H. Freeman & Co.

Two-Session Implementation

Session One

Workshop Session (1 hour, 5 minutes)

- **In Session Activity: Small-Group Discussion** (30 minutes)
  - Trainees form small groups with members who have varying levels of research experience. If all trainees are beginners, then invite more experienced trainees as guests to join the groups.
  - Discuss and answer questions on the mentor–trainee expectation agreement.
- **In Session Activity: Large-Group Discussion** (30 minutes)
  - Invite groups to share their answers to the questions on the trainee expectation worksheet.
- **Wrap-up: Trainee Pre-Assignment for Session Two** (5 minutes)
  - Trainees should schedule an appointment with their research mentor to discuss the questions and align their expectations. Send the research mentor the expectations worksheet (see trainee materials) before the meeting to allow the mentor to review the questions ahead of time.

Session Two

Workshop Session (1 hour, 10 minutes)

- **In Session Activity: Small-Group Discussion** (30 minutes)
  - Have trainees discuss how their conversation with their mentor went about expectations.
- **In Session Activity: Large-Group Discussion** (30 minutes)
  - Lead a large-group discussion to identify reasonable goals and expectations for the research experience, and to share how easy or difficult it was for trainees to have this conversation with their mentors. Example responses from trainees appear in the boxes below.
  - **Discussion Questions**
    - How well did your goals and expectations for your research experience align with your mentor's?
    - How does your project fit with the other projects that your research group is doing?
    - Was it difficult to generate an agreement with your mentor?
    - What were the most and least comfortable topics to discuss with your mentor?
    - What are you most excited about, and what are you most concerned about after talking with your mentor?

Example Responses from Trainees

How well did your goals and expectations for your research experience align with your mentor's?

- I'm still unsure about what exactly I would like to get out of the research experience, but I am happy with what the mentor defined.
- The research experience is going to be much more independent than I expected, and my concern is whether there will be enough guidance in the beginning to get started.
- We talked very little about the research questions. The conversation was focused on learning experimental techniques rather than working on any particular experiment at this early stage in project development.

Contributed by E. Frazier, C. Pfund, and A. R. Butz with information from Branchaw, Pfund, and Rediske. (2010). *Entering Research: A Facilitator's Manual.* New York: W.H. Freeman & Co.

Example Responses from Trainees

How does your project fit with the other projects that your research group is doing?

- My project is vital to only a certain part of the lab. The lab is broken into multiple projects and my project is a small part of my mentor's project.
- There are two projects going on in the group and mine is connected to only one of them.
- My project is considered a side project and not connected to the main projects in the lab.

Example Responses from Trainees

Was it difficult to generate an agreement with your mentor?

- The goals and expectations discussion helped.
- We still have a few things to settle, but most of it got done.
- My mentor didn't really want to go through the agreement, but we had an informal conversation about expectations.

- **Wrap-up** (10 minutes)
  - Summarize the main ideas generated from the large-group discussion.
  - Encourage trainees to revisit their agreements with their mentor regularly throughout the course of their relationship. Invite trainees to suggest how frequently this should occur (e.g., every 6 months, annually, at milestones during training experience—after a preliminary exam, after a conference presentation, after the publication of a paper, etc.).

Alternative Activity for Graduate Trainees (30 minutes)

- Instead of completing the "Goals and Expectations Exercise," graduate trainees can review a sample mentor–trainee agreement (compact) included in the Trainee Materials (see "Example Mentor–Graduate Trainee Agreement"). Invite trainees to circle items that they think they should discuss with their mentors (or have already discussed). Trainees may also use the questions raised in the Goals and Expectations exercise as a guide for what they wish to discuss with their mentors. Graduate students should send the questions/topics they wish to discuss to their mentor in advance of their meeting.

NOTE: Graduate trainees could incorporate a discussion about expectations into a larger discussion of their Individual Development Plan (IDP; see "Professional Development Plans" activity) with their mentor.

- **In Session Activity: Discussion Questions**
  - What items from the example were important to discuss with your mentor?
  - What, if anything, is missing from these agreements (compacts)?
  - What ideas do you have about how you establish expectations with your mentor?
  - How do you start a conversation around expectations?

NOTE: A role-playing exercise can be used to help graduate trainees practice raising difficult questions with their mentors. Trainees pair up and take turns being the trainee and the mentor.

Contributed by E. Frazier, C. Pfund, and A. R. Butz with information from Branchaw, Pfund, and Rediske. (2010). *Entering Research: A Facilitator's Manual.* New York: W.H. Freeman & Co.

ALIGNING MENTOR AND TRAINEE EXPECTATIONS

Undergraduate

Learning Objectives

Trainees will:

- ▶ Understand that their expectations of their research mentoring relationship may be different from their mentors.
- ▶ Establish an open line of communication with their mentor in order to address any differences and align their expectations.

Assignment

1. Answer the questions on the Undergraduate Research Trainee Expectations worksheet.
2. Meet with your mentor to discuss the questions and to align your goals and expectations for the research experience. Send the Research Mentor Expectations worksheet to your mentor before the meeting so that he/she can prepare.
3. Trainees and mentors should tailor their discussion to the specific needs of *their* relationship.
4. After the discussion, complete the Mentor–Trainee Expectation document together.

Contributed by Frazier, C. Pfund, and A. R. Butz with information from Branchaw, Pfund, and Rediske. (2010). *Entering Research: A Facilitator's Manual.* New York: W.H. Freeman & Co.

UNDERGRADUATE RESEARCH TRAINEE EXPECTATIONS

1. Why do you want to do research?
2. What are your career goals? How can this research experience and the mentor–trainee relationship help you achieve them?
3. What would success in this research experience look like to you?
4. How many hours per week and at what times/days do you expect to work on your mentor's research?
5. Assuming a good fit, how long do you expect to work with this research group?
6. What, if any, specific technical or communication skills do you expect to learn as part of the research experience?
7. How do you learn best (written procedure, verbal instructions, watch and repeat, etc.)? What can your mentor do to help you learn the techniques and skills in a timely manner that you need to be successful in your research lab? What can you do before you start in the lab to allow you to be successful in this research group?
8. Once you are trained in basic techniques, would you prefer to continue to work closely with others (e.g., on a team project), or independently?
9. Once you have learned the techniques and procedures used in your lab do you prefer that your mentor watch closely what you do, walking your through all the steps, or do you prefer a hands off approach to being supervised?
10. How will you document your research results? Is there a specific protocol for keeping a laboratory notebook in your research group?
11. To whom do you expect to go if you have questions about your research project? Does your mentor expect you to come solely (or first) to them, or should you feel free to ask others in the research group? If others, can your mentor identify those in the group who would be good resource people for your project?
12. Are you comfortable with the methodology used in the lab? Does it involve the use of animals, for example? Does it involve lengthy field trips where you might be isolated with other researchers for weeks at a time? How do you feel about that?
13. Will the research that you will be involved in be confidential? Are you allowed to discuss your project with other individuals outside of your laboratory?
14. What role will your mentor play in the development of your skills as a writer? Is your mentor willing to help you with your research-related writing along the way or does he/she only want to read it after it is in its final version? If your mentor only wants to read final versions of your writing, could they appoint someone in the laboratory/research team to proofread your writings?
15. Do you know all the institutional safety and ethics training that is required to work in your research project? Discuss the required training with your mentor and establish a deadline by which you should complete it.
16. If you have previous research experience, what skills do you expect to bring to your new research group?

Contributed by E. Frazier, C. Pfund, and A. R. Butz with information from Branchaw, Pfund, and Rediske. (2010). *Entering Research: A Facilitator's Manual.* New York: W.H. Freeman & Co.

RESEARCH MENTOR EXPECTATIONS

1. Why do you want to mentor an undergraduate researcher?
2. What are your career goals? How can this research experience and the mentor–trainee relationship help you achieve them?
3. What would success in this research experience look like to you?
4. How many hours per week and at what times/days do you expect your trainee to work on the research project?
5. Assuming a good fit, how long would you like your trainee to remain with the group?
6. What, if any, specific technical or communication skills do you expect your trainee to learn as part of the research experience?
7. What level of independence do you expect your trainee to achieve, once basic techniques are learned? How will you let your trainee know when they have reached this level?
8. What is your mentoring approach? Once your trainee has learned the techniques and procedures used in your lab do you prefer to watch your trainee closely, walking them through all the steps or do you prefer a more hands-off approach?
9. How will your trainee document research results? Is there a specific protocol for keeping a laboratory notebook in your research group?
10. To whom should your trainee go, if they have questions about their research project? Do you expect them to come to you solely (or first), or should they feel free to ask others in the research group? If others, who would be good resource people for their research project?
11. What are your expectations for your trainee's level of comfort with the methodology used in the lab? For example, does your research involve working with animals, lengthy field trips, or working in isolation with other researchers, and is your trainee comfortable with this?
12. Is the research that your trainee will be involved in confidential? Are they allowed to discuss your project with other individuals outside of your laboratory? What are your expectations?
13. What role will you play in the development of your trainee's skills as a writer? Are you willing to help them with research-related writing along the way or do you only want to read final versions? Is there someone else in the lab/ research team who is available to help your trainee with their writing?
14. Discuss the institutional training that is required for your trainee to work on your research project and establish a deadline by which they should complete it.
15. If a trainee has previous research experience, is there anything that you need to share about this research group that is unique and that the trainee should be aware of?

Contributed by E. Frazier, C. Pfund, and A. R. Butz with information from Branchaw, Pfund, and Rediske. (2010). *Entering Research: A Facilitator's Manual.* New York: W.H. Freeman & Co.

MENTOR–UNDERGRADUATE TRAINEE EXPECTATIONS AGREEMENT

Trainee (print) ________________________ Mentor (print) ________________________

This agreement outlines the parameters of our work together on this research project.

1. Our major goals are:
   A. proposed research project goals –
   B. trainee's personal and/or professional goals –
   C. mentor's personal and/or professional goals –
2. Our shared vision of success in this research project is:
3. We agree to work together on this project for at least __________ semesters.
4. The trainee will work at least __________ hours per week on the project during the academic year, and __________ hours per week in the summer.
5. The trainee will propose their weekly schedule to the mentor by the __________ week of the semester.

   If the trainee must deviate from this schedule (e.g., to study for an upcoming exam), then they will communicate this to the mentor at least __________ (weeks/days/hours) before the change occurs.
6. On a daily basis, our primary means of communication will be through (circle all that apply):

   face to face/phone/email/instant messaging/ __________
7. We will meet one-on-one to discuss our progress, the larger project goals, and to evaluate the trainee's performance in the lab. We will reaffirm or revise our goals and/or expectations going forward for at least __________ minutes __________ time(s) per month.
   a. (Circle one): It will be the (trainee's/mentor's) responsibility to schedule these meetings.
   b. In preparation for these meetings, the trainee will:
   c. In preparation for these meetings, the mentor will:
8. At these meetings, the mentor will provide feedback on the trainee's performance and specific suggestions for how to improve or progress to the next level of responsibility through a
   a. written evaluation
   b. a verbal evaluation
   c. other __________
9. When learning new techniques and procedures, the mentor will train the trainee using the following procedure(s) (write out directions, hands-on demonstration, verbally direct as trainee does the procedure, etc.):
10. The proper procedure for documenting research results (laboratory notebook) in our research group is:

    The notebook will be checked __________ (e.g., weekly/monthly).
11. If the trainee gets stuck while working on the project (e.g., has questions or needs help with a technique or data analysis), the procedure to follow will be:

Contributed by E. Frazier, C. Pfund, and A. R. Butz with information from Branchaw, Pfund, and Rediske. (2010). *Entering Research: A Facilitator's Manual.* New York: W.H. Freeman & Co.

12. The standard operating procedures for working in our research group, which all group members must follow and the trainee agrees to follow, include: (e.g., require institutional training to wash your own glassware, attend weekly lab meetings, reorder supplies when you use the last of something, etc.)
13. The mentor and trainee have agreed on a mentoring approach, which consists of:
14. The mentor and trainee have discussed the methodology used in the lab in detail and the trainee understands what is expected of them. To become part of the lab, the trainee must complete the following safety procedures and/or ethics training(s): ______
15. The mentor agrees to read and revise the trainees research writing according to the following procedure:
16. The trainee agrees to not present any of the research findings from this laboratory in any shape or form without the explicit consent and approval of the mentor.
17. Other issues not addressed above that are important to our work together:

By signing below, we agree to these goals, expectations, and working parameters for this research project.

Trainee's signature ______ Date: ______

Mentor's signature ______ Date: ______

Professor's signature ______ Date: ______

Trainee Materials

Contributed by E. Frazier, C. Pfund, and A. R. Butz with information from Branchaw, Pfund, and Rediske. (2010). *Entering Research: A Facilitator's Manual.* New York: W.H. Freeman & Co.

ALIGNING MENTOR AND TRAINEE EXPECTATIONS

Graduate

Learning Objectives

Trainees will:

- Understand that their expectations of their research mentoring relationship may be different from their mentors.
- Establish an open line of communication with their mentor in order to address any differences and align their expectations.

Assignment

1. Answer the questions on the Graduate Research Trainee Expectations worksheet.
2. Meet with your mentor to discuss the questions and to align your goals and expectations for the research experience. Send the Research Mentor Expectations worksheet to your mentor before the meeting so that he/she can prepare.
3. Trainees and mentors should tailor their discussion to the specific needs of their relationship.
4. After the discussion, complete the Mentor–Trainee Expectation document together.

Contributed by E. Frazier, C. Pfund, and A. R. Butz with information from Branchaw, Pfund, and Rediske. (2010). *Entering Research: A Facilitator's Manual.* New York: W.H. Freeman & Co.

GRADUATE RESEARCH TRAINEE EXPECTATIONS

1. Why do you want to do research?
2. What are your career goals? How can this research experience and the mentor–trainee relationship help you achieve them?
3. What would success in this research experience look like to you? What would you like to achieve:
   a. By the end of your 1st year?
   b. By the end of your 3rd year?
   c. By the time you complete your degree?
4. How many hours per week and at what times/days do you expect to do your research?
5. What other commitments or obligations will you have during graduate school (group meetings, teaching, family, religious, community, etc.)? How many hours will these take? How will you schedule around these commitments?
6. Assuming a good fit, how long do you expect to work with this research group?
7. What, if any, specific technical or communication skills do you expect to learn as part of the research experience?
8. How do you learn best (written procedure, verbal instructions, watch and repeat, etc.)? What can your mentor do to help you learn the techniques and skills in a timely manner that you need to be successful in your research lab? What can you do before you start in the lab to allow you to be successful in this research group?
9. Once you are trained in basic techniques, the goal should be to gain independence. What can you do to gain independence in your research? How long do you expect this transition to take?
10. What role do you want your mentor to take throughout your graduate career? For example, would you prefer that your mentor is hands-on throughout your graduate work? Or do you prefer a more hands-off approach to being supervised?
11. How will you document your research results? Is there a specific protocol for keeping a laboratory notebook in your research group?
12. To whom do you expect to go to if you have questions about your research project? Does your mentor expect you to come solely (or first) to them, or should you feel free to ask others in the research group? If others, can your mentor identify those in the group who would be good resource people for your project?
13. Are you comfortable with the methodology used in the lab? Does it involve the use of animals, for example? Does it involve lengthy field trips where you might be isolated with other researchers for weeks at a time? How do you feel about that?
14. Is the research that you will be involved in confidential? Are you allowed to discuss your project with other individuals outside of your laboratory?
15. What role will your mentor play in the development of your writing skills? Will they provide feedback and guidance on numerous drafts or will they only want to provide feedback on the final draft? If your mentor only wishes to read final drafts of writing, are there others in the research group who are willing to provide feedback on earlier drafts?
16. Do you know all the institutional safety or ethics training that is required to work in your research project? Discuss the required training with your mentor and establish a deadline by which you should complete it.
17. If you have previous research experience, what skills do you expect to bring to your new research group?

Contributed by E. Frazier, C. Pfund, and A. R. Butz with information from Branchaw, Pfund, and Rediske. (2010). *Entering Research: A Facilitator's Manual.* New York: W.H. Freeman & Co.

RESEARCH MENTOR EXPECTATIONS

1. Why do you want to mentor a graduate researcher?
2. What are your research goals? How can this research experience and the mentor–trainee relationship help you achieve them?
3. What would success for your trainee look like to you? What would you like your trainee to achieve:
   a. By the end of his/her 1st year?
   b. By the end of his/her 3rd year?
   c. By the time he/she completes his/her degree?
4. How many hours per week and at what times/days do you expect your trainee to do research?
5. Assuming a good fit, how long would you like your trainee to remain with the group?
6. What, if any, specific technical or communication skills do you expect your trainee to learn as part of the research experience?
7. What level of independence do you expect your trainee to achieve, once basic techniques are learned? What can your trainee do to gain independence in research? How long do you expect this transition to take?
8. What is your mentoring approach? Once your trainee has learned the techniques and procedures used in your lab, do you prefer to watch your trainee closely, walking them through all the steps, or do you prefer a more hands-off approach?
9. How will your trainee document research results? Is there a specific protocol for keeping a laboratory notebook in your research group?
10. To whom should your trainee go if they have questions about your research project? Do you expect them to come to you solely (or first), or should they feel free to ask others in the research group? If others, who would be good resource people for their research project?
11. What are your expectations for your trainee's level of comfort with the methodology used in the lab? For example, does your research involve working with animals, lengthy field trips, or working in isolation with other researchers, and is your trainee comfortable with this?
12. Will the research that your trainee will be involved in be confidential? Are they allowed to discuss your project with other individuals outside of your laboratory? What are your expectations?
13. What role will you play in the development of your trainee's writing skills? Will you provide feedback and guidance on numerous drafts or do you only want to provide feedback on the final draft? If you are only willing to read final drafts of writing, are there others in the lab who are willing to provide feedback on earlier drafts?
14. Discuss the institutional safety or ethics training that is required for your trainee to work on their research project and establish a deadline by which they should complete it.
15. If a trainee has previous research experience, is there anything that you need to share about this research group that is unique and that the trainee should be aware of?

Contributed by E. Frazier, C. Pfund, and A. R. Butz with information from Branchaw, Pfund, and Rediske. (2010). *Entering Research: A Facilitator's Manual.* New York: W.H. Freeman & Co.

MENTOR–GRADUATE TRAINEE EXPECTATIONS AGREEMENT

Trainee (print) ______________________ Mentor (print) ______________________

This agreement outlines the parameters of our work together on this research project.

1. Our major goals are:
   A. Proposed research project goals –
   B. Trainee's personal and/or professional goals –
   C. Mentor's personal and/or professional goals –
2. Our shared vision of success in this research project is:
3. We agree to work together on this project for at least __________ years.
4. The trainee will propose their weekly schedule to the mentor by the __________ week of the semester.

   If the trainee must deviate from this schedule (e.g., to study for an upcoming exam), they will communicate this to the mentor at least __________ (weeks/days/hours) before the change occurs.
5. On a daily basis, our primary means of communication will be through (circle all that apply):

   face to face/phone/email/instant messaging/ __________
6. We will meet one-on-one to discuss our progress on the project and to reaffirm or revise our goals for at least __________ minutes __________ time(s) per month.
   a. (Circle one): It will be the (trainee's/mentor's) responsibility to schedule these meetings.
   b. In preparation for these meetings, the trainee will:
   c. In preparation for these meetings, the mentor will:
7. At these meetings, the mentor will provide feedback on the trainee's performance and specific suggestions for how to improve or progress to the next level of responsibility through a
   a. written evaluation.
   b. a verbal evaluation.
   c. other __________.
8. The trainee is expected to participate in the following (journal club, teaching commitments, etc.).
9. When learning new techniques and procedures, the mentor will train the trainee using the following procedure(s) (write out directions, hands-on demonstration, verbally direct as the trainee does the procedure, etc.):
10. The proper procedure for documenting research results (laboratory notebook) in our research group is:

    The notebook will be checked __________ (e.g., weekly/monthly).
11. If the trainee gets stuck while working on the project (e.g., has questions or needs help with a technique or data analysis), the procedure to follow will be:
12. The standard operating procedures for working in our research group, which all group members must follow and the trainee agrees to follow, include (e.g., require institutional training, wash your own glassware, attend weekly lab meetings, reorder supplies when you use the last of something, etc.):

Contributed by E. Frazier, C. Pfund, and A. R. Butz with information from Branchaw, Pfund, and Rediske. (2010). *Entering Research: A Facilitator's Manual.* New York: W.H. Freeman & Co.

13. The mentor and trainee have agreed on a mentoring approach, which consists of:
14. The mentor and trainee have discussed the methodology used in the lab in detail and the trainee understands what is expected of him/her.
15. The mentor agrees to read and revise the trainee's research writing according to the following procedure:
16. The trainee agrees not to present any of the research findings from this laboratory in any shape or form without the explicit consent and approval of the mentor.
17. Other issues not addressed above that are important to our work together are:

By signing below, we agree to these goals, expectations, and working parameters for this research project.

Trainee's signature ______________________ Date: ____________

Mentor's signature ______________________ Date: ____________

Trainee Materials

Contributed by E. Frazier, C. Pfund, and A. R. Butz with information from Branchaw, Pfund, and Rediske. (2010). *Entering Research: A Facilitator's Manual.* New York: W.H. Freeman & Co.

EXAMPLE OF MENTOR–GRADUATE TRAINEE AGREEMENT
DR. TRINA McMAHON, UNIVERSITY OF WISCONSIN–MADISON

The broad goals of my research program

As part of my job as a professor, I am expected to write grants and initiate research that will make tangible contributions to science, to the academic community, and to society. You will be helping me carry out this research. It is imperative that we carry out good scientific method and conduct ourselves in an ethical way. We must always keep in mind that the ultimate goal of our research is publication in scientific journals. Dissemination of the knowledge we gain is critical to the advancement of our field. I also value outreach and informal science education, both in the classroom and while engaging with the public. I expect you to participate in this component of our lab mission while you are part of the lab group.

What I expect from you

Another part of my job as a professor is to train and advise students. I must contribute to your professional development and progress in your degree. I will help you set goals and hopefully achieve them. However, I cannot do the work for you. In general, I expect you to:

- Learn how to plan, design, and conduct high-quality scientific research
- Learn how to present and document your scientific findings
- Be honest, ethical, and enthusiastic
- Be engaged within the research group and at least two programs on campus
- Treat your lab mates, lab funds, equipment, and microbes with respect
- Take advantage of professional development opportunities
- Obtain your degree
- Work hard; don't give up!

- ***You will take ownership over your educational experience***

Acknowledge that you have the primary responsibility for the successful completion of your degree. This includes commitment to your work in classrooms and the laboratory. You should maintain a high level of professionalism, self-motivation, engagement, scientific curiosity, and ethical standards.

Ensure that you meet regularly with me and provide me with updates on the progress and results of your activities and experiments. Make sure that you also use this time to communicate new ideas that you have about your work and the challenges that you are facing. Remember, I cannot address or advise about issues if you do not bring them to my attention.

Be knowledgeable of the policies, deadlines, and requirements of the graduate program, the graduate school, and the university. Comply with all institutional policies, including academic program milestones, laboratory practices, and rules related to chemical safety, biosafety, and fieldwork.

Actively cultivate your professional development. UW–Madison has outstanding resources in place to support professional development for students. I expect you to take full advantage of these resources, since part of becoming a successful engineer or scientist involves doing more than just academic research. You are expected to make continued progress in your development as a teacher, as an ambassador to the general public representing the University and your discipline, with respect to your networking skills, and as an engaged member of broader professional organizations. The Graduate School has a regular seminar series related to professional development. The Delta Program offers formalized training in the integration of research, teaching, and learning. All graduate degree programs require attendance at a weekly seminar. Various organizations on campus engage in science outreach and informal education activities. Attendance at conferences and workshops will also provide professional development opportunities. When you attend a conference, I expect you to seek out these opportunities to make the most of your attendance. You should become a member of one or more professional societies such as the Water Environment Federation, the American Society for Microbiology, or the American Society for Limnology and Oceanography.

Contributed by E. Frazier, C. Pfund, and A. R. Butz with information from Branchaw, Pfund, and Rediske. (2010). *Entering Research: A Facilitator's Manual.* New York: W.H. Freeman & Co.

► ***You will be a team player***

Attend and actively participate in all group meetings, as well as seminars that are part of your educational program. Participation in group meetings does not mean only presenting your own work, but providing support to others in the lab through shared insight. You should refrain from using your computer, Blackberry, or iPhone during research meetings. Even if you are using the device to augment the discussion, it is disrespectful to the larger group to have your attention distracted by the device. Do your part to create a climate of engagement and mutual respect.

Strive to be the very best lab citizen. Take part in shared laboratory responsibilities and use laboratory resources carefully and frugally. Maintain a safe and clean laboratory space where data and research participants confidentiality are protected. Be respectful, tolerant of, and work collegially with all laboratory colleagues: Respect individual differences in values, personalities, work styles, and theoretical perspectives.

Be a good collaborator. Engage in collaborations within and beyond our lab group. Collaborations are more than just publishing papers together. They demand effective and frequent communication, mutual respect, trust, and shared goals. Effective collaboration is an extremely important component of the mission of our lab.

Leave no trace. As part of our collaborations with the Center for Limnology and other research groups, you will often be using equipment that does not belong to our lab. I ask that you respect this equipment and treat it even more carefully than our own equipment. Always return it as soon as possible in the same condition you found it. If something breaks, tell me right away so that we can arrange to fix or replace it. Don't panic over broken equipment. Mistakes happen. But it is not acceptable to return something broken or damaged without taking the steps necessary to fix it.

Acknowledge the efforts of collaborators. This includes other members of the lab as well as those outside the lab. Don't forget important individuals like Dave Harring at the CFL and Jackie Cooper at CEE.

► ***You will develop strong research skills***

Take advantage of your opportunity to work at a world-class university by developing and refining stellar research skills. I expect that you will learn how to plan, design, and conduct high-quality scientific research.

Challenge yourself by presenting your work at meetings and seminars as early as you can and by preparing scientific articles that effectively present your work to others in the field. The "currency" in science is published papers; they drive a lot of what we do and because our lab is supported by taxpayer dollars, we have an obligation to complete and disseminate our findings. I will push you to publish your research as you move through your training program, not only at the end. Students pursuing a master's degree will be expected to author or make major contributions to at least one journal paper submission. Students pursuing a doctoral degree will be expected to be lead author on at least two journal papers' submissions, preferably three or four.

Keep up with the literature so that you can have a hand in guiding your own research. Block at least one hour per week to peruse current tables of contents for journals or do literature searches. Participate in journal clubs. Better yet, organize one!

Maintain detailed, organized, and accurate laboratory records. Be aware that your notes, records, and all tangible research data are my property as the lab director. When you leave the lab, I encourage you to take copies of your data with you. But one full set of all data must stay in the lab, with appropriate and accessible documentation. Regularly backup your computer data to the Bacteriology Elizabeth McCoy server (see the wiki for more instructions).

Be responsive to advice and constructive criticism. The feedback you get from me, your colleagues, your committee members, and your course instructors is intended to improve your scientific work.

► ***You will work to meet deadlines***

Strive to meet deadlines: This is the only way to manage your progress. Deadlines can be managed in a number of ways, but I expect you to work your best to maintain these goals. We will establish mutually agreed upon deadlines for each phase of your work during one-on-one meetings at the beginning of each term. For graduate students, there is to be a balance between time spent in class and time spent on research and perhaps on outreach or teaching. As long as you are meeting expectations, you can largely set your own schedule. It is your responsibility to talk with me if you are having difficulty completing your work and I will consider your progress unsatisfactory if I need to follow up with you about completion of your lab or coursework.

Be mindful of the constraints on my time. When we set a deadline, I will block off time to read and respond to your work. If I do not receive your materials, I will move your project to the end of my queue. Allow a minimum of one week prior to submission deadlines for me to read and respond to short materials such as conference abstracts and three weeks for me to work on manuscripts or grant proposals. Please do not assume I can read materials within a day or two, especially when I am traveling.

► ***You will communicate clearly***

Remember that all of us are "new" at various points in our careers. If you feel uncertain, overwhelmed, or want additional support, please overtly ask for it. I welcome these conversations and view them as necessary.

Let me know the style of communication or schedule of meetings that you prefer. If there is something about my mentoring style that is proving difficult for you, please tell me so that you give me an opportunity to find an approach that works for you. No single style works for everyone; no one style is expected to work all the time. Do not cancel meetings with me if you feel that you have not made adequate progress on your research; these might be the most critical times to meet with a mentor.

Be prompt. Respond promptly (in most cases, within 48 hours) to emails from anyone in our lab group and show up on time and prepared for meetings. If you need time to gather information in response to an email, please acknowledge receipt of the message and indicate when you will be able to provide the requested information.

Discuss policies on work hours, sick leave, and vacation with me directly. Consult with me and notify fellow lab members in advance of any planned absences. Graduate students can expect to work an average of 50 hours per week in the lab; post-docs and staff at least 40 hours per week. I expect that most lab members will not exceed two weeks of personal travel away from the lab in any given year. Most research participants are available during University holidays, so all travel plans, even at the major holidays, must be approved by me before any firm plans are made. I believe that work–life balance and vacation time are essential for creative thinking and good health and encourage you to take regular vacations. Be aware, however, that there will necessarily be epochs—especially early in your training—when more effort will need to be devoted to work and it may not be ideal to schedule time away. This includes the field season for students/post-docs working on the lakes.

Discuss policies on authorship and attendance at professional meetings with me before beginning any projects to ensure that we are in agreement. I expect you to submit relevant research results in a timely manner. Barring unusual circumstances, it is my policy that students are first-author on all work for which they took the lead on data collection and preparation of the initial draft of the manuscript.

Help other students with their projects and mentor/train other students. This is a valuable experience! Undergraduates working in the lab should be encouraged to contribute to the writing of manuscripts. If you wish to add other individuals as authors to your papers, please discuss this with me early on and before discussing the situation with the potential coauthors.

What you should expect from me

I will work tirelessly for the good of the lab group; the success of every member of our group is my top priority, no matter their personal strengths and weaknesses, or career goals.

Contributed by E. Frazier, C. Pfund, and A. R. Butz with information from Branchaw, Pfund, and Rediske. (2010). *Entering Research: A Facilitator's Manual.* New York: W.H. Freeman & Co.

I will be available for regular meeting and informal conversations. My busy schedule requires that we plan in advance for meetings to discuss your research and any professional or personal concerns you have. Although I will try to be available as much as possible for "drop-in business," keep in mind that I am often running to teach a class or to a faculty meeting and will have limited time.

I will help you navigate your graduate program of study. As stated above, you are responsible for keeping up with deadlines and being knowledgeable about requirements for your specific program. However, I am available to help interpret these requirements, select appropriate coursework, and select committee members for your oral exams.

I will discuss data ownership and authorship policies regarding papers with you. These can create unnecessary conflict within the lab and among collaborators. It is important that we communicate openly and regularly about them. Do not hesitate to voice concerns when you have them.

I will be your advocate. If you have a problem, come and see me. I will do my best to help you solve it.

I am committed to mentoring you, even after you leave my lab. I am committed to your education and training while you are in my lab and to advising and guiding your career development—to the degree you wish—long after you leave. I will provide honest letters of evaluation for you when you request them.

I will lead by example and facilitate your training in complementary skills needed to be a successful scientist, such as oral and written communication skills, grant writing, lab management, mentoring, and scientific professionalism. I will encourage you to seek opportunities in teaching, even if not required for your degree program. I will also strongly encourage you to gain practice in mentoring undergraduate and/or high school students, and to seek formal training in this activity through the Delta program.

I will encourage you to attend scientific/professional meetings and will make an effort to fund such activities. I will not be able to cover all requests, but you can generally expect to attend at least one major conference per year, when you have material to present. Please use conferences as an opportunity to further your education, and not as a vacation. If you register for a conference, I expect you to attend the scientific sessions and participate in conference activities during the time you are there. Travel fellowships are available through the Environmental Engineering program, the Bacteriology Department, and the University if grant money is not available. I will help you identify and apply for these opportunities.

I will strive to be supportive, equitable, accessible, encouraging, and respectful. I will try my best to understand your unique situation, and mentor you accordingly. I am mindful that each student comes from a different background and has different professional goals. It will help if you keep me informed about your experiences and remember that graduate school is a job with very high expectations. I view my role as fostering your professional confidence and encouraging your critical thinking, skepticism, and creativity. If my attempts to do this are not effective for you, I am open to talking with you about other ways to achieve these goals.

Yearly Evaluation

Each year we will sit down to discuss progress and goals. At that time, you should remember to tell me if you are unhappy with any aspect of your experience as a graduate student here. Remember that I am your advocate, as well as your advisor. I will be able to help you with any problems you might have with other students, professors, or staff. Similarly, we should discuss any concerns that you have with respect to my role as your advisor. If you feel that you need more guidance, tell me. If you feel that I am interfering too much with your work, tell me. If you would like to meet with me more often, tell me. At the same time, I will tell you if I am satisfied with your progress, and if I think you are on track to graduate by your target date. It will be my responsibility to explain to you any deficiencies, so that you can take steps to fix them. This will be a good time for us to take care of any issues before they become major problems.

Contributed by E. Frazier, C. Pfund, and A. R. Butz with information from Branchaw, Pfund, and Rediske. (2010). *Entering Research: A Facilitator's Manual.* New York: W.H. Freeman & Co.

ARTICLE ORGANIZATION, COMPREHENSION, AND RECALL

Learning Objectives

Trainees will:

- ► Learn about electronic tools for organizing papers, citing papers, and taking notes on papers.
- ► Learn to use a framing funnel as a strategy to actively read papers.
- ► Learn to use guided questions to understand and evaluate scholarly papers.

Trainee Level

undergraduate or graduate trainees
intermediate or advanced trainees

Areas of Trainee Development

- ► Research Comprehension and Communication Skills
  - Develop disciplinary knowledge.
- ► Professional and Career Development Skills
  - Explore and pursue a research career.

Activity Components and Estimated Time for Completion

- ► Trainee Pre-Assignment Time: 20 minutes
- ► In Session Time: 1 hour

Total time: 1 hour, 20 minutes

When to Use This Activity

This activity is recommended for intermediate and advanced trainees who have foundational research experience. It can be implemented on its own or paired with:

- ► Research Writing 4: Research Literature Review and Publishing Process

Inclusion Considerations

Learning and reading styles will vary among trainees. Invite them to share with you any learning accommodations they need or preferences they have and be flexible when setting reading and writing assignment deadlines. Encourage them to share with the group alternative ideas about how to approach reading scientific papers, organizing information, and constructing reviews.

Contributed by A. Sokac with information from Branchaw, J. L., Pfund, C., and Rediske, R. (2010). *Entering Research: A Facilitator's Manual.* New York: W.H. Freeman & Co.

Implementation Guide

Trainee Pre-Assignment (20 minutes)

► Ask trainees to complete the "Tools for Handling Your Papers" worksheet by interviewing their primary mentor, at least one member of their research team, or a senior member of their research team. The worksheet may be distributed in the previous session or electronically.

Workshop Session (1 hour)

► **Introduction** (2–3 minutes)
- The goal of this activity is for trainees to learn how to organize, cite, and take notes on papers. Using the framing funnel and guided questions, trainees will learn to better evaluate and understand scientific papers.

► **Large-Group Discussion** (10 minutes)
- Use the following questions to lead the discussion in the large group.
  - What are the tools and combinations of tools that people use to organize their papers?
  - What tools are people using that did not appear on the worksheet?
  - Why do people like the specific tools that they are using?
  - What limitations do these tools have? Benefits?
  - Which tools are the most attractive to you and why?

► **Activity 1: *The Framing Funnel*** (20 minutes)
- In the large group, ask trainees the following questions about reading papers. Either assign a paper for this activity or instruct students to read and bring a paper from their research group. Alternatively, the facilitator can provide an article where all elements of the *Framing Funnel* are apparent and ask the students to read it before class. Record responses and/or strategies on a whiteboard or chart.
  - What are the easiest and most challenging parts of reading research articles?
  - How might you use the sections of a research article to break down complex ideas and results?
  - How might reading a textbook help you understand a research article?
  - Can you link ideas and concepts learned in courses to research articles?
- Distribute the *Framing Funnel* document. Explain that the framing funnel is a tool that can be used by readers to outline research articles. It represents the language that scientists use to think about and present their research and what they expect from other scientists. Every research paper has an underlying framing funnel. Using the framing funnel to map scientific articles can increase understanding and retention of the information that is presented and develop logical thinking skills.
- Review each element of the framing funnel with the group.

► **Activity 2: Critical Evaluation of Papers** (20 minutes)
- Brainstorm: Ask trainees to describe the function of each part of a research paper. For example:
  - What should the *Title* of a research article do?
  - What information should be in the *Abstract*?
  - In the *Introduction*?
  - In the *Methods*?
  - In the *Results*?
  - In the *Discussion*?

Contributed by A. Sokac with information from Branchaw, J. L., Pfund, C., and Rediske, R. (2010). *Entering Research: A Facilitator's Manual.* New York: W.H. Freeman & Co.

- Distribute the *Paper Parts and Evaluation* document and compare the trainees' ideas to the descriptions on the handout.
- Ask the group how confident they feel critically evaluating a paper. Individually, have them take a few minutes to look over the questions behind the descriptions of paper sections in the *Critical Evaluation of Papers* handout provided in the trainee materials. Ask them to consider which questions they think they can answer, and which would require them to read the paper more closely. For example:
  - Do the authors' methods critically or directly test their hypothesis?
  - Did the authors use a creative method to evaluate their hypothesis?
  - Do you agree with the authors' interpretation of the data or are there other interpretations?

Optional Assignment: Facilitators may assign trainees to use the *Critical Evaluation of Papers* activity to identify and review a scientific paper that is relevant to their research project. These review assignments can be assessed by peers or the facilitator using the following rubric.

Contributed by A. Sokac with information from Branchaw, J. L., Pfund, C., and Rediske, R. (2010). *Entering Research: A Facilitator's Manual.* New York: W.H. Freeman & Co.

CRITICAL EVALUATION OF PAPERS

Assessment Rubric

	0 Absent	1 Does not meet expectations	2 Meets expectations	3 Exceeds expectations
Title, Abstract, Introduction				
Articulated the title and purpose of the article.				
Provided a brief overview of the article, including hypotheses and key results.				
Methods				
Determined whether the proposed methods critically test their hypotheses.				
Identified limitations of the method.				
Examined creativity of the author's methodology.				
Examined methodological process.				
Identified methodological innovation.				
Results				
Determined whether data supports argument.				
Interpreted the data.				
Formulated own opinion based on the data.				
Examined format in which data was presented in the paper.				
Evaluated all figures and supporting documents.				
Discussion				
Determined whether the conclusion is supported by the data.				
Identified alternative interpretations of the data.				
Identified novel insights gained from the results.				
Determined if the results may be applied more generally.				
Identified author's future directions.				

Notes:

Contributed by A. Sokac with information from Branchaw, J. L., Pfund, C., and Rediske, R. (2010). *Entering Research: A Facilitator's Manual.* New York: W.H. Freeman & Co.

ARTICLE ORGANIZATION, COMPREHENSION, AND RECALL

Learning Objectives

Trainees will:

- Learn about electronic tools for organizing papers, citing papers, and taking notes on papers.
- Learn to use a framing funnel as a strategy to actively read papers.
- Learn to use guided questions to understand and evaluate scholarly papers.

Pre-Assignment: Tools for Handling Your Papers

Generate a list of electronic (and other) tools for organizing your papers, citing papers, and taking notes on your papers by interviewing your current mentor or a senior person on your research team to ask what tools and methods they use to organize, cite, and take notes on the papers. If you interview multiple people, you may find that everyone has their own system!

Based on your interview, circle the tools used below. If your interviewee suggests new tools, record them. Also ask why your interviewee likes or dislikes the tools they use.

Finding papers:

- Google Scholar
- PubMed
- Bing
- Papers (Mekentosj)
- Readcube

Organizing and note-taking for PDFs:

- Adobe Acrobat
- Papers (Mekentosj)
- OneNote
- Word
- Google Docs
- Notes Plus
- Good Note
- Evernote
- Readcube

Citing papers:

- EndNote
- Papers (Mekentosj)
- Mendeley
- Readcube

Tools at journal websites for digging deeper:

- Social media (Facebook, Twitter)
- eAlerts
- Journal-specific Apps
- Video Portals
- Podcasts
- Webinars
- Blogs
- Twitter

Mindful note-taking (plagiarism scanning software):

- TurnItIn
- iThenticate
- Doc Cop
- Grammarly

Contributed by A. Sokac with information from Branchaw, J. L., Pfund, C., and Rediske, R. (2010). *Entering Research: A Facilitator's Manual.* New York: W.H. Freeman & Co.

Consider these questions about the pre-meeting activity:

- Who did you interview to complete this worksheet?
- What are the tools and combinations of tools that people use to handle their papers?
- What tools are people using that did not appear on the worksheet?
- Why do people like the specific tools that they are using?
- What limitations do these tools have? Benefits?

Contributed by A. Sokac with information from Branchaw, J. L., Pfund, C., and Rediske, R. (2010). *Entering Research: A Facilitator's Manual.* New York: W.H. Freeman & Co.

ARTICLE ORGANIZATION, COMPREHENSION, AND RECALL

Activity 1: The Framing Funnel

Use these questions to explore strategies to improve your understanding and retention of the research presented in scholarly articles:

- What are the easiest and most challenging parts of reading research articles?
- How might you use the sections of a research article to break down complex ideas and results?
- How might reading a textbook help you understand a research article?
- Can you link ideas and concepts learned in courses to research articles?

The framing funnel is a tool that can be used by readers to outline research articles. It represents the language that researchers use to think and present their research and what they expect from other researchers. Every research paper has an underlying framing funnel. Using the framing funnel to map research articles can increase understanding and retention of the information that is presented and develop logical thinking skills.

Select a research article to read and use the framing funnel below to outline the content.

The framing funnel

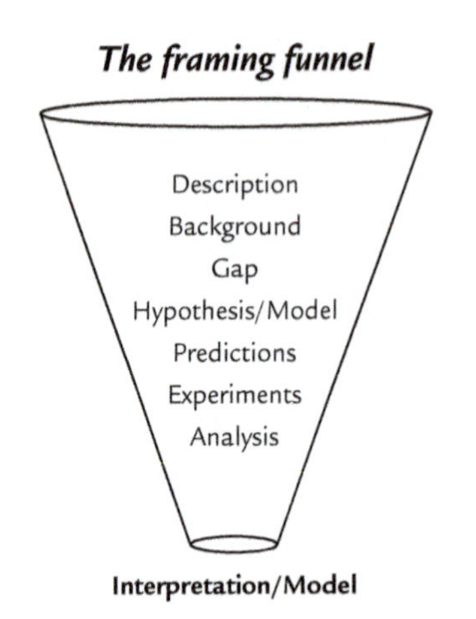

Description: In very few words, what is this paper about? What is the synopsis?

Background: What is already known?

Gap: What knowledge is still missing? What are the open questions?

Hypothesis*/Model/Research Question: What is the working hypothesis, model, or research question that these authors are testing or investigating? What is the rationale for doing this specific set of experiments or collecting the data?

Predictions: If the authors' hypothesis/model is correct, what results are expected? What if the hypothesis/model is incorrect?

Experiments/Data Collection: What are the experiments or data collection methods? What is the flow of the experiments or the collection of data? What are the strengths and weaknesses of the methods used? Are there alternative methods? What are the results?

**A hypothesis is a possible explanation that is proposed on the basis of a small amount of preliminary data. A hypothesis serves to launch and focus future research.*

Contributed by A. Sokac with information from Branchaw, J. L., Pfund, C., and Rediske, R. (2010). *Entering Research: A Facilitator's Manual.* New York: W.H. Freeman & Co.

Trainee Materials

Data Analysis: How are the data analyzed? Are the analysis techniques appropriate to address the hypothesis or research questions?

Interpretation/Model: Based on the results, what new things were learned? What do the results mean? Was the initial hypothesis/model correct or incorrect? How does the hypothesis/model need to be revised? *What are the next directions? What new **gap** opened up?*

Contributed by A. Sokac with information from Branchaw, J. L., Pfund, C., and Rediske, R. (2010). *Entering Research: A Facilitator's Manual.* New York: W.H. Freeman & Co.

ARTICLE ORGANIZATION, COMPREHENSION, AND RECALL

Activity 2: Critical Evaluation of Papers

The goal of this activity is to provide a set of questions and criterion that will help you critically evaluate research papers that you read. Research articles are typically organized in sections as outlined below. Knowing what types of information are present in each section allows you to more efficiently and effectively find information.

In your own words, describe each part of a research paper.

- What should the *Title* of a research article do?
- What information should be in the *Abstract*?
- In the *Introduction*?
- In the *Methods*?
- In the *Results*?
- In the *Discussion*?

How confident do you feel in critically evaluating a paper?

Take a few minutes to look over the questions following the descriptions of paper sections below. Among these, identify good questions to ask when you are critically evaluating a paper. For example:

- Do the authors' methods critically or directly test their hypothesis?
- Did the authors use a creative method to evaluate their hypothesis?
- Do you agree with the authors' interpretation of the data or are there other interpretations?

Title

Paper titles are usually brief, stand-alone overviews of a paper's contents. Authors make an effort to include keywords that abstracting services could use in indexing the article. Therefore, if you are new to a field and/or subject, it is useful to take note of the words used in the title as they may provide you with useful keywords to use in future literature searches.

Abstract

The purpose of the abstract is to provide the reader with a brief summary of the article. Thus, the abstract should provide information about the specific research problem being investigated, the methods used, the results obtained, and what the results of the study mean in the larger context of the research study and in some cases the field of study. This means that the abstract is a good place to look first if you are trying to decide whether or not the paper is relevant to your work.

Introduction

The introduction section provides a general overview of the research question being studied—why it is a worthy question, what work has already been done by others to address it, and what the authors may have already done in this area. Introductions are a good place to go if you are new to the subject.

- What is the main question they are interested in pursuing?
- What background research, pattern, theoretical prediction, or theoretical framework motivates this question?
- Why is this question interesting in light of the background they discuss?
- Do they offer one hypothesis or more than one?
- What assumptions are made when proposing the hypotheses?

Contributed by A. Sokac with information from Branchaw, J. L., Pfund, C., and Rediske, R. (2010). *Entering Research: A Facilitator's Manual.* New York: W.H. Freeman & Co.

Methods

The methods section will provide detailed information about experiments performed or data that was collected. Such information typically includes sources for all materials used, names of all data collection instruments, detailed descriptions of experimental or data gathering techniques, and detailed descriptions of data analysis techniques, including computer programs used.

- Do the proposed methods test the hypotheses or address the research question?
- Are any of the methods confounded?
- Did the authors use a creative method to evaluate their hypothesis or interpret their data?
- Are the methods simple and elegant or complicated and convoluted?
- Was a new technique or research approach presented that will better evaluate a problem that others have struggled with?

Results

Some articles will distinguish between "Results" and "Discussion" while others will combine this information into one section "Results and Discussion." In papers that contain two distinct sections ("Results" and "Discussion"), the data obtained from the study are introduced in the "Results" section and their interpretation is delayed until the "Discussion" section. In papers that contain one section ("Results and Discussion"), results are introduced and interpreted experiment-by-experiment.

- What does the data say about the hypotheses or research questions?
- Is there only one interpretation of the data?
- Are there any big surprises/unexpected results?

Discussion

In papers that contain a distinct "Discussion" section, the interpretations of the results are included here. The "Discussion" should also place the results in the context of the existing literature in the field of interest. Authors may also address limitations to the study or suggest future directions in this section.

Keep the following in mind:

- Does the author say that they support or reject the hypothesis?
- Do you agree with the author's interpretation of the data?
- What novel insights are gained from the results?
- What do the results imply more generally for the field of interest? For other fields?
- What will the authors do next?

Sophisticated Understanding

With experience, reading the literature in a given field will come more easily. This includes the ability to better evaluate what is being presented, and the ability to ask more sophisticated questions.

- Be critical when reading papers but also pay attention to exciting findings, novel insights, and creative ideas. It's easy to criticize, but hard to praise!
- What critical experiment would you do to evaluate the proposed hypothesis?
- What data would you collect to address the research questions?

Contributed by A. Sokac with information from Branchaw, J. L., Pfund, C., and Rediske, R. (2010). *Entering Research: A Facilitator's Manual.* New York: W.H. Freeman & Co.

- Form an opinion after looking at the data, before reading the author's interpretation and conclusions.
- Do you agree with the author's interpretation or are there others?
- If more than one hypothesis is offered, does each hypothesis propose a distinctly alternative explanation that is incompatible with the others, or could some of the hypotheses operate simultaneously?
- Are there compelling alternatives given the data?
- What assumptions are made about the effectiveness of the experiments or the accuracy of the data?

Consider these questions:

- Is a research paper comprised of "facts" or "arguments"?
- What is the difference between a result and an interpretation?

Contributed by A. Sokac with information from Branchaw, J. L., Pfund, C., and Rediske, R. (2010). *Entering Research: A Facilitator's Manual.* New York: W.H. Freeman & Co.

CASE STUDY: AUTHORSHIP

Learning Objectives

Trainees will:

- Identify the authorship guidelines for a discipline/career stage.
- Practice strategies for discussing authorship with mentors and members of the research group.

Trainee Level

undergraduate or graduate trainees
novice, intermediate, or advanced trainees

Areas of Trainee Development

- Research Ethics
  - Develop responsible and ethical research practices.
- Research Comprehension and Communication Skills
  - Develop effective interpersonal communication skills.

Activity Components and Estimated Time for Completion

- In Session Time: 20 minutes

Total time: 20 minutes

When to Use This Activity

This activity is suitable for an undergraduate or graduate trainee who has completed at least one semester of research. It can be implemented as part of a group of activities focused on ethics or as part of a discussion on aligning expectations.

Inclusion Considerations

Some trainees may come from backgrounds where humility is valued and where trainees have been acculturated not to speak of their achievements freely. Talk about why it is important to articulate one's research contributions positively and with confidence and reassure them that they can do this while remaining true to their ideals around humility.

Trainees may also resist explaining why they feel they deserve to be an author with a mentor who holds authority over them. Reassure them that their mentors are there to support their training and development and encourage them to respectfully advocate for themselves when authorship is being discussed.

Contributed by C. Pfund (2012). *Authorship*.

Implementation Guide

In Session Activity (20 minutes)

- Ask one person to read the case aloud or have individuals read the case silently to themselves.
- Ask for initial reactions to the case study, then discuss the questions as a large group. Alternatively, you can have small groups address the discussion questions.
  - How should Amy respond?
  - Whose responsibility is it to decide issues of authorship?
  - When should issues of authorship be discussed?
  - Who do you think should be an author on the paper described in this scenario?
  - How could each of the players in this scenario have avoided this situation?
  - Have you discussed authorship policies with your mentor? Have trainees discuss the authorship policies in their lab; do all mentors have the same authorship policies?

Note: Optional role-playing activity: In pairs, select one individual to play the role of the student and one to play the role of the mentor. What is each person's role in the authorship discussion?

Contributed by C. Pfund (2012). *Authorship.*

CASE STUDY: AUTHORSHIP

Undergraduate

Learning Objectives

Trainees will:

- Identify the authorship guidelines for their discipline/career stage.
- Practice strategies for discussing authorship with their mentor and members of their research group.

Amy is a sophomore and has been working on a new research project for the past 6 months. Her project is part of a larger research study led by an advanced graduate student on the research team where Amy is an undergraduate researcher. Over the past four months, Amy has attended weekly meetings for the project, read the background research, assisted with some of the experiments under the mentorship of the graduate student, and even contributed some fresh ideas for future experiments. Before one of the weekly lab meetings, Amy overhears the PI tell the graduate student that he feels the project is coming along nicely and that it should be ready to submit as a conference proposal in the next few weeks. One month later, Amy overhears the graduate student tell the PI that the proposal has been accepted. Amy is excited about the idea of having research she contributed to be presented at a national conference. She glances at a copy of the proposal on the graduate student's desk and notices that her name was not included as an author. When the graduate student sees Amy looking at the list of authors, he quickly says "Don't worry, we included your name in the acknowledgements. You will be an author soon enough." Amy is surprised and disappointed; as a member of another research team this past year, she was listed as an author on a conference proposal for this same level of contribution.

Discussion Questions

1. How should Amy respond?

2. Whose responsibility is it to decide issues of authorship?

3. When should issues of authorship be discussed?

4. Who do you think should be an author on the paper described in this scenario?

5. How could each of the players in this scenario have avoided this situation?

6. Have you discussed authorship policies with your mentor? Do all mentors have the same authorship policies?

Contributed by C. Pfund (2012). *Authorship.*

CASE STUDY: AUTHORSHIP

Graduate

Learning Objectives

Trainees will:

- Identify the authorship guidelines for a discipline/career stage.
- Practice strategies for discussing authorship with mentors and members of the research group.

Robert is a first-year graduate student and has been working on a new research project for the past 6 months. His project is part of a larger research study being led by Dr. Althia Thompson, a senior post-doc on the research team. Over the past four months, Robert has attended weekly meetings for the project, read the background research, conducted some experiments under the mentorship of Dr. Thompson, and even contributed some fresh ideas for future experiments. Before one of their weekly meetings, Robert overhears the PI, Dr. Jones, tell Dr. Thompson that he feels the manuscript is coming along nicely and should be ready to submit in the next few weeks. When Robert inquires about the manuscript, Dr. Jones says that he and Dr. Thompson are pushing the manuscript out the door as quickly as possible so that she can include the submission on her resume when she begins applying for jobs next month. Robert sees the manuscript lying on the desk and looks at the title page. He is surprised that his name has not been included as an author. When Dr. Jones sees Robert looking at the title page, he quickly says "Don't worry, we included your name in the acknowledgements. We just don't have the time to include you in the writing this time. You will be an author soon enough."

Discussion Questions

1. How should Robert respond?

2. Whose responsibility is it to decide issues of authorship?

3. When should issues of authorship be discussed?

4. Who do you think should be an author on the paper described in this scenario?

5. How could each of the players in this scenario have avoided this situation?

6. Have you discussed authorship policies with your mentor? Do all mentors have the same authorship policies?

Trainee Materials

Contributed by C. Pfund (2012). *Authorship.*

CASE STUDY: AWKWARD MENTOR

Learning Objectives

Trainees will:

- Discuss challenges that may arise in mentoring relationships.
- Develop strategies to address mentoring relationship challenges.

Trainee Level

undergraduate or graduate trainees
novice, intermediate, or advanced trainees

Areas of Trainee Development

- Equity and Inclusion Awareness and Skills
  - Develop skills to deal with personal differences in the research environment.
- Research Comprehension and Communication Skills
  - Develop effective interpersonal communication skills.

Activity Components and Estimated Time for Completion

- In Session Time: 25 minutes

Total time: 25 minutes

When to Use This Activity

This activity is suitable for undergraduate and graduate trainees at any level who have a mentor. It can be combined with any other activities addressing professional relationships including:

- Aligning Mentor and Trainee Expectations
- Research Experience Reflections 1: Entering Research?
- Three Mentors
- Prioritizing Research Mentor Roles

Inclusion Considerations

Individuals from backgrounds historically underrepresented in STEM may struggle to find others who can relate to their lived experiences and how these experiences may be impacting their research learning experience. Different mentor and trainee identities and backgrounds can create barriers to connecting and to the development of effective mentor–trainee relationships. Help trainees realize that even with different backgrounds and identities it is possible to develop positive and trusting relationships through open and honest communication. Reassure them that their mentors are there to support their training and development. Encourage them to share their experiences and to ask for what they need to be successful.

Contributed by J. Branchaw. (2018). *Awkward Mentor.*

Implementation Guide

Workshop Session (25 minutes)

► **Introduction** (5 minutes)

- The goal of this activity is for trainees to consider challenges they may face in their mentoring relationships and to discuss possible strategies to address these challenges.

► **Case Study: Awkward Mentor** (5 minutes)

- Distribute the case study included in the trainee materials. Have one trainee read the case study aloud, display the case on a projector screen, or have trainees read and consider the case study silently.

► **Discussion** (10 minutes)

- As a group, discuss the following questions. Alternatively, if there are more than 10 trainees, they can discuss the questions in small groups of three or four and report highlights from their discussion to the larger group. Example responses appear below each question.
  - How would you describe the communication between Sheneka and her mentor?

    - They are not listening to one another.
    - They are not being open to learning about the other and finding common ground.
    - Sheneka does not seem to trust her career mentor and therefore she isn't comfortable sharing her perspective.
    - The conversations sound very one sided in that the mentor is sharing her story, but Sheneka is not contributing much to the conversation.

  - With whom might Sheneka share her frustration and ask for advice about what to do?

    - Transfer advisor
    - Training program director
    - Peers in the program
    - Research mentor
    - Family members
    - Other mentors in her network who are currently serving in this capacity

  - How might Sheneka structure the meetings with her career mentor to better meet her needs?

    - Make a list of things that she'd like to discuss (e.g., current or future challenges she is facing) and ask her career mentor specifically about these things.
    - Be more open about sharing her own story to help the career mentor better understand her perspective.

► **Wrap-up** (5 minutes)

- Summarize the main ideas generated from the large-group discussion and generate an "action plan" for Sheneka. Would you be comfortable carrying out all of these action items if you were Sheneka? Why or why not?

Contributed by J. Branchaw. (2018). *Awkward Mentor.*

CASE STUDY: AWKWARD MENTOR

Learning Objectives

Trainees will:

- Discuss challenges that may arise in mentoring relationships.
- Develop strategies to address mentoring relationship challenges.

Sheneka has overcome many challenges to successfully transfer from a local community college to a research university, where she is earning a B.S. degree in biochemistry. She has developed a strong support network and learned a lot about how to be successful as a student and in life. A transfer advisor at the research university, who is part of her network, helped her to prepare a successful application to a prestigious, federally funded research training program. Through this program, she is preparing for admission to graduate school.

As a trainee in the program, Sheneka was matched with a research mentor, who provides very useful help when she needs it, but otherwise leaves her alone to work on her research project. In addition, each student in the program is matched with a career mentor. Sheneka's career mentor is nice, but offers advice and mentorship in areas where she feels that she doesn't really need it. She already has a strong network of mentors who understand where she is coming from and how to support her.

The meetings with her career mentor have begun to feel awkward. The mentor constantly talks about her own experience, which has been very different from Sheneka's experience. Everyone in her career mentor's family has a college degree and her greatest challenge growing up was waiting to hear which top-tier colleges she was admitted to. Sheneka's mentor doesn't seem to understand the kinds of challenges that Sheneka has overcome and will continue to face. Sheneka doesn't want to risk offending her career mentor by pointing this out, but has come to dread the monthly meetings that are required by the program. They feel like a waste of time. What can she do?

Discussion Questions

1. How would you describe the communication between Sheneka and her career mentor?

2. To whom might Sheneka share her frustration and ask for advice about what to do?

3. How might Sheneka structure the meetings with her career mentor to better meet her needs?

Contributed by J. Branchaw. (2018). *Awkward Mentor.*

BARRIERS TO EFFECTIVE COMMUNICATION

Learning Objectives

Trainees will:

- ► Learn strategies for improving communication (in person, at a distance, across multiple mentors, and within proper personal boundaries).

Trainee Level

undergraduate or graduate trainees
novice or intermediate trainees

Areas of Trainee Development

- ► Research Comprehension and Communication Skills
  - Develop effective interpersonal communication skills.
- ► Equity and Inclusion Awareness and Skills
  - Develop skills to deal with personal differences in the research environment.

Activity Components and Estimated Time for Completion

- ► In Session Time: 35 minutes

Total time: 35 minutes

When to Use This Activity

This activity can be used once a trainee has established a relationship with a research mentor.

Inclusion Considerations

Discuss how different backgrounds can play a role in how communication is perceived and handled. Trainees may have different ideas about what good communication looks like and how to handle communication barriers depending on their background or their mentor's background. As a facilitator, use the differences in trainee responses as an opportunity to reflect upon assumptions that may have been made. Encourage trainees to seek to understand what others are saying and not to rely on stereotypes when interpreting situations.

Contributed by A. R. Butz and C. Pfund with information from Pfund, C., Brace, C., Branchaw, J., Handelsman, J., Masters, K., and Nanney, L. (2013). *Mentor Training for Biomedical Researchers.* New York: W.H. Freeman & Co.

Implementation Guide

Good communication is a key element of any relationship, and a mentoring relationship is no exception. It is not enough to say that we know good communication when we see it. Rather, it is critical that trainees reflect upon and identify characteristics of effective communication and take time to practice communication skills.

Workshop Session (25 minutes)

- **Trainee–Mentor Communication** (10 minutes)
  - Trainees answer the questions about communication with their own mentor, then pair with a peer to compare answers and strategies.
- **Large-Group Discussion: Barriers to Effective Communication** (15 minutes)

Share the barriers to good communication that trainees discussed in pairs and use the table provided in the trainee materials to record them. Choose two or three barriers and discuss practical ways to overcome them.

  - For example, one barrier might be a lack of time to meet one-on-one. Some solutions might be more frequent email, telecoms, or setting up a time to chat by instant message each week.
  - How will you determine if communication has improved?

Note: Alternative Activity

- Trainees create a list of all the forms of communication used by them and their mentor (face-to-face meetings, email, sticky notes, phone calls, etc.).
- Organize the resulting list by forms of communication and assign each type to a group of two or three trainees.
- Each subgroup identifies the pros and cons of each form and brainstorms ideas about how each can be improved.
- At the end, each small group reports to the entire group. Record all ideas on the whiteboard or flipchart. You may want to send a compiled list to the entire group.

Contributed by A. R. Butz and C. Pfund with information from Pfund, C., Brace, C., Branchaw, J., Handelsman, J., Masters, K., and Nanney, L. (2013). *Mentor Training for Biomedical Researchers.* New York: W.H. Freeman & Co.

BARRIERS TO EFFECTIVE COMMUNICATION

Learning Objectives

Trainees will:

- Learn strategies for improving communication (in person, at a distance, across multiple mentors, and within proper personal boundaries).

Answer these questions on your own, then pair with a peer and share.

1. In what ways and how often do you communicate with your mentor?

2. How do you know when communication with your mentor is good?

3. What are the signs that communication with your mentor is not good?

4. What strategies do you use or would you like to use to improve communication with your mentor when things are not going as well as you would like?

Contributed by A. R. Butz and C. Pfund with information from Pfund, C., Brace, C., Branchaw, J., Handelsman, J., Masters, K., and Nanney, L. (2013). *Mentor Training for Biomedical Researchers.* New York: W.H. Freeman & Co.

A few examples of barriers to effective communication between mentors and trainees are listed below. Write down a solution for each barrier.

Barrier to Effective Communication	**Solutions to Overcome Barrier**
Your mentor has very little time to meet with you one-on-one.	
Your mentor only provides verbal feedback on your work, and you often fail to catch everything that he/she suggests.	
You perceive the feedback you receive from your mentor as particularly critical.	
You are anxious about asking your mentor questions during meetings.	
English is not your mentor's first language and at times you are unsure that you understand him/her.	
Other:	
Other:	
Other:	

Contributed by A. R. Butz and C. Pfund with information from Pfund, C., Brace, C., Branchaw, J., Handelsman, J., Masters, K., and Nanney, L. (2013). *Mentor Training for Biomedical Researchers.* New York: W.H. Freeman & Co.

BIAS LITERACY: *FAIR PLAY* VIDEO GAME

Learning Objectives

Trainees will:

- Learn about the different types of bias individuals encounter in research environments and in everyday life.
- Develop awareness of the potential for bias to impact all individuals' judgment and behavior, even if it is unintentional.
- Identify specific strategies to counteract bias.

Trainee Level

undergraduate or graduate trainees
intermediate or advanced undergraduate trainees
novice, intermediate, or advanced graduate trainees

Areas of Trainee Development

- Equity and Inclusion Awareness and Skills
  - Advance equity and inclusion in the research environment.

Activity Components and Estimated Time for Completion

- Trainee Pre-Assignment Time: 10 minutes
- In Session Time: 2 hours

Total time: 2 hours, 10 minutes

Resources and Further Readings for Facilitators

- Kaatz, A., Carnes, M., Gutierrez, B., . . . Pribbenow, C. M. (2017). Fair Play: A study of scientific workforce trainers' experience playing an educational video game about racial bias. *CBE Life Sciences Education, 16*(2): ar27. doi:10.1187/cbe.15-06-0140
- *Fair Play Facilitator Resource Guide.* Available from www.workshops.fairplaygame.org (free with registration).

When to Use This Activity

Before implementing this activity, trainees should read the "Bias Literacy: An Introduction" and "Bias Concepts and Definitions," found in the Trainee Materials, to develop a basic, shared understanding of the concept of bias literacy. The group of trainees should be reasonably well established and trust one another for this activity to be effective. It is ideal to implement with an already established cohort of trainees or after a course has been meeting regularly for a few weeks and they have had time to co-construct ground rules for their discussions. Displaying the ground rules on a whiteboard or flipchart throughout this session is useful.

If it is not possible to implement the entire activity in one session, facilitators may break up the activity into two sessions, where the introduction and game are completed in session 1 or as homework, and the discussion and reflection exercise are completed in session 2 (30 minutes).

Contributed by A. Kaatz, C. Pribbenow, and M. Carnes. (2018). *Bias Literacy: Fair Play Video Game.*

Inclusion Considerations

Everyone has unconscious bias. Individuals from backgrounds historically underrepresented in STEM are particularly vulnerable to bias. These biases can appear in research environments, where they may create barriers to trainees' efforts to engage in training activities or to build effective professional relationships with mentors and colleagues. While implementing this activity, facilitators should be cautious in asking or expecting individuals from historically underrepresented groups to speak for their respective identity group. To mitigate this, encourage all trainees to share in discussions about identifying and mitigating bias and stereotypes.

Contributed by A. Kaatz, C. Pribbenow, and M. Carnes. (2018). *Bias Literacy: Fair Play Video Game.*

Implementation Guide

Trainee Pre-Assignment (10 minutes)

- ► Reading: Bias Literacy: An Introduction and Bias Concepts and Definitions (see trainee materials)

Facilitator Background and Preparation

- ► Recommended technology preparations:
  - If the game will be played during the session, facilitators should reserve a computer lab or ask trainees to bring a computer.
  - Headphones or earbuds are recommended, since this game has audio.
  - For additional information on technology recommendations, read the *Fair Play Facilitator Resource Guide* included in the facilitator resources listed above.

Workshop Session (2 hours, 5 minutes)

- ► **Introduction: Review of Pre-Assignment: Bias Concepts and Definitions** [5 minutes]
  - Invite trainees to ask any questions about the reading or bias concepts that they reviewed prior to this session. Remind them that they will be revisiting these terms later in the session.
- ► **Fair Play** (1 hour–1 hour, 15 minutes)
  - Each trainee should **download the game:** https://fairplaygame.org/
  - The game has five chapters and takes approximately 60 to 75 minutes to complete.
  - *Note:* Facilitators can assign this game as an assignment if there is not enough time to complete the game as part of an in-class activity.
  - *Fair Play* is an interactive, role-playing, educational video game targeting science students and faculty to address racial bias in academic science (Kaatz et al., 2017). In *Fair Play*, trainees assume the role of Jamal Davis, a Black student working toward a graduate degree in science at a large research university. As Jamal, players experience and learn about implicit bias as they interact with other game characters and complete tasks relevant to a science graduate student, such as selecting an advisor, writing research articles, and attending professional conferences.
  - Have trainees launch the game and begin to navigate in the game. Left-click on a location to have Jamal walk there.
  - Jamal will have the opportunity to speak with people in the game. Clicking on a character will cause Jamal to walk to that character and begin a conversation.
  - Red exclamation points indicate bias in the environment. Trainees will have the option to identify each bias or ignore the bias and continue playing the game. Trainees will be able to keep track of the biases that they have successfully identified, but should also make a note of the biases encountered and any questions that arise for them during the game.
- ► **Discussion** (20 minutes)
  - Ask trainees to reflect on their experience with the game using the reflection questions in the trainee materials as a guide:
    - What are your initial reactions to your experience with this game?
    - What types of bias did you encounter as Jamal? (*Note:* Facilitators can use the table below as a guide to identifying the different types of bias presented in the game.)

TYPES OF BIAS IN THE GAME

Critical Bias: Biases on the "critical path," which means a player will always encounter them on the first play through of the game; all other biases are optional and may not be experienced by every player. Critical biases are typically conversational biases, which occur through conversations with other characters in the game.

Environmental Bias: Biases present in the environment through interacting with objects in the world or observing ambient conversations between nonplayer characters.

Conversational Bias: Biases that occur through conversations with other characters in the game. Not every conversation in the game contains a bias nor is every conversational bias also a critical bias.

Bias Construct	In-Game Examples
Attributional Rationalization	Chapter 3, Environmental Bias: Bias Encounter: You overheard people talking about Deirdre's tenure case, dismissing her contributions to an article. Bias Hint: People are talking about Deirdre's tenure case. Chapter 5, Critical Bias: Bias Encounter: Lucas unfairly presumed that your high-quality proposal would be attributed to him rather than you, perhaps because of group stereotypes about ability. Bias Hint: A graduate student talks about your proposal.
Color-Blind Racial Attitudes	Chapter 2, Critical Bias: Bias Encounter: Nick mentioned that paying attention to race is unimportant, and that it has nothing to do with graduate studies. Bias Hint: A faculty member talks about departmental fit. Chapter 2, Environmental Bias: Bias Encounter: You overheard two staff members outside Nick's office talk about how they don't see race any longer; they only see people. Bias Hint: Staff members are chatting outside Nick's office. Chapter 5, Critical Bias: Bias Encounter: Lucas mentioned that students of color are unfairly advantaged by diversity programs that target underrepresented minorities because he assumed discrimination no longer exists. Bias Hint: A graduate student believes students of color are unfairly advantaged by diversity programs.
Competency Proving	Chapter 1, Critical Bias: Bias Encounter: Morgan suggested that instead of assuming you are here because of your academic accomplishments, people may assume you are only attending the university to play basketball. Bias Hint: A researcher discusses assumptions about your admission to the university. Chapter 4, Critical Bias: Bias Encounter: Nick was surprised by the quality of your excellent paper, which he had asked you to write to prove your academic ability. Bias Hint: A professor is surprised by your academic writing ability.

(*Continued*)

Contributed by A. Kaatz, C. Pribbenow, and M. Carnes. (2018). *Bias Literacy: Fair Play Video Game.*

Failure to Differentiate	Chapter 1, Conversational Bias: Bias Encounter: Shania confused you for another Black graduate student she had met at a research lab. Bias Hint: A graduate student is hiding on the balcony to get work done. Chapter 2, Critical Bias: Bias Encounter: Shania confused you for another Black graduate student she works with in the office. Bias Hint: A graduate student is waiting for Tyrone.
Impression Management	Chapter 5, Environmental Bias: Bias Encounter: You dressed up for your presentation, and you were glad that your formal clothing made people recognize you as a conference participant instead of as support staff. Bias Hint: Check to make sure you are well-dressed for the conference.
Racial Microaggression	Chapter 2, Environmental Bias: Bias Encounter: Portraits of the past faculty in the department showed that they were all older White men. This is an example of a microinvalidation. Bias Hint: Portraits hanging in the hallway show the past faculty in the department. Chapter 3, Critical Bias: Bias Encounter: Franklin singled you out and asked you, but not others, for ID before you may enter the library. This is an example of a microinsult. Bias Hint: A library staff member asks to see your ID. Chapter 3, Critical Bias: Bias Encounter: Engaging in racial profiling, Franklin followed you around the stacks. This is an example of a microassault. Bias Hint: A library staff member keeps a close eye on patrons' activities. Chapter 3, Environmental Bias: Bias Encounter: The book collection on Black history was separated from American history in the stacks. This is an example of a microinvalidation. Bias Hint: There are perplexing disciplinary divisions among the library's collections.
Shifting Standards of Judgment	Chapter 1, Critical Bias: Bias Encounter: Lucas suggested that you apply for fellowships targeted for minorities, because he presumed that you have lower competence compared to all applicants but would be considered highly competent when compared to other minority applicants. Bias Hint: A graduate student gives you advice for funding. Chapter 5, Critical Bias: Bias Encounter: Lucas presumed there are higher expectations for his performance, ignorant of how evaluators actually require greater proof of minority group members' ability prior to confirming their competence. Bias Hint: A graduate student presumes there are higher expectations for his performance.
Status Leveling	Chapter 1, Critical Bias: Bias Encounter: Lucas assumed that you were a caterer for the incoming students' event and asked you to get more soda. Bias Hint: A graduate student notices that the buffet is low on soda.

Contributed by A. Kaatz, C. Pribbenow, and M. Carnes. (2018). *Bias Literacy: Fair Play Video Game.*

Stereotype Threat	Chapter 5, Conversational Bias: Bias Encounter: When Crystal found out she was the only female presenter on a panel, she became nervous, and it compromised her performance. Bias Hint: A fellow student shares her experience of being a minority at a previous conference.
Tokenism	Chapter 1, Critical Bias Bias Encounter: Morgan wondered if the lack of minorities makes the campus less appealing to other Black students and asked for your opinion. Bias Hint: A researcher is concerned with the lack of minorities on campus. Chapter 2, Critical Bias: Bias Encounter: Morgan asked you to speak to his class about the Black graduate student experience. Bias Hint: A researcher asks you to speak to his class. Chapter 4, Critical Bias: Bias Encounter: Nick suggested that you may have been admitted to the university through a diversity program as a token minority. Bias Hint: A professor discusses your admission to the university. Chapter 5, Critical Bias: Bias Encounter: Lucas implied that indicating a fake minority status would give him an advantage in an application process. Bias Hint: A graduate student discusses the application process.

► **Reflection Exercise** (10 minutes)

- After game play have trainees review the bias reducing strategies provided in the trainee materials and reflect on how each strategy can be implemented in their own life.

► **Wrap-up** (10 minutes)

- Summarize the main ideas generated from the large-group discussion.
- Remind trainees that the reflections they provided in the previous exercise can serve as their commitment to action in the research environment. Ask trainees to consider the one thing that they can do to address bias in their environment and interactions with others. Invite them to share their ideas out loud or to write them down on a sticky note or index card so that suggestions can be compiled and shared with the class anonymously.

Portions of this facilitator guide were reproduced and adapted with permission from the Fair Play Facilitator Resource Guide. https://workshops.fairplaygame.org

Contributed by A. Kaatz, C. Pribbenow, and M. Carnes. (2018). *Bias Literacy: Fair Play Video Game.*

BIAS LITERACY: *FAIR PLAY* VIDEO GAME

Learning Objectives

Trainees will:

- Learn about the different types of bias individuals encounter in research environments and in everyday life.
- Develop awareness of the potential for bias to impact all individuals' judgment and behavior, even if it is unintentional.
- Identify specific strategies to counteract bias.

Activity: Video Game

Fair Play is an interactive, role-playing, educational video game to address racial bias in academic science (Kaatz et al., 2017). In *Fair Play*, you will assume the role of Jamal Davis, a Black student working toward a graduate degree in science at a large research university. As Jamal, you will experience and learn about implicit bias as you interact with other game characters and complete tasks relevant to a science graduate student, such as selecting an advisor, writing research articles, and attending professional conferences.

Download the game here: https://fairplaygame.org/

The game has five chapters and takes approximately 60 to 75 minutes to complete.

Contributed by A. Kaatz, C. Pribbenow, and M. Carnes. (2018). *Bias Literacy: Fair Play Video Game.*

POST-GAME REFLECTION: BIAS-REDUCING STRATEGIES

Read through the bias-reducing strategies below and write your response to each reflection in the space provided below.

Perspective-Taking: Adopt the perspective (in the first person) of a member of a stigmatized group (Galinsky and Moskowitz, 2000; Todd et al., 2011; Gutierrez et al., 2014; Carnes et al., 2012, 2015; Kaatz et al., 2017).

How does it feel to be Jamal? Or how would it feel to be treated like you were not competent in science just because of your race or gender? Think about this in detail and write it out.

Stereotype Replacement: Recognize when you have stereotypic thoughts. Recognize stereotypic portrayals in society. Label the characterization as stereotypical. Identify precipitating factors. Challenge the fairness of the portrayal and replace it with a nonstereotypic response (Devine et al., 2012; Carnes et al., 2012, 2015).

Write about a time when stereotypes may have impacted your appraisal of another person and explain why assumptions you made were unfair. What would happen if you challenged your assumptions?

Counter-Stereotypic Imaging: Help regulate your response by imaging a counter-stereotypic woman or racial/ethnic minority in detail (Blair et al., 2001; Carnes et al., 2012, 2015).

Building off your previous answer, describe racial/ethnic minorities and/or women who do not fit stereotypic portrayals.

Contributed by A. Kaatz, C. Pribbenow, and M. Carnes. (2018). *Bias Literacy: Fair Play Video Game.*

Individuating (vs. Generalizing): Avoid making a snap decision based on a stereotype. Obtain more information on specific qualifications, past experiences, etc., before making a decision. Practice making situational attributions rather than dispositional attributions (Heilman, 1984, 2001; Carnes et al., 2012, 2015).

Write about how, upon meeting or interviewing people, you can purposely focus on getting to know them as a way to prevent the influence of stereotypic assumptions about their ability.

Increase Opportunities for Contact: Seek out opportunities for greater interaction with counter-stereotypic women and racial/ethnic minorities (Alloport, 1979; Carnes et al., 2012, 2015).

Building off of your responses above, write out some ways that you can improve your contact with members of groups outside of your own.

Commitment to action: Given what you have learned today, what is one thing you can do to address bias in your environment or in your interactions with others? Write down something that feels easy for you to commit to on a regular basis.

Contributed by A. Kaatz, C. Pribbenow, and M. Carnes. (2018). *Bias Literacy: Fair Play Video Game.*

BIAS LITERACY: AN INTRODUCTION

The National Academies of Sciences (NAS) investigated barriers to diversity in STEM fields, and concluded that women and racial/ethnic minorities are similarly talented, and committed to careers in STEM; however, bias that arises from stereotypes depicting women and racial/ethnic minorities as deficient in the skills and traits associated with competence in STEM operates in personal interactions, evaluation processes, and institutional cultures to subtly, yet systematically disadvantage them as they work to succeed. Stereotype-based bias is particularly problematic because people are usually unaware that it can impact their judgment, decision making, and behavior. For that reason, it is commonly called implicit or unconscious bias. Studies show that such bias can impact a person's decision making and behavior regardless of their gender, race/ethnicity, and extent to which they consciously hold egalitarian beliefs. It can also impact self-perceptions about ability. Common forms of implicit gender and racial bias are reviewed on the next page.

Although fully addressing stereotype-based bias requires interventions at multiple levels, individuals can learn how to recognize and reduce the impact of such bias on their decision making and behavior. One of the most effective strategies involves becoming "bias literate," that is, learning a vocabulary about bias; a skill set to recognize, understand, and converse about bias; and intentionally practicing cognitive and behavioral strategies to mitigate the impact of group stereotypes on judgment and decision making (Sevo and Chubin, 2008). Bias literacy has been used to successfully reduce implicit stereotype-based bias in students (Devine et al., 2012) and to reduce biased behavior and improve the department's climate for faculty (Carnes et al., 2012, 2015) in academic science fields.

References

Carnes, M., Devine, P. G., Baier Manwell, L., Byars-Winston, A., Fine, E., Ford, C. E., Forscher, P., Issac, C., Kaatz, A., Magua, W., et al. (2015). The effect of an intervention to break the gender bias habit for faculty at one institution: a cluster randomized, controlled trial. *Acad Med.*, *90*: 221–230.

Devine, P. G., Forscher, P. S., Austin, A. J., and Cox, W. T. L. (2012). Long-term reduction in implicit race prejudice: A prejudice habit-breaking intervention. *J Exp Soc Psychol.*, *48*: 1267–1278.

Carnes, M., Devine, P. G., Isaac, C., et al. (2012). Promoting institutional change through bias literacy. *J Divers High Educ.*, *5*: 63–77.

Sevo, R., and Chubin, D. E. (2008). *Bias Literacy: A Review of Concepts in Research on Discrimination.* Washington, DC: American Association for the Advancement of Science Center for Advancing Science & Engineering Capacity.

Contributed by A. Kaatz, C. Pribbenow, and M. Carnes. (2018). *Bias Literacy: Fair Play Video Game.*

BIAS CONCEPTS AND DEFINITIONS

Attributional Rationalization: Group stereotypes may lead to assumptions that people from underrepresented groups are less competent than their majority peers. As a result, they may not receive credit for their accomplishments, which are often incorrectly attributed to those in the majority or to factors other than their efforts (e.g., luck).

Color-Blind Racial Attitudes: Color-blind racial attitudes reflect the belief that discrimination no longer exists. Though based on the positive premise that we should all be treated equally, a color-blind approach discounts the experiences of members of minority groups and can backfire to promote bias.

Competency Proving: To counter common assumptions about their presumed incompetence, members of minority groups have to frequently and repeatedly demonstrate that they are indeed qualified, capable, and/or competent.

Failure to Differentiate: Members of a particular minority group may sometimes be mistaken for one another by a person of a different group. All groups share this unintentional recognition bias, but research suggests the effect is most pronounced for White individuals when viewing racial and ethnic minorities.

Impression Management: People from historically low status or underrepresented groups must often pay more conscious attention to how they behave (e.g., a Black student may consciously modulate his/her tone of voice or volume of speech to prevent activating the racial stereotype of being angry or aggressive) or how they dress in order to reinforce their professional role. A casual appearance may elicit prevailing negative images of their group.

Racial Microaggression: Microaggressions are brief and subtle comments, behaviors, or environmental cues that intentionally or unintentionally communicate hostile, derogatory, or unwelcoming messages toward members of underrepresented groups. When accumulated, these seemingly minor messages lead to harmful isolation and alienation. There are three types of microaggressions: microassaults, microinsults, and microinvalidations.

Shifting Standards of Judgment: The presumed incompetence of members of underrepresented groups causes well-qualified, underrepresented individuals to be judged as highly competent when compared to members of their group. However, they are held to even higher standards and require greater proof of competence than comparable members of the majority group.

Status Leveling: Based on stereotypes about the lower social standing of minority groups, status leveling occurs when a person from an underrepresented group is assumed to belong to a lower social category or position.

Stereotype Threat: Stereotype threat occurs when awareness of negative stereotypes about one's own group induces stress and anxiety about confirming the stereotype. Situations that consciously or unconsciously trigger a stereotype threat can lead members of minority groups to underperform relative to their actual ability.

Tokenism: Tokenism is treating members of minority groups as representative of their entire group rather than as individuals, especially when they are a numeric minority or the only person from that group present (solo status).

Contributed by A. Kaatz, C. Pribbenow, and M. Carnes. (2018). *Bias Literacy : Fair Play Video Game.*

CHALLENGES FACING DIVERSE TEAMS

Learning Objectives

Trainees will:

- ► Learn about three common challenges that individuals working in diverse teams may face.
- ► Identify practical strategies for recognizing and mitigating challenges to create more inclusive research groups.

Trainee Level

undergraduate or graduate trainees, novice trainees

Activity Components and Estimated Time for Completion

- ► In Session Time: 1 hour

Total time: 1 hour

When to Use This Activity

This activity is ideal for novice researchers early in their training (e.g., undergraduate research program or graduate school orientation) and works best coupled with a discussion about why diversity is important for STEM.

Inclusion Considerations

In discussions, be cautious about asking or expecting individuals from groups historically underrepresented in research fields to speak for their respective identity group. Instead, encourage all trainees to share in identifying challenges and barriers to creating more inclusive teams and diversifying STEM careers.

Areas of Trainee Development

- ► Equity and Inclusion Awareness and Skills
  - Develop skills to deal with personal differences in the research environment.
  - Advance equity and inclusion in the research environment.

Contributed by A. O'Connell, and J. Harrell. (2018). *Challenges Facing Diverse Teams.*

Implementation Guide

Workshop Session (1 hour)

► **Introduction** (2 minutes)
- There is a lot of evidence that diverse teams produce better results.
- It is the responsibility of every member of a research team, not just the team leader, to build a welcoming environment for individuals from all backgrounds. To be successful, trainees must learn to appreciate and value the ideas, experiences, and perspectives of everyone on the team. These skills will help them to build and lead successful teams as they advance in their careers.

► **Brainstorm: Challenges Facing Diverse Teams** (10 minutes)
Diverse research teams can face challenges that make it difficult to incorporate multiple perspectives. For example, someone who observes different religious holidays than everyone else on the team could miss important meetings or opportunities if they are out of work regularly while everyone else is in the office. This person could also be viewed as a burden if others must cover for them while they are out.
- **What are other challenges that diversity might pose in the research environment?** Examples trainees might come up with could include:

> - isolation if a research group member is not a native English language speaker or if a research group member is the only one of their race/ethnicity in the research group.
> - awkward research group celebrations if some research group members like to drink alcohol and others don't.
> - challenges for research group members who are parents with time constraints versus others who aren't parents.
> - research group members who have difficulty making it to the field station to collect data due to transportation issues (e.g., no or limited access to reliable transportation).

► **Activity: Challenges Facing Diverse Teams** (30 minutes)
- (10 minutes) **Small-Group Discussions:** Distribute the "Challenges Facing Diverse Teams" handout included in the trainee materials. Divide trainees into groups of three or four. Assign each group a number (1, 2, or 3) corresponding to a section on the handout. Distribute the corresponding reading for each topic (see "Trainee Materials" below for brief readings on "Imposter Syndrome, Stereotype Threat, and Implicit Bias"). Instruct groups to assign a scribe to take notes, and a reporter who will speak for the group at the end. Each group should:
  - review the reading on their topic;
  - prepare to briefly define their topic and provide examples of how their topic impacts research groups; and
  - generate strategies for how to address the challenge presented in their examples.
- (20 minutes) **Large-Group Discussion:** Groups explain their topic and their example.
  - What is (Imposter Syndrome? Stereotype Threat? Implicit Bias?)
    - Has anyone ever experienced this or known someone who may have experienced it?
  - What are the strategies that you came up with to address each challenge?
  - If members of a team don't feel a sense of true belonging and equal status, what impact might this have on their contributions to the research team?

Contributed by A. O'Connell, and J. Harrell. (2018). *Challenges Facing Diverse Teams.*

Additional Discussion Question (if time permits):

- What are other examples of how this challenge might present itself in the research environment?

Note: During these discussions, it may be helpful to specifically highlight or list biases and stereotypes that exist in science/academia/research. Examples might include:

- ► *Women are less capable in math/engineering (the Google memo!).*
- ► *Minority students were admitted to address affirmative action efforts, but aren't as qualified.*
- ► *Students who are parents cannot put in enough time to be successful in research*
- ► *Asian students are extremely smart and driven ('the model minority').*
- ► *Scientists/academics are liberal.*

► **Activity: Practical Strategies for Building Inclusive Teams** (15 minutes)

- Invite trainees to brainstorm some additional practical strategies to address the three types of challenges. Generate a list. Examples are provided in the boxes below.

Reducing Implicit Bias:

- Educate yourself and decision makers on bias. Raising awareness can reduce reliance on stereotypes.
- Exposure to individuals from different backgrounds helps reduce reliance on stereotypes.

Reducing Stereotype Threat:

- Practice, build confidence in ability to succeed.
- Cultivate a growth mindset (e.g., I can improve my skills if I keep working on them).
- When receiving feedback, remember that it is being given in order to help you develop—not because you have a deficiency.
- If you are in a position to give feedback, focus on the standards for success and the individual's ability to reach the goal.

Reducing Imposter Syndrome:

- Learn to internalize external validation.
- Talk about your fears—to a friend, partner, therapist, academic advisor, or someone in your department.
- Develop a more realistic view of yourself (your abilities and your habits) and others (No one is perfect!).
- Write down your successes; keep a record.

► **Wrap-up** (3 minutes)

- Summarize the key points raised in the discussion.
- Encourage trainees to continue to consider the ideas shared during the discussion and to try a few of the strategies. Invite them to learn more using the additional readings and resources.

Contributed by A. O'Connell, and J. Harrell. (2018). *Challenges Facing Diverse Teams.*

CHALLENGES FACING DIVERSE TEAMS

Learning Objectives

Trainees will:

- ▶ Learn about three common challenges facing individuals working in diverse teams may face.
- ▶ Identify practical strategies for recognizing and mitigating challenges to create more inclusive research groups.

Instructions: Use the provided readings on Imposter Syndrome, Stereotype Threat, and Implicit Bias to write a definition of each of the types of challenges listed below. Work in groups to review your assigned research environment example, offer another example based on the experience of group members or those known to group members, and then brainstorm a list of strategies one could use to mitigate each of the challenges.

1. IMPOSTER SYNDROME

RESEARCH ENVIRONMENT EXAMPLE: Carl is a third-year graduate student in the research group of a world-renowned chemistry professor. The recipient of a prestigious graduate research fellowship, Carl has published several articles with his famous mentor, presented at national conferences, and is already making a name for himself in the field. Yet, Carl still faces his own self-doubts about whether he is good enough to pursue a career in research. Sometimes, Carl finds himself wondering if it was a fluke that he was admitted to his graduate program.

YOUR RESEARCH ENVIRONMENT EXAMPLE:

STRATEGIES TO ADDRESS IMPOSTER SYNDROME:

2. STEREOTYPE THREAT

RESEARCH ENVIRONMENT EXAMPLE: Elena is a first-year student at a large research university who wants to major in Biology and go on to study medicine. On her first day of classes, she takes a seat among 300 other students in her introductory Biology class and waits for Professor Jones to begin her lecture and go over the course. As Elena scans the room, she sees that there are few other Latinx students in the class. Professor Jones emphasizes to students that this course is designed to weed out those students who "won't cut it" in science. The first test is worth 30% of her final grade, and Professor Jones encouraged all students who don't make at least a B to drop the class because if they don't understand the material by that point in the course, they will never get it.

YOUR RESEARCH ENVIRONMENT EXAMPLE:

STRATEGIES TO ADDRESS STEREOTYPE THREAT:

Contributed by A. O'Connell, and J. Harrell. (2018). *Challenges Facing Diverse Teams.*

3. IMPLICIT BIAS

RESEARCH ENVIRONMENT EXAMPLE: John and Naomi are both recent Ph.D. graduates applying for laboratory manager positions. Both have impressive research and publication records. They are invited to interview on campus at a research-intensive university. When John, a man who identifies as White and is originally from the region, shows up to the front desk, he is greeted with a smile and asked to take a seat. Naomi, a woman who identifies as Chicana and African American, enters the office next. Before she can introduce herself, a faculty member passing through the office makes eye contact and asks with a smile "are you lost?"

YOUR RESEARCH ENVIRONMENT EXAMPLE:

STRATEGIES TO ADDRESS IMPLICIT BIAS:

Contributed by A. O'Connell, and J. Harrell. (2018). *Challenges Facing Diverse Teams.*

Imposter Syndrome

Imposter syndrome is a persistent feeling of self-doubt and non-belonging, despite evidence to the contrary. The phenomenon was first described in the context of high-achieving women (Clance and Imes, 1978). Since then, it's become clear that a person of any gender can experience imposter syndrome. Individuals who experience imposter feelings often harbor fears that they were 'accidentally' chosen for their current position, or that they are advancing only due to luck and soon will be 'found out' and 'exposed as a fraud'. These feelings are common among high achieving individuals and have been reported by professionals in many settings including academia, entertainment, business, and others.

Citation: Clance, Pauline Rose, and Ament Imes, Suzanne. (1978). The imposter phenomenon in high achieving women: Dynamics and therapeutic intervention. *Psychotherapy: Theory, Research & Practice*, *15*(10): 241–247. 10.

Stereotype Threat

Anna Kaatz, Ph.D.

Stereotype threat is the psychological experience of anxiety about performing in a way that reinforces a negative stereotype about your group (e.g., girls are bad at math). It was first discovered by Dr. Claude Steele while he was at the University of Michigan. He noticed a troubling trend where White students would start to outperform Black students by their sophomore year, despite having similar ability and credentials when they started college. He attempted to re-create this performance gap in his lab, and after several failed attempts he identified the causal mechanism (Steele & Aronson, 1995). He gave two groups of similarly qualified White and Black students a portion of the verbal GRE. Prior to taking the test, he told the first group that it was a test of their ability; he told the second group that it was a problem-solving task (Steele & Aronson, 1995). Black students performed significantly worse than White students when they were told the exam was a test of ability. By comparison, Black and White students performed similarly when told the exam was a problem-solving task. These results remained even after controlling for students' prior standardized test scores.

Hundreds of studies and decades later, we now know that stereotype threat can be triggered by subtle cues that make membership of a negatively stereotyped-group salient, and that it undermines performance by causing anxiety. In Dr. Steele's study, simply saying the exam was a test of ability made salient the stereotype that Blacks have low academic competence. This led Black students to experience test anxiety, which took away some of their working memory, led them to underperform, and subsequently reinforced a negative stereotype about ability for their social group.

In academia, there are strong negative stereotypes that women, racial/ethnic minorities, and first-generation college students lack the intrinsic ability to succeed in science, technology, engineering, and mathematics (STEM) fields, which makes members of these group highly vulnerable to stereotype threat. Stereotype threat can be activated by essentially any means that makes stereotype-group membership salient. This includes emphasizing that a test is indicative of ability (Steele & Aronson, 1995), stating that performance disparities between certain groups (e.g., men vs. women, Whites vs. Blacks; Spencer, Steele, & Quinn, 1999), providing demographic information before a test (e.g., gender, race/ethnicity; Steele & Aronson, 1995),

Contributed by A. O'Connell, and J. Harrell. (2018). *Challenges Facing Diverse Teams.*

being the only member of your social group in a classroom (e.g., only woman, only Black student) in a field where there are stereotypes about ability (Sekaquaptewa, Waldman, & Thompson, 2007), or any other interaction or experience that makes a negatively stereotyped identity salient (e.g., pictures of prior faculty that only include White men).

References

Sekaquaptewa, D., Waldman, A., and Thompson, M. (2007). Solo status and self-construal: Being distinctive influences racial self-construal and performance apprehension in African American women. *Cultural Diversity and Ethnic Minority Psychology, 13*: 321–327.

Spencer, S. J., Steele, C. M., and Quinn, D. M. (1999). Stereotype threat and women's math performance. *Journal of Experimental Social Psychology, 35*: 4–28.

Steele, C. M., and Aronson, J. (1995). Stereotype threat and the intellectual test performance of African-Americans. *Journal of Personality and Social Psychology, 69*: 797–811.

Bias Literacy: An Introduction

Anna Kaatz, Ph.D.

The National Academies of Sciences (NAS) investigated barriers to diversity in STEM fields, and concluded that women and racial/ethnic minorities are similarly talented, and committed to careers in STEM; however, bias that arises from stereotypes depicting women and racial/ethnic minorities as deficient in the skills and traits associated with competence in STEM operates in personal interactions, evaluation processes, and institutional cultures to subtly, yet systematically disadvantage them as they work to succeed. Stereotype-based bias is particularly problematic because people are usually unaware that it can impact their judgment, decision making, and behavior. For that reason, it is commonly called implicit or unconscious bias. Studies show that such bias can impact peoples' decision making and behavior regardless of their gender, race/ethnicity, and extent to which they consciously hold egalitarian beliefs. It can also impact self-perceptions about ability. Common forms of implicit gender and racial bias are reviewed on the next page.

Although fully addressing stereotype-based bias requires interventions at multiple levels, individuals can learn how to recognize and reduce the impact of such bias on their decision making and behavior. One of the most effective strategies involves becoming "bias literate"—i.e., learning a vocabulary about bias; a skill set to recognize, understand, and converse about bias; and intentionally practicing cognitive and behavioral strategies to mitigate the impact of group stereotypes on judgment and decision making (Sevo & Chubin, 2008). Bias literacy has been used to successfully reduce implicit stereotype-based bias in students (Devine et al., 2012) and to reduce biased behavior and improve department climate for faculty (Carnes et al., 2012, 2015) in academic science fields.

Contributed by A. O'Connell, and J. Harrell. (2018). *Challenges Facing Diverse Teams.*

References

Carnes, M., Devine, P. G., Baier Manwell, L., Byars-Winston, A., Fine, E., Ford, C. E., Forscher, P., Issac, C., Kaatz, A., Magua, W., et al. (2015). The effect of an intervention to break the gender bias habit for faculty at one institution: a cluster randomized, controlled trial. *Acad Med.*, *90*: 221–230.

Devine, P. G., Forscher, P. S., Austin, A. J., and Cox, W. T. L. (2012). Long-term reduction in implicit race prejudice: A prejudice habit-breaking intervention. *J Exp Soc Psychol.*, *48*: 1267–1278.

Carnes, M., Devine, P. G., Isaac, C., et al. (2012). Promoting institutional change through bias literacy. *J Divers High Educ.*, *5*: 63–77.

Kaatz, A., Carnes, M., Gutierrez, B., et al. (2017). Fair play: A study of scientific workforce trainers' experience playing an educational video game about racial bias. Kenneth, G., ed. *CBE Life Sciences Education*, *16*(2): ar27. doi:10.1187/cbe.15-06-0140

Sevo, R., and Chubin, D. E. (2008). Bias literacy: A review of concepts in research on discrimination. Washington, DC: American Association for the Advancement of Science Center for Advancing Science & Engineering Capacity.

Contributed by A. O'Connell, and J. Harrell. (2018). *Challenges Facing Diverse Teams.*

COMMUNICATING RESEARCH FINDINGS 1: POSTER PRESENTATIONS

Learning Objectives

Trainees will:

- Identify the characteristics of effective research posters.
- Learn how to give a poster presentation of research findings.
- Recognize that research presentations are opportunities for discussing and networking with colleagues about research, not only for reporting research.

Trainee Level

undergraduate and graduate trainees
novice or intermediate trainees

Areas of Trainee Development

- Research Comprehension and Communication Skills
  - Develop research communication skills.
  - Develop disciplinary knowledge.
  - Develop logical/critical thinking skills.

Activity Components and Estimated Time for Completion

- Trainee Pre-Assignment Time: 30 minutes
- In Session Time: 1 hour

Total time: 1 hour, 30 minutes

When to Use This Activity

This activity should be implemented before or concurrently with the "Communicating Research Findings 2: Oral Presentations" activity. It can be used when trainees have been working in their research groups for several months and are preparing to present their results or thinking about how scientists communicate progress on their research.

Inclusion Considerations

Novice trainees, especially those who have traveled nontraditional academic pathways, may have had limited opportunity to give scientific or research presentations. Though all trainees can be nervous about giving presentations, these trainees may be especially nervous and would benefit from extra practice sessions and encouragement. Consider using a "compliment sandwich" approach: Start with what was done well, insert what can be improved, and close with a positive comment. Reassure trainees that their presentation skills will improve with practice.

Some trainees may come from backgrounds where humility is valued, and they have been acculturated not to speak of their achievements freely. Talk about why it is important to speak about one's research positively and with confidence and reassure trainees that they can do this while remaining true to their ideals around humility.

Contributed by A. Bramson and K. Eskine with information from Branchaw, J. L., Pfund, C., and Rediske, R. (2010). *Entering Research: A Facilitator's Manual.* New York: W.H. Freeman & Co.

Implementation Guide

Trainee Pre-Assignment (30 minutes)

- Trainees should complete the Scientific Poster Hunt and Poster Presentation Video assignments (included in trainee materials)

Workshop Session (1 hour)

- **Introduction** (5 minutes)
  - Ask trainees to share if they have ever given a research presentation, either a poster or oral presentation. If so, ask them to describe the experience in one word.

> - Stressful
> - Terrifying
> - Compact—you're taking something you've been working on for months or years and condensing it into a single poster or a short talk.

  - Ask trainees: What is/are the goal(s) of a research presentation?

> - To share the status of a project with colleagues.
> - To meet the scientists in your field and put faces to names from papers you've read.
> - To get comments from others in your field about how to further your research.

- **Elements of Effective Posters: Results of the Scientific Poster Hunt** (15–20 minutes)
  - Put trainees into pairs or small groups to share their notes about the posters they viewed in preparation for class. Have each group discuss the questions below and contribute to a shared list of positive and negative poster features that can be generated on a board, a large piece of paper, or a shared online document that is projected.
    - *What characteristics of posters make them most effective?*

> - Eye catching
> - Clear, concise title
> - Lots of figures and/or images, not too wordy
> - Large font size
> - Clear statement of hypothesis or research question
> - Diagrams/images used to explain experimental techniques
> - Sections are easy to follow

Contributed by A. Bramson and K. Eskine with information from Branchaw, J. L., Pfund, C., and Rediske, R. (2010). *Entering Research: A Facilitator's Manual.* New York: W.H. Freeman & Co.

- *What were the characteristics that made you list a poster as your least favorite?*

  - Plain
  - Lengthy, confusing title
  - Few images and/or figures, lots of text
  - Very small font size
  - Hard to know at a glance what the poster is about
  - No diagrams/images used to explain experimental techniques
  - Hard to follow the flow of the poster

- Bring the students together as a large group to review the lists of positive and negative features. The facilitator can capture this list and distribute it to the trainees as a guide to use when making posters.

► **Poster Presentation Tips** (15 minutes)

- Ask trainees to view the short video "Giving an Effective Poster Presentation" and complete the worksheet before class. The video provides content on what makes a good poster presentation and highlights some of the common mistakes.
- **Discussion Questions** (15 minutes)
  - What are the six tips identified in the video?
    - 1. Don't read your poster.
    - 2. Be prepared.
    - 3. Handouts are good—two sided also might be even better.
    - 4. Make viewers responsible—give them your information (on a card or on the handouts).
    - 5. Admit when you don't know.
    - 6. Put the viewer first—could be a judge or possible collaboration, this is not the time to chit chat with friends.
  - Do you agree with all of the tips? Why or why not?
  - What kinds of things would you bring as supplemental materials to a poster session?
- Ask trainees, if given the choice, would they prefer to present a poster during a 3-hour poster session or give a 15-minute oral presentation. Which would they choose and why? What factors influenced their answers?

  - I would choose a poster if my research was still in the beginning stages because it would be more useful to have in-depth conversations with other scientists about my project.
  - I would choose a presentation because I'm applying for graduate schools, so it would be a good advertisement to get my name out there.

- Optional (5–10 minutes): Facilitators may identify a few examples from their own work and/or by asking colleagues a few days before class for examples of a poster and oral presentation presented on the same topic. Using printouts of posters and corresponding slides from oral presentations on the same project can help trainees compare formats. Trainees can spend a few minutes looking through the examples while thinking about their experiences either giving or being in the audience for a talk, and how this compares/contrasts with the posters they found in the Poster Hunt assignment.

Contributed by A. Bramson and K. Eskine with information from Branchaw, J. L., Pfund, C., and Rediske, R. (2010). *Entering Research: A Facilitator's Manual.* New York: W.H. Freeman & Co.

Wrap-up (5 minutes)

► Links for more information/tips on making posters:
- Making a good poster: http://colinpurrington.com/tips/poster-design
- Online tutorials: http://www.kumc.edu/SAH/OTEd/jradel/effective.html
- Baylor's College of Medicine Beyond the Beakers, Chapters 10–12: https://media.bcm.edu/documents/2015/37/beyond-the-beakers.pdf

Contributed by A. Bramson and K. Eskine with information from Branchaw, J. L., Pfund, C., and Rediske, R. (2010). *Entering Research: A Facilitator's Manual.* New York: W.H. Freeman & Co.

COMMUNICATING RESEARCH FINDINGS 1: POSTER PRESENTATIONS

Learning Objectives

Trainees will:

- Identify the characteristics of effective research posters.
- Learn how to give a poster presentation of research findings.
- Recognize that research presentations are opportunities for discussing and networking with colleagues about research, not only for reporting research.

Scientific Poster Hunt

Explore the halls in the building where your research group resides (or another research building) to find scientific posters hanging on the walls. Select one favorite, and one least favorite poster.

Identify the characteristics of each poster that make it your favorite or least favorite.

1. Favorite Poster Title:

   What characteristics make this poster your favorite?

2. Least Favorite Poster Title:

   What characteristics make this poster your least favorite?

Contributed by A. Bramson and K. Eskine with information from Branchaw, J. L., Pfund, C., and Rediske, R. (2010). *Entering Research: A Facilitator's Manual.* New York: W.H. Freeman & Co.

COMMUNICATING RESEARCH FINDINGS 1: POSTER PRESENTATIONS

Poster Presentation Tips

Watch this video to answer the questions https://www.youtube.com/watch?v=vMSaFUrk-FA

1. What are the six tips for giving a poster presented in the video?

2. Do you agree with all of the tips? Why? Why not?

3. What would you put on the second page of your handout?

Contributed by A. Bramson and K. Eskine with information from Branchaw, J. L., Pfund, C., and Rediske, R. (2010). *Entering Research: A Facilitator's Manual.* New York: W.H. Freeman & Co.

COMMUNICATING RESEARCH FINDINGS 2: ORAL PRESENTATIONS

Learning Objectives

Trainees will:

- Identify the characteristics of effective oral presentations.
- Learn how to give an oral presentation of research findings.
- Recognize that research presentations are opportunities for discussing and networking with colleagues about research, not only for reporting research.

Trainee Level

undergraduate or graduate trainees
novice or intermediate trainees

Areas of Trainee Development

- Research Comprehension and Communication Skills
  - Develop research communication skills.
  - Develop disciplinary knowledge.
  - Develop logical/critical thinking skills.

Activity Components and Estimated Time for Completion

- In Session Time: 1 hour

Total time: 1 hour

When to Use This Activity

This activity should be implemented following or concurrently with the "Communicating Research Findings 1: Poster Presentations" activity. It can be used when trainees have been working in their research labs for several months and are preparing to present their results or thinking about how scientists communicate progress on their research.

Inclusion Considerations

Novice trainees, especially those who have traveled nontraditional academic pathways, may have had limited opportunity to give scientific or research presentations. Though all trainees can be nervous about giving presentations, these trainees may be especially nervous and would benefit from extra practice sessions and encouragement. Consider using a "compliment sandwich" approach: Start with what was done well, insert what can be improved, and close with a positive comment. Reassure trainees that their presentation skills will improve with practice.

Some trainees may come from backgrounds where humility is valued, and they have been acculturated not to speak of their achievements freely. Talk about why it is important to speak about one's research positively and with confidence and reassure trainees that they can do this while remaining true to their ideals around humility.

Contributed by A. Bramson and K. Eskine with information from Branchaw, J. L., Pfund, C., and Rediske, R. (2010). *Entering Research: A Facilitator's Manual.* New York: W.H. Freeman & Co.

Implementation Guide

Workshop Session (50 minutes)

- Invite an experienced researcher (e.g., graduate student, post-doc, scientist) to give a 10- to 15-minute oral presentation about their research.
- BEFORE the presentation (15 minutes)
  - Allow students to see the presentation materials and consider the following questions (best if speaker is not present):
    - What information should be included in the presentation slides?
    - What information should be communicated verbally?
    - What should the take home message be?
  - Distribute the worksheet included in trainee materials and encourage trainees to take notes in response to each of the questions posed on the worksheet.
- **Guest Speaker Research Presentation** (15 minutes)
- AFTER presentation (15 minutes)
  - Invite trainees to ask questions of the presenter.
  - Revisit the discussion questions:
    - What pieces of information or explanation did the presenter add to the information given on the poster/slide?

      - details about the experimental design, graphs, figures, and results.
      - personal stories about doing the research, including setbacks and how they were overcome (troubleshooting strategies).

    - How did the oral presentation of the research add to your understanding of the research poster or slides being presented?

      - opportunity to ask questions when something on the poster or in the talk was unclear.
      - opportunity to discuss the connection of my own research to the research presented in the poster or talk.
      - opportunity to meet the author.

  - Highlight that the speaker focused on the hypothesis and findings and not the details of the methods.
  - Discuss attire, posture, question fielding techniques, and sophistication of language.
  - For more advanced undergraduate and graduate trainees, also discuss the depth and clarity of the introduction, and how to relate research to listening (i.e., networking).

Contributed by A. Bramson and K. Eskine with information from Branchaw, J. L., Pfund, C., and Rediske, R. (2010). *Entering Research: A Facilitator's Manual.* New York: W.H. Freeman & Co.

► **Discussion:** Compare/Contrast Poster and Oral Presentation Formats (15–20 minutes)

- What features on the lists we generated for the best and worst features of poster presentations apply to oral presentations as well?

> - Both presentation styles need lots of figures.
> - A poster may need to be more eye catching and have visual appeal since people can often choose whether or not to stop and look at it.

- What is different about oral presentations?

> - Slides for a talk should contain less words/complete sentences since you are there talking your audience through it. Posters may need to stand alone if you won't be standing by it the whole time it is up.
> - Don't have the slides full of text. Explain in pictures.

- How does the presentation format affect the presenter's interaction with the audience?

> - For a talk there is more pressure to perform since you have to operate slides, project confidence, and be able to answer unanticipated questions on the spot in front of a lot of people. A poster is more conversational, so you generally just want to prepare a 2- or 3-minute concise summary for people passing by.
> - Having the right pacing so you don't lose or bore your audience.
> - Good body language: make eye contact, use hand gestures.

- Ask trainees, if given the choice, would they prefer to present a poster during a 3-hour poster session or give a 15-minute oral presentation. What factors influenced their choice?

> - I would choose a poster if my research was still in the beginning stages because it would be more useful to have in-depth conversations with other scientists about my project.
> - I would choose a presentation because I'm applying for graduate schools, so it would be good advertisement to get my name out there.

Contributed by A. Bramson and K. Eskine with information from Branchaw, J. L., Pfund, C., and Rediske, R. (2010). *Entering Research: A Facilitator's Manual.* New York: W.H. Freeman & Co.

COMMUNICATING RESEARCH FINDINGS 2: ORAL PRESENTATIONS

Learning Objectives

Trainees will:

- Identify the characteristics of effective oral presentations.
- Learn how to give an oral presentation of research findings.
- Recognize that research presentations are opportunities for discussing and networking with colleagues about research, not only for reporting research.

As you listen to the presentation today, write down your responses to each of the questions:

1. What pieces of information or explanation did the presenter add to the information given on the slide?

2. How did the oral presentation of the research add to your understanding of the research slides being presented?

3. What did the researcher leave out of the presentation? Did it make the presentation better or worse?

4. Identify one thing the presenter did that you thought was very effective.

5. What was the most interesting part? Why?

Contributed by A. Bramson and K. Eskine with information from Branchaw, J. L., Pfund, C., and Rediske, R. (2010). *Entering Research: A Facilitator's Manual.* New York: W.H. Freeman & Co.

COMMUNICATING RESEARCH FINDINGS 3: DEVELOPING YOUR PRESENTATION

Learning Objectives

Trainees will:

- Prepare and confidently present a research poster or talk summarizing research findings.
- Recognize that scientific presentations are opportunities for discussing and networking with colleagues about research, not only for reporting research.

Trainee Level

undergraduate or graduate trainees
novice or intermediate trainees

Areas of Trainee Development

- Research Comprehension and Communication Skills
  - Develop research communication skills.
  - Develop disciplinary knowledge.
  - Develop logical/critical thinking skills.
  - Develop effective interpersonal communication skills.
- Researcher Confidence and Independence
  - Develop confidence as a researcher.
  - Develop independence as a researcher.

Activity Components and Estimated Time for Completion

- Trainee Pre-Assignment Time: Varies
- In Session Time: Multiple sessions. *Time to implement this activity will vary depending upon the number of sessions dedicated to this topic, the number of students that will present, and the amount of time outside of the session that trainees are expected to work on their poster/presentation.*

Total time: Varies

When to Use This Activity

This activity is best suited for novice trainees; however, it can be used at any time during a trainee's career, including when he/she has the opportunity to present research locally or nationally or would like to gain experience in constructing and presenting a poster/research talk. A research group may also wish to use this activity as a way to involve members at all levels in the development of their presentation skills. If used as part of an *Entering Research* (or similar) course, this activity can be used at the end of the course as a way for trainees to present their research to the rest of the class.

Prior to beginning work on their research poster or oral presentation, facilitators may wish to have trainees complete:

- Communicating Research Findings 1: Poster Presentations
- Communicating Research Findings 2: Oral Presentations
- Mentor Interview About Making Research Posters

This activity also links to sites that may be useful for trainees to consult when preparing a poster or oral presentation.

Contributed by C. Barta and A. R. Butz with information from Branchaw, J. L., Pfund, C., and Rediske, R. (2010). *Entering Research: A Facilitator's Manual.* New York: W.H. Freeman & Co.

Inclusion Considerations

Novice trainees, especially those who have traveled nontraditional academic pathways, may have had limited opportunity to give scientific or research presentations. Though all trainees can be nervous about giving presentations, these trainees may be especially nervous and would benefit from extra practice sessions and encouragement. Consider using a "compliment sandwich" approach: Start with what was done well, insert what can be improved, and close with a positive comment. Reassure trainees that their presentation skills will improve with practice.

Some trainees may come from backgrounds where humility is valued and they have been acculturated not to speak of their achievements freely. Talk about why it is important to speak about one's research positively and with confidence and reassure trainees that they can do this while remaining true to their ideals around humility.

Contributed by C. Barta and A. R. Butz with information from Branchaw, J. L., Pfund, C., and Rediske, R. (2010). *Entering Research: A Facilitator's Manual.* New York: W.H. Freeman & Co.

Implementation Guide

Note: This activity has three phases: development, peer review, and presentation, which should be spread over the course of several weeks leading up to a trainee's presentation. Facilitators may wish to use part or all of these activities, depending on the individual needs, time constraints, and career stages of their trainees.

Rubrics and peer-review forms for posters and oral presentations are included in the trainee materials for this activity.

Trainee Pre-Assignment

► **Phase 1: Presentation Development**

- Guidelines and worksheets for developing poster and oral presentations are available in the Trainee Materials. These resources provide helpful information on the key components of an effective presentation.
- If trainees created a research project outline, they should use this to generate the first draft of their presentation.
- You may wish to provide additional guidance and resources based on any time or design constraints for the trainee's presentation.
- The work on the presentation or the poster can be conducted individually outside of the session, but facilitators are encouraged to provide opportunities for trainees to talk about their progress and ask questions as they develop their presentations.

Workshop Sessions

► **Phase 2: Review** (1–3 sessions, depending on the number of trainees and the number of drafts that are reviewed)

- Reviews can take place in one session, or, if time permits, you may wish to offer a few opportunities for trainees to receive feedback on their presentations as they iteratively revise and finalize them. Below are a few different options for structuring peer-reviewed feedback sessions.
- ***Option 1: In-class presentations with peer review.*** This implementation is best for sessions with a small number of trainees, a session with a larger group that can be broken into smaller groups, or a session that has enough time to allow everyone to present a draft of their presentation. Trainees project their poster or talk on a screen and practice their presentation as they would either at a poster session or when giving a formal research talk. Before each presentation, presenters should share with the audience one thing they would like to improve. Trainees who are not presenting can use the Poster or Presentation Review Rubric and Form (see Trainee Materials) to evaluate each presentation and provide feedback. Before presentations begin, ask trainees to identify the following for each presentation they review (in addition to completing the rubric and corresponding form, if used):
  - What is one thing the presenter did well?
  - What is one thing that the presenter could improve for his/her next presentation? What specific suggestions can you offer for improvement?
  - Comment on the one thing the presenter indicated they would like to improve.
- ***Option 2: Pair or small-group peer reviews.*** This implementation works well for any group size, large or small. Assign pairs or groups when you introduce the poster/presentation activity and ask trainees to send their poster/presentation to their partner or group members at least one session before you plan to do peer reviews. Students should pair up (or get into groups) with their peer-review partners. Before students exchange reviews, ask the reviewers to identify the best part of the presentation, one specific suggestion for how to improve the presentation, and how they would suggest improving the one thing the author identified they would like to improve. The pairs or small-group peer-review discussions can be followed by a large-group discussion if time allows.
  - ***Large-group Discussion.*** The goal of this discussion is to give students the opportunity to process the feedback they received and to identify specific strategies they will use to improve their presentation.

Contributed by C. Barta and A. R. Butz with information from Branchaw, J. L., Pfund, C., and Rediske, R. (2010). *Entering Research: A Facilitator's Manual.* New York: W.H. Freeman & Co.

Discussion questions:

- What are the major issues you need to address in your presentation?
- Specifically, how will you address these issues in your second draft?
- To whom, if anyone, will you go for help in revising?
- If another round of review is needed, provide trainees with a week or two to revise their presentations. During the next review session, you may repeat the peer-review process above and assign different pairs/groups, or invite a group of outside reviewers to provide feedback on the second draft (see below).

- ***Option 3: Outside Reviews of Poster/Presentation.*** Invite a few experienced presenters (e.g., graduate students, postdoctoral researchers, scientists, or professors) to attend your review session. You should invite enough reviewers so that each trainee has the opportunity to get feedback from several individuals. Ask the trainees and outside reviewers to introduce themselves. Trainees should share the area of their research and the research group with which they work. Outside reviewers should share their position (e.g., graduate student, postdoctoral scholar, professor), their department, and their research area of expertise.
  - Trainees sit around the room, at individual tables with mini-copies (8½" × 11") of their poster or printed slides from their oral presentation in front of them. Reviewers rotate around the room to visit each student individually (or in pairs). If there are a large number of trainees, it may be useful to invite more reviewers, and split the group in half for rotations. Trainees present a fresh mini-copy of their presentation to each reviewer, on which they and their reviewers can make notes about the feedback they receive. At the end of the session, trainees will have several copies of their presentation with specific feedback from each reviewer. The review session can be followed by a large-group discussion.

- ***Discussion of Revision Plans.*** The goal of this discussion is to give students the opportunity to process the feedback they receive and to identify specific strategies they will use to improve their presentation.

Discussion questions:

- What are the major issues you need to address in your presentation?
- Specifically, how will you address these issues in your final draft?
- To whom, if anyone, will you go for help in revising?

► Phase 3: Poster/Oral Presentation

- If your campus sponsors a formal event at which trainees can present their research results, or if trainees intend to present at a local or national conference, either would be an ideal venue in which to have trainees present their posters or talks. Alternatively (or in addition), the facilitator(s) can organize a mini-symposium for the trainees and invite faculty, scientists, post-docs, graduate, and undergraduate students to attend.
- If this activity is used as part of a course, a mini-poster session or research conference can be organized within the course. Each trainee presents their poster or research talk to the class, just as they would present their poster at a conference to an interested group of scientists. Posters can either be projected on the wall or printed and attached to the wall. If the class is presenting research talks, their talks can be projected followed by a short question/answer session after each talk.
- ***Optional activity: New Presentation Evaluation Exercise.*** If there are multiple sections of the course (e.g., *Entering Research*) being offered on the campus, then trainees can be assigned to visit the poster/presentation of trainees from another section of the course. In addition to the informal conversation that the trainees will have about the work, they can complete a review rubric (see trainee materials) about the presentation as an assignment. The reviews can then be shared with the presenter. If the presentations are part of a larger symposium, students can also be assigned to review presentations given by students in other sections of the course.

Contributed by C. Barta and A. R. Butz with information from Branchaw, J. L., Pfund, C., and Rediske, R. (2010). *Entering Research: A Facilitator's Manual.* New York: W.H. Freeman & Co.

COMMUNICATING RESEARCH FINDINGS 3: POSTER PRESENTATION

Assessment Rubric

	0	1	2	3
Title and Authors	Absent	Title is lengthy and unclear; Authors and/ or affiliations are not included	Title is lengthy, but clear; Authors and affiliations may be included.	Title is concise and clear; Authors and affiliations are included and clear.
Abstract (brief summary; may be optional)	Absent	Does not summarize the research and results. Provides only basic background information.	Summarizes research and results, but provides too much information, is not concise or uses jargon without definitions.	Concise descriptive summary of research and results; provides only relevant information with limited jargon.
Introduction (Background, context and relevance**)**	Absent	The background information lacks the content needed to understand the scientific basis of the hypothesis or research question.	Relevant background information and broader significance of the research is presented, but poorly organized. Therefore, the hypothesis or research question does not follow logically.	Relevant background information and broader significance of the research is presented and organized such that the hypothesis or research question follows logically.
Hypothesis or Research Question (purpose or aim/ goal)	Absent	A statement is made, but it is neither a hypothesis nor a research question.	A hypothesis or research question statement is made, but it is neither concise nor follows logically from the background information.	A clear and concise hypothesis or research question statement is made that follows logically from the background information.
Research Methods (How was it done?)	Absent	Experiments are listed but lack necessary detail and are not connected to the stated hypothesis or research question.	Experiments are listed, and either well explained or connected to the stated hypothesis or research question, but not both.	Experiments are listed, well explained, and connected to the stated hypothesis or research question.
(Expected) Results (What did you learn?)	Absent	Results are described, but lack a figure to represent them or a statement of whether they support the stated hypothesis or research question.	Results are described, but lack either a figure to represent them or a statement of whether they support the stated hypothesis or research question.	Results are presented in a figure and a statement about whether they support the stated hypothesis or research question made.

Contributed by C. Barta and A. R. Butz with information from Branchaw, J. L., Pfund, C., and Rediske, R. (2010). *Entering Research: A Facilitator's Manual.* New York: W.H. Freeman & Co.

	0	1	2	3
Conclusions and Future Directions (What does it mean and where do we go from here?)	Absent	Some obvious conclusions are not identified, or are not connected to the results. There is no mention of future hypotheses to test or research questions to investigate.	The conclusions are not well connected to the results. Possible future hypotheses to test or research questions to investigate are presented, but not well formulated or connected to the results.	Clear and relevant conclusions that follow logically from the results. Possible future hypotheses to test or research questions to investigate are presented, and are well formulated and connected to the results.
References and Acknowledgments	Absent	Few or not credible references are listed.	Credible references are listed, but not cited in text.	Credible references are listed and cited in text.
Poster (Is it visually appealing?)	Absent	Figures/Images are absent; text is too small, lengthy and detailed.	Figures are present, but are cluttered or unclear; text is large enough, but too lengthy.	Clear figures with concise text outlining the most important points.
Presentation of Poster	Absent	Speaking style was choppy, there was poor eye contact, and transitions between slides were not made.	Speaker was clear, but had poor eye contact and made only weak connections between poster sections.	Speaker was clear, made eye contact, and effectively transitioned between poster sections.

Contributed by C. Barta and A. R. Butz with information from Branchaw, J. L., Pfund, C., and Rediske, R. (2010). *Entering Research: A Facilitator's Manual.* New York: W.H. Freeman & Co.

COMMUNICATING RESEARCH FINDINGS 3: ORAL PRESENTATION

Assessment Rubric

	0	1	2	3
Title and Authors	Absent	Title is lengthy and unclear.	Title is lengthy, but clear.	Title is concise and clear.
Overview (Brief, usually 1–2 sentence summary)	Absent	Did not present the context or motivation for which the research was conducted.	Summarized the content of the talk, but did not present the context or motivation for which the research was conducted.	Summarized the content of the talk and presented the context or motivation for which the research was conducted.
Introduction (Background, context, and relevance)	Absent	The background information presented lacked the content needed to understand the scientific basis of the hypothesis or research question.	Relevant background information and broader significance of the research was presented, but poorly organized. Therefore, the hypothesis or research question did not follow logically.	Relevant background information and broader significance of the research was presented and organized such that the hypothesis or research question followed logically.
Hypothesis or Research Question (Purpose or aim/goal)	Absent	A statement was made, but it was neither a hypothesis nor a research question.	A hypothesis or research question statement was made, but it was neither concise nor followed logically from the background information.	A clear and concise hypothesis or research question statement was made that followed logically from the background information.
Research Methods (How was it done?)	Absent	Experiments were described but lacked detail and were not connected to the stated hypothesis or research question.	Experiments were described and were either well explained or connected to the stated hypothesis or research question, but not both.	Experiments were well explained and connected to the stated hypothesis or research question.
(Expected) Results (What did you learn?)	Absent	Results were orally presented, but not in a clear format, nor was a statement of whether they supported the stated hypothesis or research question.	Results were described but were not clearly depicted or a statement of whether they supported the stated hypothesis or research question.	Results were presented in a clear manner and a statement about whether they support the stated hypothesis or research question was made.

Contributed by C. Barta and A. R. Butz with information from Branchaw, J. L., Pfund, C., and Rediske, R. (2010). *Entering Research: A Facilitator's Manual.* New York: W.H. Freeman & Co.

	0	1	2	3
Conclusions and Future Directions (What does it mean and where do we go from here?)	Absent	Some obvious conclusions are not identified, or are not connected to the results. There is no mention of future hypotheses to test or research questions to investigate.	The conclusions are not well connected to the results. Possible future hypotheses to test or research questions to investigate are presented, but not well formulated or connected to the results.	Clear and relevant conclusions that follow logically from the results. Possible future hypotheses to test or research questions to investigate are presented, and are well formulated and connected to the results.
References and Acknowledgments	Absent	Few or not credible references were listed.	Credible references were listed.	Credible references were listed and connected to the work presented.
Slides (Is it visually appealing?)	Absent	Figures/Images are absent; text is too small, lengthy and detailed.	Figures are present, but are cluttered or unclear; text is large enough, but too lengthy.	Clear figures with concise text outlining the most important points.
Presentation	Absent	Speaking style was choppy; there was poor eye contact, and transitions between slides were not made.	Speaker was clear, but had poor eye contact and made only weak connections between slides.	Speaker was clear, made eye contact, and effective transitions between slides.

Contributed by C. Barta and A. R. Butz with information from Branchaw, J. L., Pfund, C., and Rediske, R. (2010). *Entering Research: A Facilitator's Manual.* New York: W.H. Freeman & Co.

COMMUNICATING RESEARCH FINDINGS 3: DEVELOPING YOUR PRESENTATION

Learning Objectives

Trainees will:

- Prepare and confidently present a research poster or talk summarizing research findings.
- Recognize that scientific presentations are opportunities for discussing and networking with colleagues about research, not only for reporting research.

Use your research project outline assignment (if you did this assignment) to generate the first draft of a research poster or oral presentation. Use the peer-reviewed rubric and any resources provided for advice on how to best construct your presentation.

Contributed by C. Barta and A. R. Butz with information from Branchaw, J. L., Pfund, C., and Rediske, R. (2010). *Entering Research: A Facilitator's Manual.* New York: W.H. Freeman & Co.

POSTER PRESENTATION PEER REVIEW

Presenter: ______________________________

Presentation Title: ______________________________

Use the following scale to rate the presenter:

1 = strongly disagree, 2 = disagree, 3 = somewhat agree, 4 = agree, 5 = strongly agree

1. The presenter clearly explained his/her research question.	1	2	3	4	5
2. The presenter clearly explained the importance of the research within the first minute of his/her presentation.	1	2	3	4	5
3. The presenter included the appropriate amount of background information so I could easily follow the presentation.	1	2	3	4	5
4. The presenter explained the methods section clearly.	1	2	3	4	5
5. If results were included, they were clearly articulated.	1	2	3	4	5
6. The presenter had a clear conclusion that gave an appropriate overview of his/her talk.	1	2	3	4	5
7. The presentation was well organized and followed a logical order.	1	2	3	4	5
8. The presenter explained unfamiliar terms/vocabulary.	1	2	3	4	5
9. The poster design and layout contributed to the effectiveness of the presentation (there were few or no distracting items).	1	2	3	4	5
10. The delivery of the presentation was good (the presenter was loud and had a good, consistent pace).	1	2	3	4	5
11. The presenter was enthusiastic about his/her research.	1	2	3	4	5
12. The presenter was confident in the material he/she was presenting.	1	2	3	4	5
13. The presenter maintained good eye contact with the audience.	1	2	3	4	5
14. The presenter helped me to understand his/her research project better.	1	2	3	4	5
15. Overall, I would give the speaker a grade of:	C	BC	B	AB	A

Comments:

Contributed by C. Barta and A. R. Butz with information from Branchaw, J. L., Pfund, C., and Rediske, R. (2010). *Entering Research: A Facilitator's Manual.* New York: W.H. Freeman & Co.

ORAL PRESENTATION PEER REVIEW

Presenter: ____

Presentation Title: ______________________________________

Use the following scale to rate the presenter:

1 = strongly disagree, 2 = disagree, 3 = somewhat agree, 4 = agree, 5 = strongly agree

1. The presenter clearly explained his/her hypothesis or research question.	1	2	3	4	5
2. The presenter included the appropriate amount of background information so I could easily follow the presentation.	1	2	3	4	5
3. The presenter clearly explained his/her results.	1	2	3	4	5
4. The presenter had a clear conclusion slide that gave an appropriate overview of his/her talk.	1	2	3	4	5
5. The presentation was well organized and followed a logical order.	1	2	3	4	5
6. The presenter explained unfamiliar terms/vocabulary.	1	2	3	4	5
7. Slide design and layout contributed to the effectiveness of the presentation (there were few or no distracting items).	1	2	3	4	5
8. The delivery of the presentation was good (the presenter was loud and clear, and had a good, consistent pace).	1	2	3	4	5
9. The presenter was enthusiastic about his/her research.	1	2	3	4	5
10. The presenter was confident in the material he/she was presenting.	1	2	3	4	5
11. The presenter maintained good eye contact with the audience.	1	2	3	4	5
12. The presenter helped me to understand his/her research project better.	1	2	3	4	5
13. Overall, I would give the speaker a grade of:	C	BC	B	AB	A

Comments:

Contributed by C. Barta and A. R. Butz with information from Branchaw, J. L., Pfund, C., and Rediske, R. (2010). *Entering Research: A Facilitator's Manual.* New York: W.H. Freeman & Co.

GUIDELINES FOR PREPARING A RESEARCH POSTER

Some general guidelines about how to prepare a poster are presented below, followed by a worksheet to help you organize your poster presentation.

Posters come in all shapes and sizes and can be generated using many different computer programs. A common computer program used to construct posters is Microsoft PowerPoint. In PowerPoint, the dimensions of a single slide can be altered to fit the desired dimensions of a poster (for example, 3 ft by 4 ft). Thus, you will only need to work with this one slide, which will become your entire poster when printed. **Pay careful attention to the conference or class recommendations for poster sizes because some conferences will not allow you to present posters that are larger than their specifications.**

When designing your poster, consider your audience. Will your poster be displayed at a conference that is open to the public, to scientists of all fields, or only to researchers in your discipline? Depending on the audience, you may need to alter the amount of technical language/jargon you use. In addition, consider whether you will always be standing by your poster when it is displayed, or will your poster be displayed at times when you are not there? If you are not there to talk someone through your poster, you may want to include more details or carefully state major conclusions, so any viewer can understand your poster without a verbal explanation. Alternatively, if you will be standing by your poster the entire time, you may not need to include as many details because you will be able to articulate these during your presentation.

Title, Authors, and Affiliations: Include a short, concise title at the top of your poster. Often titles are written in font sizes of 72 pt. or larger. The list of authors should include yourself, your mentors, and any other contributors to the project. Authors are often listed under the title in slightly smaller font. The affiliations of the authors are generally listed under the author list (example: Department of Chemistry, University of Wisconsin–Madison). If the authors have multiple affiliations or are from different institutions/organizations, use symbols such as *, †, ‡ to denote which authors are from what institutions/organizations. Ask your research mentor to help you determine the appropriate order in which the authors should be listed.

Background: This section needs to include a summary of the research you are conducting, the importance of the research to both your field and to the broader scientific community, relevant background that explains the topic and also identifies any "holes" in the existing knowledge, and your research question/hypothesis—keep it brief and pertinent! Be sure to explicitly state what question **your** research is asking. Sometimes it is helpful to even bold your research question/hypothesis, so viewers immediately know what you are studying.

Experimental Design/Methods: No matter who your audience is (your lab, people in the same field, people in diverse fields, nonscientists), they appreciate an explanation of how the experiment was set up and why those experiments were chosen. Remember that words should be kept to a minimum on posters. Instead of using sentences or paragraphs to describe your methods, consider using schematics (vector diagram, protocol sketch) or a written flowchart. Keep things simple; you can elaborate verbally, but you don't want the poster to be too overwhelming that the main point of the research is lost!

Data: Present data clearly. Every chart, graph, figure, table, scheme, etc., should be properly labeled with descriptive titles that convey the "bottom line" or the main conclusion from the presented results. Be sure to think about how to most effectively present your data so that the most important findings/conclusions are highlighted. If you have mostly numerical data, ask yourself if a table would be the best fit, or maybe you can make your point more effectively with a graph or histogram? If you have a LOT of numbers, will you present ALL the data? Or can it be summarized into numbers (i.e., percentages) that allow comparison? Are your data significant? Do you need to show

Contributed by C. Barta and A. R. Butz with information from Branchaw, J. L., Pfund, C., and Rediske, R. (2010). *Entering Research: A Facilitator's Manual.* New York: W.H. Freeman & Co.

all spectra to make your point, or can you show just one or two well-chosen examples? Also make sure the axes in the graphs, headings in tables, lanes in gel photographs, lines in spectra, etc., are clearly labeled so that the audience can read and follow your explanations.

Conclusions and/or Future Research: Bring your research full circle by highlighting the major findings from your work and explain how these results relate to the initial objective of your work. You may also propose future experiments or alternative directions based on your results.

Acknowledgments: Science is a collaborative process. Give credit to people in your research group who contributed intellectually or taught you techniques needed to complete this project. It also is important to list sources of funding that supported the research.

References: References are usually included at the bottom of a poster using the appropriate referencing style for your discipline.

Presenting: Practice! Stand in front of your poster and practice explaining the main parts of the poster. Have a 1-minute version, a 3-minute version, and a 5-minute version ready for various audience members who might want a quick snapshot or would like a detailed explanation of your work. The more times you present a poster, the easier it gets!

Other considerations

- Use a simple background color/image for your poster. Many eye-catching posters have light color, solid backgrounds (such as White, pale yellow, or pale blue) with dark text. Add additional color sparingly so as not to distract from the information on your poster.
- Make sure the overall layout of the poster has a logical flow so your audience can easily find relevant information.
- Try to use 800 words or less on posters. Instead, use carefully chosen visuals such as flowcharts, descriptive graphics, tables, figures, and schemes.
- Define unfamiliar terms, especially acronyms, for your audience.
- Keep text legible (the body text of a poster should not be smaller than 24 pt.; Subheadings should be 36 pt. or larger; authors would be 56 pt. or larger; title should be 72 pt. or larger).
- Use high-resolution pictures (at least 150 dpi). Low-resolution pictures may look fine on your computer screen, but might not look so great when enlarged on your printed poster.
- Proofread, Proofread, Proofread!!! Have a friend proofread for you!!
- When presenting, always stand to the side of your poster so you do not obstruct the view. Remember, face your audience when presenting, not your poster.
- Help your audience through all graphics on your poster. You've seen them before, they haven't (define axis, point out control vs. experimental, tell your audience the main conclusions, etc.).
- Show enthusiasm and confidence!
- Look around your campus, department, or on the Internet for examples of posters. Identify styles or elements that you like and try to incorporate these styles/elements in your poster.
- Some additional links for more information/tips on posters:
  - Making a good poster: http://colinpurrington.com/tips/poster-design
  - Online tutorials: http://www.kumc.edu/SAH/OTEd/jradel/effective.html
  - Baylor's College of Medicine Beyond the Beakers, Chapters 10–12 (readings included in original ERv1 instructor materials, and can be found online here: https://media.bcm.edu/documents/2015/37/beyond-the-beakers.pdf)

Readings

Contributed by C. Barta and A. R. Butz with information from Branchaw, J. L., Pfund, C., and Rediske, R. (2010). *Entering Research: A Facilitator's Manual.* New York: W.H. Freeman & Co.

GUIDELINES FOR PREPARING AN ORAL PRESENTATION OF RESEARCH

Some general guidelines about how to prepare an oral presentation are presented below, followed by a worksheet to help you organize your presentation. An oral presentation should have approximately 1 slide/minute. (i.e., a 10-minute oral presentation should have ***approximately 10–12 slides***).

Objectives: Start by telling your audience what type of research you have conducted and **why** this research is important. This often is followed by relevant background that explains the topic and also identifies any "holes" in the existing knowledge—keep it brief and pertinent! Be sure to explicitly state what question **your** research is asking. For a 10-minute talk, the first two or three slides are usually spent on introductory material.

Experimental design: No matter who your audience is (your lab, people in the same field, people in diverse fields, non-scientists), they appreciate an explanation of how the experiments were conducted. Visuals help depict any experimental design! Consider using schematics (vector diagram, protocol sketch, synthetic schemes, etc.) or a written flowchart. Careful: Keep these simple—you can elaborate verbally, but you don't want the schematic to be too complex! For a 10-minute talk, the experimental design is often summarized in 1–3 slides.

Data: Present data clearly. Use the slide title to convey the "bottom line," or the main conclusion from the presented results. Think about how to most effectively present your data. If it is numerical, is a table OK, or will you make your point more effectively with a graph or histogram? If you have a LOT of numbers, will you present ALL the data? Or can it be summarized into numbers (i.e., percentages) that allow comparison? Are your data significant? Do you need to show all spectra to make your point? Or can you show just one or two well-chosen examples? Make sure the axes in graphs, headings in tables, lanes in gel photographs, lines in spectra, etc., are labeled so that the audience can read and follow your explanation. Depending on how many results you have, the results portion of a 10-minute talk may take 1–3 slides.

Make conclusions when discussing your results: In 10 minutes, your listeners might not be able to analyze your data completely. TELL them what you believe the data shows. Don't overstate your conclusions—if these experiments need to be replicated or if additional controls need to be run, say so. These conclusions should be made on the slides where the results are presented. Remember do not include too many words on each individual slide—consider if you need to write down your conclusions on each slide or simply state verbally.

Summarize your results and put them in context: Bring the talk full circle by explaining how your results relate to the initial objective of your work. This is often the time to propose future experiments or alternative directions the project might take. This is often called a "Conclusion" slide or a "Future Work" or "Future Directions" slide. Make sure to allocate 1-slide to this in a 10-minute talk.

Acknowledgments: Science is a collaborative process. Give credit to people in your research group who contributed intellectually or taught you techniques needed to do this project. Also, thank your faculty mentors. Because you may be a little nervous, **make a slide** identifying those you wish to thank so you don't forget! Many scientific presenters will put a picture of their research group on the acknowledgments slide and point out the researchers that specifically helped with the project. On the acknowledgment slide, it is important to also list any sources of funding that made the research possible.

References: It is important to include relevant references in a scientific talk. Unlike a scientific paper where references are usually listed at the end, most scientific presentations will list the relevant references at the bottom of the slide where the reference was discussed. It is better to have your references embedded throughout the talk instead of having a slide at the very end that lists all the references because it is easier for your audience to follow.

Contributed by C. Barta and A. R. Butz with information from Branchaw, J. L., Pfund, C., and Rediske, R. (2010). *Entering Research: A Facilitator's Manual.* New York: W.H. Freeman & Co.

Practice: ***Do it aloud, on your feet, and pointing at your slides.*** Sometimes it helps to review that first line over and over so you can get started! After that, your visuals will help cue you.

Other considerations

- Use a simple background color/image for your slides. The background should not distract from the information on your slide.
- Avoid using distracting slide transitions or unnecessary "clip art."
- Define unfamiliar terms, especially acronyms, for your audience.
- Keep text on the slides and on graphs, tables, schemes, etc., legible (at least 24 pt. type). Proofread!
- Not too many slides—usually no more than 1/minute!
- Stand clear of your slides and point so you can return to make eye contact with the audience.
- Help your audience through each graphic—you've seen them before, they haven't (define axis, point out control vs. experimental, etc.).
- Show enthusiasm and confidence!

Some WORDS to help you get started . . .

Why?
Our research QUESTION is . . . Previous studies showed . . .
We used the following EXPERIMENTS to investigate our hypothesis . . .
Our RESULT showed . . . This suggests . . .
In summary, it was determined that . . .
Our next steps will be . . .
I would like to thank . . .

Contributed by C. Barta and A. R. Butz with information from Branchaw, J. L., Pfund, C., and Rediske, R. (2010). *Entering Research: A Facilitator's Manual.* New York: W.H. Freeman & Co.

POSTER WORKSHEET

Title, Authors and Affiliations:

Title:

Authors including yourself, your mentors, and any other contributing researchers: In which department and college did the authors complete the research?

Background:

What is the general area of research in which you worked? How does your research fit into and contribute to the field?

What were the most important experiments (done by others) that led to the generation of your experimental hypothesis/research question? How do the results of these experiments lead to your research? Is there a summary diagram that you could use to present this information?

What "gaps" in knowledge still exist in the field? How does your research help to address some of these "gaps"?

Hypothesis/Research Question:

State your hypothesis(es)/research question(s). This should follow directly from your background information.

Experimental Approach/Methods:

Summarize your experimental approach in a visual such as a flow diagram or a synthetic scheme (if possible).

What experimental techniques did you use to test your hypothesis/research question? Why did you use these techniques? Identify what kinds of information you get from each kind of technique.

Results/Discussion:

Write a statement/title summarizing each of your experimental results.

Result #1:

Result #2:

Result #3:

Result #4:

How will you present each of the experimental results (graphs, gel photos, spectrum, etc.)?

Result #1:

Result #2:

Result #3:

Result #4:

Contributed by C. Barta and A. R. Butz with information from Branchaw, J. L., Pfund, C., and Rediske, R. (2010). *Entering Research: A Facilitator's Manual.* New York: W.H. Freeman & Co.

How does each result support or not support your hypothesis/research question?

Result #1:

Result #2:

Result #3:

Result #4:

Conclusions/Future Works:

Restate your research question/hypothesis and summarize the major results and conclusions from your project.

How has your project contributed to the field?

What future experiments could be done to expand on your work?

Acknowledgments and References:

Identify the individuals and/or groups who helped you with your research project. Identify any sources of funding that contributed to your research.

List any references that were used throughout your poster.

Contributed by C. Barta and A. R. Butz with information from Branchaw, J. L., Pfund, C., and Rediske, R. (2010). *Entering Research: A Facilitator's Manual.* New York: W.H. Freeman & Co.

ORAL PRESENTATION WORKSHEET

Title Slide

Title of your talk: Mentors' names and titles:

In which department and college did you do your research?

Background Slide(s)

What is the general area of research in which you worked?

How does your research fit into and contribute to the field?

What were the most important experiments (done by others) that led to the generation of your experimental hypothesis? How do the results found in these studies lead to your experiments? Is there a summary diagram that you could use to present this information?

What "gaps" in knowledge still exists in the field? How does your research help to address some of the "gaps" in your existing knowledge?

Hypothesis Slide(s)

State your hypothesis(es). This should follow directly from your background information.

Experimental Approach Slide(s)

Summarize your experimental approach in a visual such as a flow diagram or a synthetic scheme (if possible).

What experimental techniques did you use to test your hypothesis? Why did you use these techniques? Identify what kinds of information you get from each kind of technique.

Results/Discussion Slides

Write a statement/title summarizing each of your experimental results.

Result #1:

Result #2:

Result #3:

Result #4:

How will you present each of the experimental results (e.g., graphs, tables, spectra, gel photos, photomicrographs)?

Result #1:

Result #2:

Result #3:

Result #4:

Contributed by C. Barta and A. R. Butz with information from Branchaw, J. L., Pfund, C., and Rediske, R. (2010). *Entering Research: A Facilitator's Manual.* New York: W.H. Freeman & Co.

How does each result support or not support your hypothesis?

Result #1:

Result #2:

Result #3:

Result #4:

Summary Slide

Restate your research question/hypothesis and summarize the major results and conclusions from your project.

How has your project contributed to the field?

What future experiments could be done to expand on your work?

Acknowledgments Slide

Identify the individuals and/or groups who helped you with your summer research project. Identify any sources of funding that contributed to your research.

Contributed by C. Barta and A. R. Butz with information from Branchaw, J. L., Pfund, C., and Rediske, R. (2010). *Entering Research: A Facilitator's Manual.* New York: W.H. Freeman & Co.

COMMUNICATING RESEARCH TO THE GENERAL PUBLIC

Learning Objectives

Trainees will:

- Understand the importance of communicating research in a way that is accessible.
- Define the "general public" and consider researchers' responsibility to communicate with this audience.
- Compare and contrast ways in which researchers communicate with each other and with non-researcher audiences.

Trainee Level

undergraduate or graduate trainees
novice or intermediate trainees

Areas of Trainee Development

- Research Comprehension and Communication Skills
  - Develop research communication skills.

Activity Components and Estimated Time for Completion

- Trainee Pre-Assignment Time: 10–20 minutes
- In Session Time: 55 minutes

Total time: 55 minutes–1 hour, 15 minutes

When to Use This Activity

This activity may be used at any time during the research experience, but is designed as an introductory activity for a unit on research communication skill building. It can serve as a precursor to:

- General Public Abstract
- Three-Minute Research Story

Inclusion Considerations

Consider how trainees from different backgrounds might present their research to family members and to the general public. Trainees who are the first in their family to attend college or graduate school may need to address concerns that their involvement in research does not align with familial notions of success (i.e., working during the summer and getting a job right after college vs. participating in a summer program and going to graduate school). Invite trainees to discuss strategies to deal with this challenge.

Some trainees may come from backgrounds where humility is valued, and they have been acculturated not to speak of their achievements freely. Talk about why it is important to speak about one's research positively and with confidence and reassure trainees that they can do this while remaining true to their ideals around humility.

Contributed by L. Adams with information from Branchaw, J. L, Pfund, C., and Rediske, R. (2010). *Entering Research: A Facilitator's Manual.* New York: W.H. Freeman & Co.

Implementation Guide

Trainee Pre-Assignment (optional, 10–20 minutes):

- In advance of the discussion, trainees should view results from a Pew Research survey conducted in 2014 on 2,002 U.S. adults and 3,748 AAAS affiliated U.S.-based scientists on public and scientists' views on science and society (http://www.pewinternet.org/interactives/public-scientists-opinion-gap/) and come to the session prepared to discuss their reactions.
- Alternatively, the facilitator can reference figures from the study to open the discussion.

Workshop Session (1 hour)

Note: This activity is designed to be completed in a standard 1-hour session, but can be condensed to fit the amount of time available by limiting discussions to the large group. Alternatively, the discussion questions can be posted online for trainees to answer and comment on each other's responses.

When using this activity with graduate students, one may quickly address questions 1 and 2, or leave them out entirely, and focus on question 3 as a segue into *how* to communicate research effectively to the general public.

- **Introduction** (5 minutes)
  - This activity is designed to raise student awareness of *why* the ability to communicate research to multiple audiences is an important skill to develop as a researcher. Trainees also start to consider ways to effectively communicate research to various audiences using different approaches.
- **Discussion #1:** The importance of communicating research to the public (15 minutes). Trainees form small groups of two or three and generate talking points related to three questions. Spend approximately 10 minutes in small groups and 5 minutes sharing ideas as a large group.
  - The general public and researchers can differ in opinion on key issues. Why does this gap in opinion exist?

    > - People don't trust researchers.
    > - People don't understand research.
    > - People have not had opportunities to engage with research.
    > - People don't think researchers agree.

  - How might this opinion gap directly impact researchers?

    > - Research is often publicly funded so if people don't see it as useful to their daily lives, they are less likely to support funding for research.
    > - The impact of research is lost if most people don't understand it.

  - Why is it important for scientists to communicate with the general public?

    > - Communicating research can help people make informed decisions.
    > - If researchers make more of an effort to communicate their research, it can also make research accessible to audiences that traditionally have been excluded from the process of research, making research more diverse and inclusive.
    > - Communicating about research may increase the support for and funding of research.

Contributed by L. Adams with information from Branchaw, J. L, Pfund, C., and Rediske, R. (2010). *Entering Research: A Facilitator's Manual.* New York: W.H. Freeman & Co.

Additional discussion topics:

- **How is academic research in the United States typically funded?** The facilitator can briefly describe how public and private funds are used to fund research.
- **Some research projects have been identified as wasteful government spending.** Facilitators can provide an overview of the following resources or invite students to visit these sites as part of an in-class activity or as a follow-up assignment:
  - Proximire's Golden Fleece Awards: http://www.wisconsinhistory.org/turningpoints/search.asp?id=1742
  - NSF "Under the Microscope" report: https://www.youtube.com/watch?v=qSLvgCb_eiw
  - Optional follow-up assignment: Students can be directed to read https://www.whitehouse.gov/blog/2015/06/02/value-basic-research and check out #basicresearch on Twitter. Students should come to class with an example of a basic research discovery that has led to an unexpected insight or application or post the example on a class blog.

► **Discussion #2:** What should the general public know about research? What does it mean to be research literate? (20 minutes)

- Give trainees 3 minutes to reflect on this question and write their responses on a notecard.
- What should the general public know about research? What does it mean to be research literate?

> - The public should appreciate how research is conducted.
> - The public should know that research is not just about memorizing facts.
> - We shouldn't force everybody to love research.

- After writing responses, watch the Neil deGrasse Tyson video in which he discusses his views on science literacy (3 minutes). Explain that Neil deGrasse Tyson is an astrophysicist who has also become a world-renowned science communicator
  - https://www.youtube.com/watch?v=gFLYe_YAQYQ
  - shorter version: https://www.youtube.com/watch?v=5gK2EEwzjPQ
- **Large-Group Discussion** (15 minutes)
  - What were the main points raised by Neil deGrasse Tyson? What aspects of Neil deGrasse Tyson's opinion align with your own? Where do you differ in opinion?

> - Scientific literacy doesn't mean you have memorized a lot of scientific facts.
> - Being scientifically literate is a way of thinking about how the world works.
> - We don't want everybody in the world to be scientists but being scientifically literate is important for everybody because science is at the foundation of a lot of issues that confront society.

Additional Discussion Questions

- Whose job is it to communicate science to the public? Should all scientists be required to communicate their research to the public?
- What role does the popular media (e.g., news, radio, social media) play in communicating science?

Contributed by L. Adams with information from Branchaw, J. L, Pfund, C., and Rediske, R. (2010). *Entering Research: A Facilitator's Manual.* New York: W.H. Freeman & Co.

► **Activity:** Compare and contrast approaches to communicating research to different audiences. (10 minutes)

- Researchers communicate all the time! Brainstorm all the ways that researchers use written, oral, and visual forms of communication to communicate with each other. (5 minutes)

> - Written—proposals, primary research papers, review articles
> - Oral—Journal clubs, Lab meetings, Conference presentations, Seminars, interpersonal communication
> - Visual—figures for papers, poster presentations, papers, and seminars

- How do researchers communicate their research to the public? Brainstorm different modes of communication and venues to communicate with the public. (5 minutes)

> - Written—Popular science magazine article, blog, social media
> - Oral—TV news, radio, Science Festival booth, public seminar, K-12 school visit
> - Visual—Museum exhibit, infographic, art exhibit

Additional Questions

- Is there such a thing as a "general audience"? Brainstorm a list of as many types of audiences (elementary school kids, local rotary club, citizen scientists) as you can think of.
- How would your communication style need to be adapted to reach a specific audience? In a specific venue?

► **Wrap-up** (5 minutes)

- Encourage trainees to meet with their mentor to discuss topics raised in class. Has the mentor had experience engaging with public audiences to talk about their research? What was the experience like? What were the challenges and benefits of the experience?
- Important follow-ups to this in-class discussion are additional assigned exercises designed to help students develop skill in *how* to effectively communicate research to a public audience in written (general public abstract) and oral (3-minute research story) formats.

Contributed by L. Adams with information from Branchaw, J. L, Pfund, C., and Rediske, R. (2010). *Entering Research: A Facilitator's Manual.* New York: W.H. Freeman & Co.

COPING EFFICACY

Learning Objectives

Trainees will:

- ► Be able to distinguish between emotion and problem-focused coping strategies.
- ► Identify when it is appropriate to apply emotion and problem-focused coping strategies.

Trainee Level

undergraduate or graduate
novice, intermediate, or advanced trainees

Areas of Trainee Development

- ► Researcher Confidence and Independence
  - Develop confidence as a researcher.
- ► Equity and Inclusion Awareness and Skills
  - Advance equity and inclusion in the research environment.

Activity Components and Estimated Time for Completion

- ► In Session Time: 55 minutes

Total time: 55 minutes

When to Use This Activity

This activity can be used with undergraduate or graduate trainees at any level. When implementing, it is important that facilitators make time to establish trust in the group. Therefore, it may be beneficial to implement with an already established cohort of trainees or after a course has been meeting regularly for a few weeks. Ground rules should have been established and be displayed throughout the session to remind participants of the guidelines that they co-constructed. Other activities that may be used with this activity include:

- ► Fostering Your Own Research Self-Efficacy

Inclusion Considerations

Using the strategies presented in this activity to cope with stress requires trainees to identify the source of the stress. This can be difficult in complicated situations or overwhelming if they feel like their peers are not experiencing the same stresses. For example, generally feeling like one does not belong in the research environment because they are "different" from everyone else can cause stress. Encourage trainees to find trusted mentors, colleagues, and friends with whom they can talk about their stressful situation. In particular, those individuals who are outside the research environment who they know well and respect (e.g., clergy, parents, hometown community members) can be particularly insightful and supportive when trainees are trying to identify and understand what is causing them stress.

Contributed by A. Kaatz. (2018). *Coping Efficacy.*

Implementation Guide

Workshop Session (55 minutes)

► **Introduction** (10 minutes)

- The goal of this activity is to help trainees identify effective coping strategies for challenges they may face in their research experiences.
- Most trainees experience psychological stress in school, particularly when learning new, challenging material or adjusting to new environments. When you feel "stressed" your resources for coping are surpassed (Bandura, 1997; Lent, Brown, & Hackett, 2000). Developing *academic coping self-efficacy*—confidence to cope effectively with academic stress—is one thing that trainees can do to help lower their stress level as they work to persist and succeed in school (Lent, Brown, & Hackett, 2000).
- Refer trainees to the first page of the trainee materials, which includes a flowchart for coping strategies. Walk them through the flowchart using the notes below as a guide.
  - Recognizing that you are experiencing stress is the first step toward combating it. Next, it is important to assess whether the situation you are in is ***changeable*** or ***unchangeable***. Identifying whether the situation you are in is changeable determines what style of coping will help most.
  - Most challenging situations in academia are changeable. For example, having a bad advisor is a situation that can be changed, as is starting out with poor grades in a class. Importantly, these changeable situations require ***problem-focused coping*** (i.e., engaging in actions that can help to address problems that cause stress; Chesney et al., 2006), such as choosing a new advisor from the multiple other advisors available in the former case, and engaging in activities, such as tutoring or working with a study group help to improve your performance in the latter case.
  - Unchangeable situations are less common, but can be impactful. For example, being expelled from school is an unchangeable situation. The death of a classmate or professor is another example of something that you cannot change. For these unchangeable situations, it is most advantageous to apply ***emotion-focused coping*** (i.e., working to adjust your emotional response to a stressor; Chesney et al., 2006), such as seeking social support from friends and loved ones, or engaging in therapy to gain psychological support for a change or loss.
- Review the examples of changeable and unchangeable situations and appropriate problem-focused and emotion-focused coping strategies outlined in the handout.
  - Often stressful situations require both types of coping. For example, if you experience stress from performing poorly in a class, it would help you to seek out some social support (i.e., to engage in emotion-focused coping), even though problem-focused coping, such as learning how to study more effectively, is the type of coping that would ultimately be able to help you change your stressful situation.
  - Trainees get into trouble when they, usually inadvertently, use problem-focused coping to address unchangeable situations, or emotion-focused coping to address changeable situations—this mismatch of coping styles is called ***maladaptive coping***.
    - As an example, consider a trainee who starts to fail a science class (i.e., changeable situation) and only seeks out social support (i.e., emotion-focused coping). Leveraging emotion-focused coping will not help the trainee learn how to improve their performance, and consequently they may begin to doubt their ability to succeed in science. In this case, seeking social support would likely help the trainee feel valued, but problem-focused coping, such as talking with the teacher about ways to improve performance, or being tutored, would be most helpful for addressing the underlying cause of stress.

Contributed by A. Kaatz. (2018). *Coping Efficacy.*

- Often, when trainees solely apply emotion-focused coping to changeable situations, such as seeking out friends to make them feel better about poor academic performance, they begin to believe that they don't have the skills or ability to succeed. This can lead to loss of interest in science and attrition from science tracks. Thus, being able to recognize if the situation you are in is changeable is an important and powerful skill that will help trainees succeed, even when faced with great challenges.

► **Activity: Case Studies** (40 minutes)

- Case studies are provided in the trainee materials. Choose the case studies based on the amount of time available and the topics that will be most relevant to your trainees. Allow approximately 10 minutes per case study. For each case study, ask trainees what they would do if they were in this situation. Suggested points for discussion are included in the Case Study Discussion Guide. (20 minutes)

► **Wrap-up** (5 minutes)

- Summarize the main ideas generated from the large-group discussion.
- Facilitators may want to follow up with trainees in a future session and ask them to share how they have implemented problem-focused and emotion-focused coping strategies in the context of their research experience.

COPING EFFICACY: CASE STUDY DISCUSSION GUIDE

► **Case Study #1:** In math class today, your professor handed back the first midterm exam. In high school, you had done well in math and you felt that you'd studied hard for this first exam. To your surprise, you see that you got a C– on the test. You feel embarrassed, and quickly turn over your exam so your classmates don't see your low grade. You start thinking that you might not be cut out for a career in science. What should you do?

- This is a changeable situation, and the student should use **problem-based coping**. The problem is that this student needs to learn how to study for university exams and could engage in a variety of activities to grow that skill set. For example, the student could go talk to their professor about ways to improve performance on future exams or join a study group.

► **Case Study #2:** You are close to finishing your biology degree and are waiting to hear from graduate schools. Last week at your annual physical health exam, your primary care doctor noticed that you'd described an interesting cluster of symptoms. So, she ran a few tests. She called you to come back into her office to discuss the results. She informed you that you have a disease that will increasingly begin to cause extreme fatigue and impair your mobility. There are drugs that can help, but no cure. You are shocked. You had your life planned out . . . you were going get a Ph.D. in biology and become faculty at a research institution . . . you don't see how you could still do that if you begin to experience the symptoms your doctor described. What should you do?

- This situation, at least the diagnosis, is unchangeable, and the student would benefit from drawing on emotion-focused coping techniques, such as seeking support from friends and family. This student will have to come to terms with the diagnosis—thus, social and psychological support might help the most.
- Though the diagnosis is unchangeable, there are many things that the student could change that would enable them to continue on the path to grad school and a research career. For example, there could be forms of physical therapy that might help to develop physical strength and endurance. Graduate schools may have special resources for students with similar conditions or disabilities that would make it easier to attend class and participate in research.
- This case shows how often we need **both emotion-based and problem-based coping** skills to navigate challenging situations—some type of mental adjustment is usually needed to move through challenges or roadblocks, even if the situation is changeable and engaging in problem-based coping would be most beneficial.

► **Case Study #3:** You've been working extra hours in the lab to get enough data to earn authorship on the research team's next paper. However, it is unclear how much data will be enough and there is no guarantee that your mentor will include you as an author, even if you contribute a lot of data. Worrying about this has created a lot of stress for you. Everyone says you need to be an author on a peer-reviewed paper to get into the good graduate training programs. On top of this, you're starting to notice that spending a lot of time in the lab is making it difficult to keep up with your coursework. Now you're beginning to worry about your grades as well. What should you do?

Contributed by A. Kaatz. (2018). *Coping Efficacy.*

- This situation is changeable, so a **problem-focused coping** strategy would be most effective. For example, you could have a conversation with your mentor to better understand the expectations for an authorship role and whether you are meeting those expectations. You can also think about the amount of time you are dedicating to research and your coursework and determine if your current commitments are sustainable. If it is not sustainable, it may be useful to have a conversation with your mentor to better understand the expectations of the research group and to talk about balancing coursework and research with your mentor and academic advisor to determine the best course of action.

► **Case Study #4:** You've finished all of your coursework, passed your preliminary exam, and now must write your thesis proposal. Up to this point, everything you've been required to do as a graduate student has been familiar: taking classes, passing exams, and carrying out experiments. These are all things that you did as an undergraduate student and you've been able to handle it. Writing a thesis proposal, on the other hand, is brand-new territory. You have to be independent. How will you come up with your own project? You're completely stressed, and it feels like your brain is frozen. You don't know where to begin. What should you do?

- This problem could be addressed using **both problem- and emotion-focused coping** strategies. For example, talking with your mentor and other students in the research group about strategies and best practices might be helpful in devising your own plan. Identifying or creating a support group for graduate students at similar stages in the thesis writing process would provide emotion-focused coping in the form of social support from peers as well as a means for collectively identifying problem-focused coping strategies.

► **Case Study #5:** Micah, a postdoctoral scholar on your research team, has been a good friend and mentor to you since you began graduate school four years ago. You and your partner frequently go camping and hiking with him and his wife, Lila, and you have become part of their circle of friends. Lately, you've noticed that Micah is making a lot of mistakes when doing experiments. You ask him if everything is OK and learn that he and Lila have been having trouble and are contemplating divorce. He begins telling you stories about Lila, who you consider a good friend, and urges you to take his side in their disputes. This has been going on now for over 6 months and it is creating a stressful environment for you in the lab. What should you do?

- Given your close relationship with Micah, it might be best to adopt a **problem-focused coping** strategy by telling Micah how these conversations make you feel and work together to find a solution that makes you both comfortable.

► **Case Study #6**: You have been working with the Nikita research group for one semester and are feeling good about your progress in understanding the research context and have gained confidence in your ability to carry out experiments independently. Recently you have noticed that your mentor seems aloof and does not want to talk to you or answer your questions. You have overheard her making remarks to another graduate student about how little you know and how much time it takes to mentor you. You feel caught because you want to do a good job, but sometimes you don't know what you are doing and need to ask questions, but she won't answer them. In the coming week you are supposed to analyze your data and you need your mentor's help. When you arrive in her office to discuss the data analysis, she sees a mistake, tells you to fix it, and puts on her headphones while you make changes to the data file. The uncomfortable working environment is stressful and has diminished your confidence to do research. You're considering leaving the lab. What should you do?

- In this case, you could employ **both problem- and emotion-focused coping** strategies. Confiding in a trusted friend or peer about your feelings may help you process your feelings and help you find support as you consider your options. Talking with your mentor to identify strategies to help you gain more independence and better understand the expectations of your mentor could also help address the inconsistency between you and your mentor's assessment of your capabilities.

► **Case Study #7**

- Invite trainees to reflect on a time when they were in a challenging academic situation. Was it a changeable situation? Was it an unchangeable situation? Was it a little of both? How did they handle it? Did they use the most beneficial coping strategies? If not, what would they do differently? (5 minutes)
- In pairs or triads, have trainees share their experiences and their coping strategies. Have trainees identify the effective and ineffective coping strategies in their examples. (5 minutes)
- As a large group discuss the following: (10 minutes)
  - What were the effective strategies that you identified?

Additional resources for facilitators on coping self-efficacy:

► Bandura, A. (1997). *Self-efficacy: The exercise of control.* New York, NY: W. H. Freeman.

► Chesney, M. A., Neilands, T. B., Chambers, D. B., Taylor, J. M., and Folkman, S. (2006, Sept. 1). A validity and reliability study of the coping self-efficacy scale. *British Journal of Health Psychology, 11*(3): 421–437.

► Lent, R. W., Brown, S. D., and Hackett, G. (2000, Jan.). Contextual supports and barriers to career choice: A social cognitive analysis. *Journal of Counseling Psychology, 47*(1): 36.

Contributed by A. Kaatz. (2018). *Coping Efficacy.*

COPING EFFICACY

Learning Objectives

Trainees will:

- Be able to distinguish between emotion- and problem-focused coping strategies.
- Identify when it is appropriate to apply emotion- and problem-focused coping strategies.

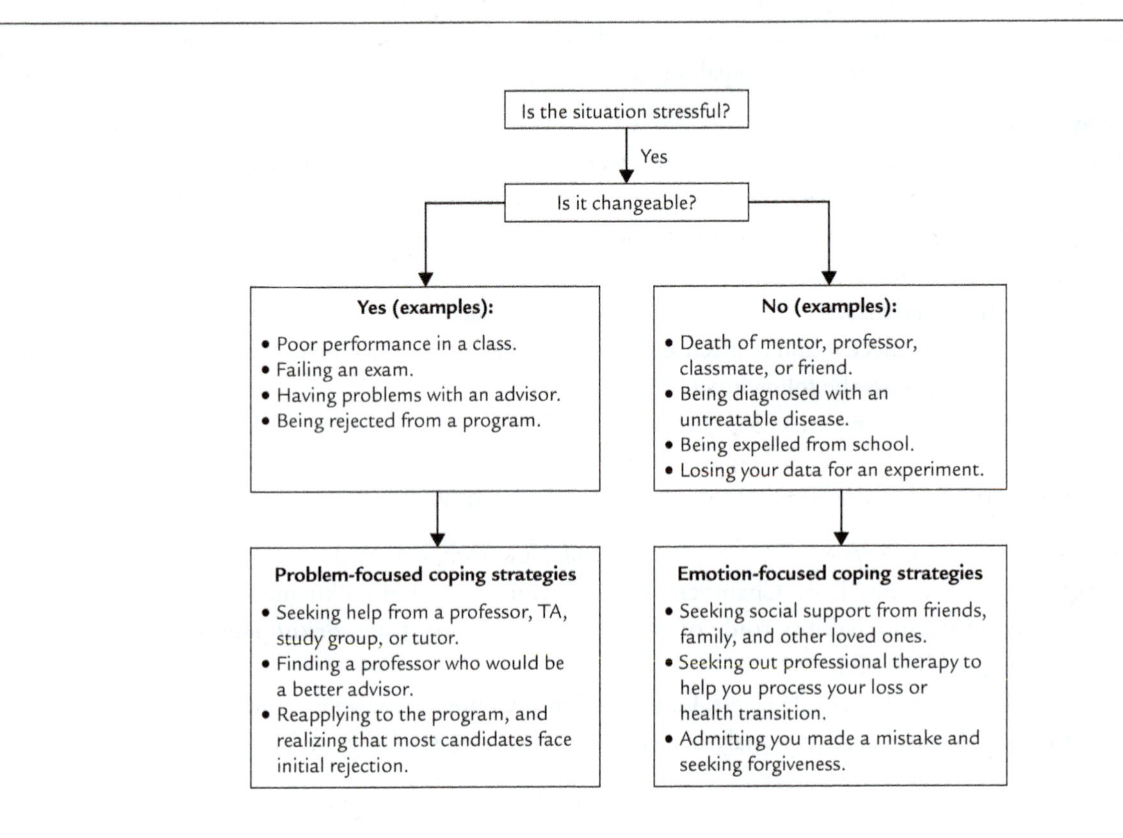

**Note:* Most situations require both types of coping. For example, being expelled from school is not something you can change, so it will require some emotion-focused coping; but you can also seek out a new school that might be a better fit, which would be a problem-based coping strategy.

Case Studies

Case Study #1: In math class today, your professor handed back the first midterm exam. In high school, you had done well in math and you felt that you'd studied hard for this first exam. To your surprise, you see that you got a C– on the test. You feel embarrassed, and quickly turn over your exam so your classmates don't see your low grade. You start thinking that you might not be cut out for a career in science. What should you do?

Contributed by A. Kaatz. (2018). *Coping Efficacy.*

Case Study #2: You are close to finishing your biology degree and are waiting to hear from graduate schools. Last week at your annual physical health exam, your primary care doctor noticed that you'd described an interesting cluster of symptoms. So, she ran a few tests. She called you to come back into her office to discuss the results. She informed you that you have a disease that will increasingly begin to cause extreme fatigue and impair your mobility. There are drugs that can help, but no cure. You are shocked. You had your life planned out . . . you were going get a Ph.D. in biology and become faculty at a research institution . . . you don't see how you could still do that if you begin to experience the symptoms your doctor described. What should you do?

Case Study #3: You've been working extra hours in the lab to get enough data to earn authorship on the research team's next paper. However, it is unclear how much data will be enough and there is no guarantee that your mentor will include you as an author, even if you contribute a lot of data. Worrying about this has created a lot of stress for you. Everyone says you need to be an author on a peer-reviewed paper to get into the good graduate-training programs. On top of this, you're starting to notice that spending a lot of time in the lab is making it difficult to keep up with your coursework. Now you're beginning to worry about your grades as well. What should you do?

Case Study #4: You've finished all of your coursework, passed your preliminary exam, and now must write your thesis proposal. Up to this point, everything you've been required to do as a graduate student has been familiar: taking classes, passing exams, and carrying out experiments. These are all things that you did as an undergraduate student and you've been able to handle it. Writing a thesis proposal, on the other hand, is brand-new territory. You have to be independent. How will you come up with your own project? You're completely stressed, and it feels like your brain is frozen. You don't know where to begin. What should you do?

Case Study #5: Micah, a postdoctoral scholar on your research team, has been a good friend and mentor to you since you began graduate school four years ago. You and your partner frequently go camping and hiking with him and his wife, Lila, and you have become part of their circle of friends. Lately, you've noticed that Micah is making a lot of mistakes when doing experiments. You ask him if everything is OK and learn that he and Lila have been having trouble and are contemplating divorce. He begins telling you stories about Lila, who you consider a good friend, and urges you to take his side in their disputes. This has been going on now for over 6 months, and it is creating a stressful environment for you in the lab. What should you do?

Case Study #6: You have been working with the Nikita research group for one semester and are feeling good about your progress in understanding the research context and have gained confidence in your ability to carry out experiments independently. Recently you have noticed that your mentor seems aloof and does not want to talk to you or answer your questions. You have overheard her making remarks to another graduate student about how little you know and how much time it takes to mentor you. You feel caught because you want to do a good job, but sometimes you don't know what you are doing and need to ask questions, but she won't answer them. In the coming week you are supposed to analyze your data and you need your mentor's help. When you arrive in her office to discuss the data analysis, she sees a mistake, tells you to fix it, and puts on her headphones while you make changes to the data file. The uncomfortable working environment is stressful and has diminished your confidence to do research. You're considering leaving the lab. What should you do?

Case Study #7: Now it is your turn. Reflect on a time when you were in a challenging academic situation. Was it a changeable situation? Was it an unchangeable situation? Was it a little of both? How did you handle it? Did you use the most beneficial coping strategies? If not, what would you do differently?

Contributed by A. Kaatz. (2018). *Coping Efficacy.*

COUNTER-STORYTELLING

Learning Objectives

Trainees will:

- Recognize that "double consciousness" can generate stress and issues of identity convergence for nonmajority people.
- Imagine scenarios of equity and mutual respect in research and other professional interactions by reframing and reenvisioning responses to microaggressions, stereotype threat, and messages of inferiority.
- Become more aware of and challenge negative stereotypes.
- Reframe racial/ethnic narratives to become more empowered.

Trainee Level

undergraduate or graduate
novice, intermediate, or advanced trainees

Areas of Trainee Development

- Equity and Inclusion Awareness and Skills
  - Advance equity and inclusion in the research environment.
- Researcher Confidence and Independence
  - Develop confidence as a researcher.
- Professional and Career Development Skills
  - Develop confidence in pursuing a research career.

Activity Components and Estimated Time for Completion

- Session 1 Time: 1 hour
- Session 2 Time: 1 hour

Total time: 2 hours

When to Use This Activity

This activity is suitable when the trainees are primarily from marginalized racial/ethnic groups, but it can be adapted for other marginalized groups (LGBT, SES, ability, veteran). It is designed to provide core skills for students to address the daily hassles they experience by being silenced or disrespected in research environments. This activity builds community through story-sharing (i.e., counter-storytelling).

This activity requires facilitation of discussions involving critical race theory, which may be unfamiliar to some facilitators. The articles listed below are recommended reading for those who wish to learn about this theory, beyond the information provided in this facilitator guide. Additional resources are listed at the end of the facilitator notes.

- Solórzano, D. G., and Yosso, T. J. (2002). A critical race counterstory of race, racism, and affirmative action. *Equity & Excellence in Education, 35*(2): 155–168.
- Terry, C. L. (2011). Mathematical counterstory and African American male students: Urban mathematics education from a Critical Race Theory perspective. *Journal of Urban Mathematics Education, 4*(1): 23–49.

Contributed by C. Saetermoe and T. Buenavista. (2018). *Counter-storytelling.*

Inclusion Considerations

Some trainees, especially those from underrepresented backgrounds, may shift from being excited by the new revelations and understandings around double consciousness to being reflective and quiet. Expect both reactions and let trainees know that this is a difficult conversation and will take some time for them to process all that it means to them.

Contributed by C. Saetermoe and T. Buenavista. (2018). *Counter-storytelling.*

Implementation Guide

Workshop Session (1 hour)

Session 1: Introduction to Counternarratives (1 hour)

- Facilitators should review the example story and counter story with annotations at the end of the facilitator guide and familiarize themselves with the terms and readings provided in both sessions.
- Introduce the concept of "**double consciousness**" (DuBois). Whites are often unaware that nonmajority individuals must negotiate two worlds—their own, indigenous knowledge from their culture, family, and environment and the majority consciousness that allows them to survive in the broader world. Ethnic minority students likely experience this, but may not know what to call it.
  - https://www.brainyquote.com/quotes/quotes/w/webdubo392715.html: "It is a peculiar sensation, this double-consciousness, this sense of always looking at one's self through the eyes of others, of measuring one's soul by the tape of a world that looks on in amused contempt and pity."
- In this session, three concepts are discussed to help trainees identify ways that can allow them to reframe their experience by recognizing patterns of thinking and behaving that can generate inequity and bad feelings in their life:
  - majoritarian narrative
  - microaggressions
  - stereotype threat
- **Activity #1: Definitions and Terms** (15 minutes)
  - List each of the terms on a whiteboard or flipchart with a brief definition (see examples listed below). Ask trainees to brainstorm some possible examples of each term. Some suggested definitions and discussion points are provided below.
  - **Majoritarian narrative:** the belief that everyone has the same opportunities to be successful in life (meritocracy).
    - Majoritarian narratives reflect the underlying belief in meritocracy and natural order. The idea is that we believe that the current state of affairs is the "natural order" of things—that students of color really ARE inferior, and that is why things don't change. The majoritarian narrative ignores the historical and structural reasons why inequity doesn't change—economic advantages of Whites from the inception of the country, racist laws, and policies like Jim Crow that forced inequality, and subtle, more insidious forms of racism that are often unconscious and unintentional, but harmful, nonetheless. (Majoritarian thinking leads people to assign individual blame for social conditions rather than recognizing the privilege of those who generated the narrative.) For further reading, see Solorzano, D. G., and Yosso, T. J. (2002). A critical race counterstory of race, racism, and affirmative action. *Equity & Excellence in Education, 35*(2): 155–168.
    - Questions for discussion:
      - When you think of the "average" researcher at your university, what are they like and what do they believe?
      - Are all voices equally powerful in research groups?
      - Who wrote the narrative that determines what is valued in your field in terms of research?
      - Who wrote the narrative that determines, expectations, decisions, and what is valued in your research group?

- **Microaggressions** are social slights, insults, or gestures that remind the recipient of bias against them as a member of marginalized groups.
  - For facilitators and students who are less familiar with the concept of microaggressions, the video "What Kind of Asian Are You?" introduces this concept through a brief encounter between two joggers. (https://www.youtube.com/watch?v=DWynJkN5HbQ)
  - Microaggressions, such as crossing the street when seeing a person of color walking up the sidewalk, holding one's purse tight in an elevator with a person of color, and complimenting a person of color for being articulate, add up to cumulative health disparities through the stressful experiences of being a person of color above and beyond economic, educational, or other resources. Microaggressions can be likened to "daily hassles" (Kanner, 1981), known to have a steady and eroding effect on cardiac health.
    - Sue, D., Capodilupo, C., Nadal, K., and Torino, G. (2008). Racial microaggression and the power to define reality. *The American Psychologist, 63*(4): 277–279.
    - http://www.microaggressions.com/
  - **Questions for Discussion:**
    - What are some examples of microaggressions that can occur in the research environment?

> - A sign that says "English only" posted in the research space.
> - The PI consistently calling upon White/majority research group members to lead projects.
> - Automatically suspecting research group members from historically underrepresented racial/ethnic groups for lost or damaged research equipment.

- **Stereotype threat** is a reaction to the low expectations of others that leads to performance levels that would typically be higher without the experience of cues to low expectations.
  - Claude Steele's book, *Whistling Vivaldi,* is a great resource to better understand the clear and evident influence of low expectations of one's marginalized group create sufficient anxiety to suppress naturally occurring positive qualities, especially in a cued high-value task such as the SAT, GRE, or final examinations.
    - Steele, C. M. (1997). A threat in the air: How stereotypes shape intellectual identity and performance. *The American Psychologist, 52*(6): 613–629.
  - Examples of how stereotype threat might be activated in research groups:
    - Asking research group members to report race/ethnicity and other demographic information on the beginning of a lab/research group application form.
    - Being the only one of a particular identity group present in a research group.

► **In-Class Activity: Writing a Counterstory** (30 minutes)

- Distribute the activity "Writing a Counterstory: Dr. Neuman's Lab."
- (5–10 minutes) Ask trainees to read the story and identify issues and concepts (middle column on the table). If trainees struggle, working in pairs may help.
- (10 minutes) Bring students together to share and discuss the issues and concepts they identified.
- (20 minutes) Ask trainees to write a counternarrative, based on the issues and concepts they identified. If time is limited, trainees may be assigned specific sections to work on.

Contributed by C. Saetermoe and T. Buenavista. (2018). *Counter-storytelling.*

Session 2: *Your* Story and Counterstory

Workshop Session (1 hour)

► **Activity: Writing and Rewriting *Your* Story**

- (15 minutes) Distribute trainee materials handout and ask trainees to respond to the following prompt: Thinking back a month or two, remember a troublesome incident that happened at your university when some aspect of your identity (e.g., race/ethnicity, gender identity, ability status, socioeconomic status, religion, sexual orientation, first-generation status) was relevant. Replay that incident in your mind a few times, writing down details in chronological order.
- (10 minutes) Invite trainees to pair up and read a peer's story. They should draw a star or an asterisk at points when their peer seemed to feel less than powerful. Ask trainees to note choice points where majoritarian narratives, microaggressions, or stereotype threat may have been relevant.
- (5 minutes) Trainees should meet with their partner to share their reflections.
- (15 minutes) Ask trainees to rewrite their story. This counternarrative should result in a different outcome that challenges the underlying assumptions that led to their first narrative. Encourage trainees to write themselves as fearless, strong, and to transform the outcome to be one where there is equity and justice.

► **Discussion:** Describe your experience with this activity. (15 minutes)

- What was easy about this assignment?
- What was challenging about this assignment?
- What are the key differences between your first and second narratives?
- How can you use counternarratives to improve your research experience? Your relationship with your mentor?

Additional Resources

Bell, D. A. (1992). *Faces at the bottom of the well.* New York, NY: Basic Books.

Carlson, J. R. (2012). Encounters with counterstories: Reading the past critically with non-fiction books for young adults. *Wisconsin English Journal, 54*(1): 52–65.

Delgado, R. (1989). Storytelling for oppositionists and others: A plea for narrative. *Michigan Law Review, 87*: 2411–2441.

Hernandez, E. (2013). Using LatCrit, autoethnography, and counterstory to teach about and resist Latina/o nihilism. In M. S. Plakhotnik & S. M. Nielsen (Eds.), Proceedings of the 12th Annual South Florida Education Research Conference (pp. 102–109). Miami: Florida International University. Retrieved from http://education.fiu.edu/research_conference/

Kanner, A. D., Coyne, J. C., Schaefer, C., and Lazarus, R. S. (1981). Comparison of two modes of stress measurement: Daily hassles and uplifts versus major life events. *Journal of Behavioral Medicine, 4*(1): 1–39.

Solórzano, D., and Delgado Bernal, D. (2001). Examining transformational resistance through a critical race and LatCrit theory framework. *Urban Education, 36*(3): 308–342.

Steele, C. M. (2010). *Whistling Vivaldi: How stereotypes affect us and what we can do.* New York, NY: W.H. Norton & Company, Inc.

Contributed by C. Saetermoe and T. Buenavista. (2018). *Counter-storytelling.*

WRITING A COUNTERSTORY: DR. NEUMAN'S LAB ANNOTATED STORY AND COUNTERSTORY

The following table takes the example story provided in the trainee materials, highlights the key topics raised in the original story, and reframes them in the counterstory to empower the student.

Original Story	Issues/Concepts	Counterstory
I had just started in Dr. Neuman's lab. He showed me the equipment, introduced me to the grad students in the lab, and talked about some of the studies that they are conducting. I was so scared!	Framing new experiences as exciting and a privilege AS WELL AS a new, scary situation, recognizing one's own strengths (social cultural wealth–Yosso). Majoritarian narrative = I start out at a disadvantage Empowered narrative = I start out ready to learn	I had just started in Dr. Neuman's lab. He showed me the equipment, introduced me to the grad students in the lab, and talked about some of the studies that they are conducting. I was so scared, but I also know that I am a good student and I'm ready to work hard and learn, so it's pretty exciting stuff!
I was worried I wouldn't remember it all, so I was taking notes. I can remember a grad student snickered when he saw me taking notes and I was super embarrassed, so I stopped taking notes.	Recognizing social cultural strengths and being grounded in them even when challenged. Reaction to *microaggressions* can be to simply be good at what you do OR to confront. She chooses to ignore him. She could also say something like, "I'm guessing you took notes when you started the lab, eh?" Majoritarian narrative = conform Empowered narrative = trust your skills and history	Knowing myself, I took the strategy of taking notes while I took the tour. I think of myself as responsible and a good student. When the graduate student laughed, I knew that was his bad trip and that I had to take notes so that I could be useful in the lab.
Dr. Neuman was going to take the grad students to dinner, and he asked me to stay back and purify DNA. Everyone left, and I burst into tears. I don't know how to purify DNA properly. So I tried, but I was fairly sure I didn't do it right.	Strengths of reaching out, knowing when you need help, making connections Majoritarian narrative = hierarchy, assumptions Empowered narrative = solidarity with others, problem solving, active response	When Dr. Neuman left with the graduate students and left me behind to purify DNA, I wasn't quite sure how to do it. I looked around the hallway, and found a grad student—she was willing to teach me, and we became friends. I asked for help because I knew I didn't know how to purify DNA.

Contributed by C. Saetermoe and T. Buenavista. (2018). *Counter-storytelling.*

I waited for Dr. Neuman to come back to the lab, and he told me that I had not purified the DNA properly. He asked me why I didn't know how to purify DNA and said, "this is why I don't usually take students in programs like this."	Dr. Neuman is engaged in stereotyping our student. He has negative expectations that she has to overcome. Majoritarian narrative = She succumbs to *stereotype threat* and doubts herself. Empowered narrative = She challenges Dr. Neuman's expectations through problem solving.	When Dr. Neuman came back to the lab, he was surprised that I had purified the DNA properly. He told me that he was surprised. I told him that I was prepared to learn and that my program and my prior education as well as my personal skills made me ready to be in his lab.
I apologized and asked if he could show me what I did wrong. He told me to ask a graduate student the next day, that he needed to do his own work.	One reason mentors do not grow is that they are rarely challenged. There are ways to approach difficult topics in a respectful manner. Majoritarian narrative = hierarchy and "sink or swim" This will continue unless it is openly confronted.	I told Dr. Neuman that it bothered me that he didn't expect me to purify DNA properly and that I needed to be treated with respect by him and by the graduate students. We discussed ways that our lab could have better communication for everyone.
The next day, I did learn to purify DNA, but I never felt quite the same about working in a lab and I may study psychology instead. Maybe biology isn't for me.	Science push-outs are common. All students need a sense of belonging. The majoritarian narrative is that folks who are in science are the ones who belong. It's tough to break in, and it takes strength beyond what is required of more privileged students to negotiate these low expectations and disrespect. This strength should not only be acknowledged, it should become unnecessary.	I decided to move to another lab because my interests changed. I became interested in biochemistry and found a lab with a really great culture of curiosity and support. I have been admitted to a doctoral program at UC Berkeley and I am excited to become a biochemist.

Contributed by C. Saetermoe and T. Buenavista. (2018). *Counter-storytelling.*

COUNTER-STORYTELLING

Learning Objectives

Trainees will:

- Recognize that "double consciousness" can generate stress and issues of identity convergence for nonmajority people.
- Imagine scenarios of equity and mutual respect in research and other professional interactions by reframing and reenvisioning responses to microaggressions, stereotype threat, and messages of inferiority.
- Become more aware of and challenge negative stereotypes.
- Reframe racial/ethnic narratives to become more empowered.

Activity #1

Writing a Counterstory: Dr. Neuman's Lab

Instructions: Read the story below, drawing a star or asterisk at points when the student may feel less than powerful. Use the table on the next page to identify where majoritarian narratives, microaggressions, or stereotype threat may have been relevant and to rewrite the narrative with a different outcome. The new outcome should challenge the underlying assumptions presented in the original narrative. Write the student to be fearless and strong, and transform the outcome to be one where there is equity and justice.

I had just started in Dr. Neuman's lab. He showed me the equipment, introduced me to the grad students in the lab, and talked about some of the studies that they are conducting. I was so scared! I was worried I wouldn't remember it all, so I was taking notes—I can remember a grad student snickered when he saw me taking notes and I was super embarrassed, so I stopped taking notes. It smelled so strange in the lab! I wondered if it was healthy to be there because I didn't see any ventilation. By then, I'd had enough embarrassment, so I didn't say anything. Dr. Neuman was going to take the grad students to dinner, and he asked me to stay back and purify DNA. Everyone left, and I burst into tears. I don't know how to purify DNA properly. So I tried, but I was fairly sure I didn't do it right. I waited for Dr. Neuman to come back to the lab, and he told me that I had not purified the DNA properly. He asked me why I didn't know how to purify DNA and said, "This is why I don't usually take students in programs like this." I apologized and asked if he could show me what I did wrong. He told me to ask a graduate student the next day, that he needed to do his own work. The next day, I did learn to purify DNA, but I never felt quite the same about working in a lab and I may study psychology instead. Maybe biology isn't for me.

Trainee Materials

Contributed by C. Saetermoe and T. Buenavista. (2018). *Counter-storytelling.*

Original Story	What issues and concepts are present?	Rewrite the story: Counterstory
I had just started in Dr. Neuman's lab. He showed me the equipment, introduced me to the grad students in the lab, and talked about some of the studies that they are conducting. I was so scared!		
I was worried I wouldn't remember it all, so I was taking notes. I can remember a grad student who snickered when he saw me taking notes and I was super embarrassed, so I stopped taking notes.		
Dr. Neuman was going to take the grad students to dinner, and he asked me to stay back and wash the DNA. Everyone left, and I burst into tears. I don't know how to wash DNA properly. So I tried, but I was fairly sure I didn't do it right.		
I waited for Dr. Neuman to come back to the lab, and he told me that I had not washed the DNA properly. He asked me why I didn't know how to wash DNA and said, "This is why I don't usually take students in programs like this."		
I apologized and asked if he could show me what I did wrong. He told me to ask a graduate student the next day, that he needed to do his own work.		
The next day, I did learn to wash DNA, but I never felt quite the same about working in a lab and I may study psychology instead. Maybe biology isn't for me.		

Contributed by C. Saetermoe and T. Buenavista. (2018). *Counter-storytelling.*

ACTIVITY #2 WRITING YOUR STORY

Step 1: Tell your story. Thinking back a month or two, remember a troublesome incident that happened in your research team when some aspect of your identity (e.g., race/ethnicity, gender identity, ability status, socioeconomic status, religion, sexual orientation, first-generation status) was relevant. Replay that incident in your mind a few times, writing down details in chronological order. Write your reflection below as a narrative story. Your narrative should allow your reader to put herself/himself in your shoes. Your reader should be able to see, feel, hear, smell, touch, and relate everything that you were experiencing at the time.

Your Story:

Step 2: Pair and Read. Pair with a peer and read one another's stories drawing a star or asterisk at points when you felt your peer was less than powerful. Note choice points where majoritarian narratives, microaggressions, or stereotype threat may be relevant.

Step 3: Tell Your Counterstory. Rewrite your narrative with a different outcome in a way that challenges the underlying assumptions that led to your first narrative. In your narrative, write yourself to be fearless, strong, and transform the outcome to be one where there is equity and justice.

Your Counterstory:

Contributed by C. Saetermoe and T. Buenavista. (2018). *Counter-storytelling.*

CASE STUDY: CREDIT WHERE CREDIT IS DUE

Learning Objectives

Trainees will:

- Discuss and practice ethical research decision making.

Trainee Level

undergraduate or graduate trainees
novice trainees

Areas of Trainee Development

- Research Ethics
  - Develop responsible and ethical research practices.
- Research Comprehension and Communication Skills
  - Develop an understanding of the research environment.

Activity Components and Estimated Time for Completion

- In Session Time: 25 minutes

Total time: 25 minutes

When to Use This Activity

This activity is suitable for undergraduate and graduate level novice trainees who have chosen a research mentor and who have been conducting research for at least one month. It can be combined with any of the following activities as part of a larger unit on ethics in research:

- Case Study: The Sharing of Research Materials
- Case Study: Selection of Data
- Ethics Case Discussion with Mentor
- Truth and Consequences Article

Inclusion Considerations

Discuss with trainees how understanding of ethical behavior may be different based on differences in cultural backgrounds or across generations. Facilitators can ask trainees to consider the case study from different cultural or generational perspectives. Emphasize that it can sometimes be as difficult to determine whether behavior is unethical as it is to decide how to deal with that behavior in a sensitive and respectful manner. Encourage trainees to seek input from others, in particular those who can offer different perspectives, when dealing with potentially unethical situations in the research environment.

Contributed by A. R. Butz and A. Smith with information from Branchaw, J. L., Pfund, C., and Rediske, R. (2010). *Entering Research: A Facilitator's Manual.* New York: W.H. Freeman & Co.

Implementation Guide

Workshop Session (25 minutes)

- ► **Introduction** (5 minutes)
  - The goal of this activity is for trainees to consider ethical dilemmas they may encounter in research and to discuss possible strategies to address these types of challenges.

Workshop Session (20 minutes)

- ► **Case Study** (5 minutes)
  - Distribute the case study and ask trainees to read and individually write down answers to the discussion questions.
  - Does Bea have any way of receiving credit for her work?

> • It is unlikely that Bea will be credited after the fact. However, Bea could try to have her technique published elsewhere, so that other researchers may credit her work in the future.

  - Should she contact Dr. Freeman in an effort to have her work recognized?

> • Bea could reach out to Dr. Freeman as a follow-up to congratulate him/her on the article. She could mention that she is working to publish the technique and would appreciate it if Dr. Freeman cites her poster until the paper comes out.

  - Is Bea's faculty advisor mistaken in encouraging his students to be open about their work?

> • Different mentors have different philosophies on sharing data, methods, and research. There are also different approaches to how ideas are cited and recognized in presentations and publications. These can vary by discipline and by mentor.

  - What could Bea have done to prevent this from happening?

> • Bea could have mentioned to Dr. Freeman that she intended to publish the technique.

- ► Alternatively, trainees can discuss these questions in small groups and report highlights from their discussion to the larger group.
- ► **Large-Group Discussion** (15 minutes)
  - Bring the entire group together to discuss the case.
- ► **Wrap-up** (5 minutes)
  - Summarize the main ideas generated from the large-group discussion.
  - Encourage trainees to talk with their mentors about their policies and philosophies regarding sharing work with others outside of the research group.

Contributed by A. R. Butz and A. Smith with information from Branchaw, J. L., Pfund, C., and Rediske, R. (2010). *Entering Research: A Facilitator's Manual.* New York: W.H. Freeman & Co.

CASE STUDY: CREDIT WHERE CREDIT IS DUE

Learning Objectives

Trainees will:

- Discuss and practice ethical research decision making.

Modified from: "On Being a Scientist: Responsible Conduct in Research," 2nd ed., National Academy Press, 1995

Bea was working on a research project that focused on developing a new experimental technique. She prepared a poster outlining the new technique and presented it at a national conference. During the poster session, Bea was surprised and pleased when Dr. Freeman, a leading researcher in her discipline, engaged her in a conversation. Dr. Freeman asked extensively about the new technique, and she described it fully, happy to be confidently discussing her work with a fellow scientist. Bea's faculty advisor had encouraged his students to openly share their research with other researchers, and Bea was flattered that Dr. Freeman was so interested in her work.

Six months later Bea was leafing through a journal when she noticed an article by Dr. Freeman. The article described an experiment that clearly depended on the technique that Bea had developed. She did not mind, in fact, she was somewhat flattered that her technique so strongly influenced Dr. Freeman's work. She turned to the citations, expecting to see a reference to her abstract or poster; however, her name was nowhere to be found.

Discussion Questions

1. Does Bea have any way of receiving credit for her work?

2. Should she contact Dr. Freeman in an effort to have her work recognized?

3. Is Bea's faculty advisor mistaken in encouraging his students to be open about their work?

4. What could Bea have done to prevent this from happening?

Contributed by A. R. Butz and A. Smith with information from Branchaw, J. L., Pfund, C., and Rediske, R. (2010). *Entering Research: A Facilitator's Manual.* New York: W.H. Freeman & Co.

DEVELOPING A CURRICULUM VITAE

Learning Objectives

Trainees will:

- Learn the differences between a resume and curriculum vitae (CV).
- Draft a CV.

Trainee Level

undergraduate or graduate trainees
novice trainees

Areas of Trainee Development

- Researcher Identity
  - Develop identity as a researcher.
- Professional and Career Development Skills
  - Explore and pursue a research career.

Activity Components and Estimated Time for Completion:

- In Session Time 1: 1 hour
- Pre-Assignment for Session 2: 1–2 hours
- In Session Time 2: 1 hour

Total time: 3–4 hours

When to Use This Activity

This activity may be used at any point in a trainee's career. However, it is particularly useful for undergraduate students preparing to apply to summer research internships or graduate training programs. For graduate students, this activity can be useful at the beginning of their program and then updated regularly (at least every 6 months).

Inclusion Considerations

Trainees from diverse backgrounds, in particular those who are the first in their family to go to college or graduate school, may wonder whether their previous experiences are "good enough" for an academic CV. To help them articulate the value of their experiences, ask them to identify and validate other models of leadership and success in their lives outside of research. For example, responsibility at a job or leadership within their family, or religious or other organizations. Discuss how these types of experiences are equally valuable and provide evidence of the kind of problem solving, work ethic, and commitment needed to do research successfully. If available, provide examples of academic CVs that include nonacademic experiences.

Some trainees may come from backgrounds where humility is valued, and they have been acculturated not to speak of their achievements freely. Talk about why it is important to present one's work positively and with confidence on their CV and reassure trainees that they can do this while remaining true to their ideals around humility.

Contributed by A. Bramson. (2018). *Developing a Curriculum Vitae.* Additional information from Purdue Online Writing Lab. (n.d.). *Writing the Curriculum Vitae.* https://owl.purdue.edu/owl/job_search_writing/resumes_and_vitas/writing_the_cv.html.

Implementation Guide

This activity includes two sessions, between which trainees draft a CV. In the first session, trainees are introduced to CVs through a large-group discussion and an in-class activity. In the second session, they share their CV drafts, discuss the content and formatting choices they made, and offer one another feedback for improvement.

BEFORE the first session, facilitators need to gather resumes and CVs from a variety of disciplines to be used as examples in the discussion. Trainees could be asked to bring a copy of their mentor's CV and/or resume.

Workshop Session One (1 hour)

► **Large-Group Discussion** (20 minutes)

- Distribute examples of a resume and CV. If the trainees represent various disciplines, then several examples from different disciplines can be distributed. Distribute the "Comparing Resumes and Curriculum Vitae" worksheet from the trainee materials. Pair trainees to review the examples of resumes and CVs and ask them to identify the similarities and differences between the documents using the worksheet. Use the questions below to guide the discussion.
- **How does a CV compare to a resume?**
  - What differences do you note on your worksheet?
  - What are the components of each and what purposes do you think each serves?
- **How might a CV change throughout one's career?**
  Distribute examples of CVs from people at different stages of their careers (undergraduate intern, graduate student, post-doc, research scientist in academy/industry/government position, assistant professor, full professor, etc.).
  - CVs are expanded over time and information is generally not deleted. However, as individuals become more advanced in their career, it may be appropriate to remove some content to eliminate less relevant items. For example:
    - By the time trainees reach the mid-point of their graduate training, they will have likely removed small/odd-end jobs completed in high school or college from their CVs.
    - As trainees progress in their research career, they will give many presentations. Shortening the presentation list to only the most recent or those that are most important may be advised.
  - Some items, such as the summary of a trainee's formal education and training experiences and peer-reviewed publications, should never be removed.
  - Sometimes a shortened (2-page) condensed CV is requested (e.g., grant proposal applications). In this situation, one should identify and retain only the information that is most relevant to the specific request.

► **Brainstorm** (15 minutes)

- **What should be on *my* CV at this stage in my career?**
  There are no specific rules for what to include on a CV. It depends on what trainees want to highlight and how the CV will be used. At the beginning stages of a career, it is best to include all academic, professional, and educational activities. Later, trainees can customize it to meet career needs. For example, undergraduate trainees might include the following:
  - **Educational Experiences and Academic Achievements:** awards and scholarships, research experiences, research publications or presentations, relevant coursework.

Contributed by A. Bramson. (2018). *Developing a Curriculum Vitae.* Additional information from Purdue Online Writing Lab. (n.d.). *Writing the Curriculum Vitae.* https://owl.purdue.edu/owl/job_search_writing/resumes_and_vitas/writing_the_cv.html.

  - **Professional Experiences:** jobs, internships, memberships in professional organizations (i.e., American Astronomical Society).
  - **Leadership and Service Experiences:** clubs, extracurricular activities, leadership activities on campus and beyond campus.
  - **Teaching, Mentoring, and Outreach Experiences:** tutoring, organizing study groups, undergraduate TA or preceptorship, organizing public outreach events, mentoring underclassmen.
- **How should I format my CV?**
  Be concise, yet descriptive. Strategies to use:
  - **Gapping:** Use incomplete sentences to present information succinctly. For example, instead of writing, "I was a guest lecturer and grader for Geology 101, an undergraduate-level course, during the Fall 2015 semester," I wrote "Geology 101 (Fall 2015): Gave guest lectures and graded exams for this undergraduate-level course."
  - **Parallelism:** Use the same structure of phrases (especially verb tenses) and visual aspects throughout the document. Likewise, list information in the same chronological order in all sections (i.e., most recent to oldest, or vice versa).

► **Activity: Developing a Curriculum Vitae Worksheet** (10 minutes)
- Distribute and introduce the assignment. Trainees should use the worksheet to list the academic and professional activities and experiences that they may want to include on their CV.
- Point out that as demonstrated by the examples shown in class, CVs do not have one standard format and do not need to be organized into the categories listed on that worksheet. They should use examples from their discipline (e.g., their mentor's CV) to decide how to organize it.

Alternative Implementations:

► *As a 1-hour workshop:* Trainees can complete the worksheet, with the recommendation to create a formatted version on their own after they leave. The discussion and worksheet should be sufficient, if combined with enough examples of formatted CVs to provide them with the knowledge they need to be able to make a formatted version on their own.

► *As a half-day workshop:* This activity could be done in a computer lab. After trainees fill out the worksheet, they help each other decide what information should go on their CVs. Trainees brainstorm together in real time as they work on the computers to create their first CVs and find the best organization and formatting to use in their documents.

Pre-Assignment for Session Two (1–2 hours)

► Assign trainees to create a formatted CV based on their completed worksheet and bring two or three copies to the next session.

Workshop Session Two (1 hour)

► **Check-in** (10 minutes): Share CV creating experiences.
- How difficult was it to draft your CV?
- How did you decide what to include? Were there items you included that were not on the worksheet? Were there items from the worksheet you did not include?
- How did you format your CV? What influenced this decision?
- What advice would you give to trainees making a CV for the first time?

Contributed by A. Bramson. (2018). *Developing a Curriculum Vitae.* Additional information from Purdue Online Writing Lab. (n.d.). *Writing the Curriculum Vitae.* https://owl.purdue.edu/owl/job_search_writing/resumes_and_vitas/writing_the_cv.html.

- **Pair and Review** (40 minutes)
  - Trainees exchange their CV with a peer and review using the guiding questions below. Depending on time, 2 to 4 rounds of pairing and reviewing should be possible.
    - Is the information clear, concise, and formatted consistently?
    - Is the information provided relevant? Is anything missing?
    - Is the format logical and easy to follow?
    - What suggestions do you have to improve the CV? Offer at least two suggestions for each CV reviewed.
- **Reflection and Next Steps** (10 minutes)
  - (5 minutes) Trainees reflect on what they have learned from the feedback they received and from reviewing their peers' CVs and write down at least three ways in which they plan to modify their CV.
  - (5 minutes) The facilitator asks for volunteers to share what they learned and their plans for modification.

Alternative Implementations (Advanced Trainees)

- This activity can be implemented with advanced trainees who have already developed a CV by only implementing the Pair and Review component of Session Two. If trainees are actively seeking jobs or advanced training, it may be useful for them to bring in a copy of a job advertisement for their partner to review in conjunction with their CV.

Contributed by A. Bramson. (2018). *Developing a Curriculum Vitae.* Additional information from Purdue Online Writing Lab. (n.d.). *Writing the Curriculum Vitae.* https://owl.purdue.edu/owl/job_search_writing/resumes_and_vitas/writing_the_cv.html.

DEVELOPING A CURRICULUM VITAE

Assessment Rubric

This rubric can be used to assess the content of curriculum vitae.

Name of trainee:

	Absent	Developing	Mature	Comments
Heading				
Name at the top of the page				
Current contact information: address, telephone number, and email address				
Education				
Education (institution, years attended and [expected] graduation date, major, GPA, and/or class rank)				
Start and end dates for each institution				
Provided information about degree classification				
Brief details of your main project/dissertation				
List of Relevant Courses (courses applicable to your career interests or a position you are applying for)				
Started with your most recent course first and then worked backward				
Professional and Work Experience				
Listed from most to least recent experience				
Provided start and end dates				
Provided company/organization name				
Job title is included				
Provided brief descriptions of main duties and responsibilities				
Skills (computer, research, instrumentation, languages spoken)				
Work, Professional and/or Research Experiences (boss/advisor/mentor, years worked there)				
Teaching Experiences				
Mentoring Experiences				
Outreach Experiences				

Contributed by A. Bramson. (2018). *Developing a Curriculum Vitae.* Additional information from Purdue Online Writing Lab. (n.d.). *Writing the Curriculum Vitae.* https://owl.purdue.edu/owl/job_search_writing/resumes_and_vitas/writing_the_cv.html.

Presentations and Publications (list authors; title of talk or paper; locations, dates [conference name and city] and format [poster or oral] of presentations)				
Honors, Hobbies, and Interests				
Concentrated on relevant interest rather than listing				
Provided short bullet point explanations				
Listed additional skills employer might be looking for				
Leadership Activities (involvement timeline and short description of your role)				
Honors, Awards, Scholarships and Fellowships (name of award and date awarded)				
Additional Information				
Additional information relevant to job/career/position				
References				
Either listed or noted as available upon request				
Layout				
Attractive and draws attention to the important information				
Sans Serif Font				
Clean appropriate line spacing and margins				
Font size 11 or 12				
Free of long paragraphs of text				
Free of typographical errors				

Notes and constructive feedback for improvement:

Contributed by A. Bramson. (2018). *Developing a Curriculum Vitae.* Additional information from Purdue Online Writing Lab. (n.d.). *Writing the Curriculum Vitae.* https://owl.purdue.edu/owl/job_search_writing/resumes_and_vitas/writing_the_cv.html.

DEVELOPING A CURRICULUM VITAE

Learning Objectives

Trainees will:

- ► Learn the differences between a resume and curriculum vitae (CV).
- ► Draft a CV.

Comparing Resumes and Curriculum Vitae

	Resume	Curriculum Vitae
Describe the length and overall format of each document.		
List the types of information that are included in each document.		
For what purposes would each type of document be used?		
What other similarities or differences do you notice?		

Contributed by A. Bramson. (2018). *Developing a Curriculum Vitae.* Additional information from Purdue Online Writing Lab. (n.d.). *Writing the Curriculum Vitae.* https://owl.purdue.edu/owl/job_search_writing/resumes_and_vitas/writing_the_cv.html.

CURRICULUM VITAE ASSIGNMENT

Create a draft of your curriculum vitae (CV) by filling in the information below and reformatting it into a concise document. It is not necessary to use all of the information nor to keep the categories the same. Use an example CV from your discipline to guide your decisions about what content and format to use.

Use these strategies to format your CV:

1. **Gapping:** Use incomplete sentences to present information succinctly. For example, instead of writing, "I was a guest lecturer and grader for Geology 101, an undergraduate level course, during the Fall 2015 semester." write "Geology 101 (Fall 2015): gave guest lectures and graded exams for this undergraduate-level course."
2. **Parallelism:** Use the same structure of phrases (especially verb tenses) and visual aspects throughout the document. Likewise, list information in the same chronological order in all sections (i.e., most recent to oldest, or vice versa).

Use these questions to identify the information to include on your CV:

1. Name:

2. Contact Information (address, email address, phone number):

3. Education (school, years attended and [expected] graduation date, major, GPA and/or class rank):

4. List of Relevant Courses (courses applicable to your career interests or a position you are applying for):

5. Honors, Awards, Scholarships, and Fellowships (name of award and date awarded):

6. Skills (computer, research, instrumentation, languages spoken):

7. Work, Professional, and/or Research Experiences (advisor/boss, years worked there):

8. Presentations and Publications (list authors; title of talk or paper; locations, dates [conference name and city] and format [poster or oral] of presentations):

9. Leadership Activities (involvement timeline and a short description of your role):

10. Teaching/Mentoring/Outreach Experiences:

Contributed by A. Bramson. (2018). *Developing a Curriculum Vitae.* Additional information from Purdue Online Writing Lab. (n.d.). *Writing the Curriculum Vitae.* https://owl.purdue.edu/owl/job_search_writing/resumes_and_vitas/writing_the_cv.html.

DISCUSSION OF THE NATURE OF SCIENCE

Learning Objectives

Trainees will:

- Explain and debate the role of science and the relationship between research and knowledge.

Trainee Level

undergraduate or graduate trainees
novice trainees

Areas of Trainee Development

- Research Comprehension and Communication Skills
  - Develop disciplinary knowledge.
  - Develop logical/critical thinking skills.

Activity Components and Estimated Time for Completion

- In Session Time: 20–30 minutes

Total time: 20–30 minutes

When to Use This Activity

This activity can be used at any trainee stage; however, it may be most beneficial for novice trainees who are beginning their first research experience. A trainee does not need to have a mentor in order to participate in this activity. Other activities that may be implemented with this activity include:

- Science and Society
- Science Literacy Test
- Science or Pseudoscience
- What Happens to Research Results?

Inclusion Considerations

Encourage respect of various religious, cultural or other backgrounds of trainees as they relate to understanding the history and thoughts of science including discoveries and their impact on society.

Contributed by A. R. Butz with information from Branchaw, J. L., Pfund, C., and Rediske, R. (2010). *Entering Research: A Facilitator's Manual.* New York: W.H. Freeman & Co.

Implementation Guide

Workshop Session (20–30 minutes)

- **Introduction** (5 minutes)
  - The goal of this activity is to share ideas about the nature of science and research. What is science, and what is not science?
- Give trainees approximately 10 minutes to individually answer the "Discussion of the Nature of Science" worksheet questions.
- **Discussion**
  - With a small number of participants, the discussion can be facilitated as one group. With larger numbers of participants (i.e., more than 8–10), trainees should work in pairs or trios to come up with answers to share with the group.

> Ideas that have emerged in response to these questions include:
>
> - **What is science?**
>   - Science is using empirical methods to explain and understand things that go on around us.
>   - Science is acquiring knowledge through methods like observation, problem solving, and experimentation.
>   - Science is the discovery of the physical world through observation and experimentation.
> - **What kinds of questions can science answer?**
>   - Science can answer any question you can collect data on.
>   - Science can answer anything that is testable.
>   - Science cannot attempt to answer questions that require a value judgment.
> - **How are research and knowledge related?**
>   - Research is usually done in order to gain knowledge or understanding.
>   - Understanding assists in designing research to gain new knowledge.
>   - Research is a process for gathering knowledge.

Some additional questions to stimulate discussion:

- What does it mean to study the natural world?
- Who does science? Does it require a certain type of specialized training?
- Is there only one way to do science? Can you think of ways different from "the scientific method"?
- Why might others' views of what science is be different?

Contributed by A. R. Butz with information from Branchaw, J. L., Pfund, C., and Rediske, R. (2010). *Entering Research: A Facilitator's Manual.* New York: W.H. Freeman & Co.

DISCUSSION OF THE NATURE OF SCIENCE

Learning Objectives

Trainees will:

- ► Explain and debate the role of science and the relationship between research and knowledge.

Answer each of the following questions:

1. What is science?

2. What kinds of questions can science answer?

3. How are research and knowledge related?

Contributed by A. R. Butz with information from Branchaw, J. L., Pfund, C., and Rediske, R. (2010). *Entering Research: A Facilitator's Manual.* New York: W.H. Freeman & Co.

Implementation Guide

DISCUSSION WITH EXPERIENCED UNDERGRADUATE RESEARCHERS

Learning Objectives

Trainees will:

- Refine and expand a definition of research.
- Increase confidence about how to identify a research mentor.

Trainee Level

undergraduate trainees
novice trainees

Areas of Trainee Development

- Professional and Career Development Skills
    - Develop confidence in pursuing a research career.
- Researcher Identity
    - Develop identity as a researcher.
- Research Comprehension and Communication Skills
    - Develop an understanding of the research environment.

Activity Components and Estimated Time for Completion:

- In Session Time: 30 minutes–1 hour

Total time: 30 minutes–1 hour

When to Use This Activity

This activity can be used to introduce novice undergraduate trainees to research through interactions with more experienced undergraduate trainees. Additionally, the activity serves as an opportunity for the experienced trainees to practice their presentation and communication skills. Recruiting experienced trainees from diverse disciplines will provide novice trainees with the opportunity to broaden their view of the types of research available to pursue.

Inclusion Considerations

Recruit experienced trainees for the panel who come from diverse backgrounds. Encourage panelists to share their experiences openly, including any personal differences they have encountered and how they have navigated those differences. Novice trainees who have had little or no exposure to research may have misconceptions or unrealistic expectations about what doing research involves.

Contributed by B. L. Montgomery with information from Branchaw, J. L., Pfund, C., and Rediske, R. (2010). *Entering Research: A Facilitator's Manual.* New York: W.H. Freeman & Co.

Implementation Guide

Workshop Session (30 minutes–1 hour)

► **Introduction** (5 minutes)

- Briefly introduce the experienced undergraduate trainees.

► **Discussion** (30–55 minutes, depending upon the number of presenters):

- Each experienced trainee should prepare brief answers to the questions listed below in preparation for the session. Facilitators may either ask each trainee to give a brief presentation or structure the session as a panel discussion.
  - How long have you been doing research?
  - How did you find your research group/mentor?
  - What is your "best piece of advice" about how to find a research mentor?
  - How frequently do you meet with your mentor?
  - What topic(s) do you discuss in these meetings?
- After the presentations, the facilitator should moderate a question-and-answer session. Alternatively, novice trainees may be invited to ask questions immediately after each presentation. However, if this strategy is used, the facilitator must diligently manage the time spent on each question period to ensure that all experienced trainees have time to present.

Contributed by B. L. Montgomery with information from Branchaw, J. L., Pfund, C., and Rediske, R. (2010). *Entering Research: A Facilitator's Manual.* New York: W.H. Freeman & Co.

DISCUSSION WITH EXPERIENCED UNDERGRADUATE RESEARCHERS

Learning Objectives

Trainees will:

- Refine and expand a definition of research.
- Increase confidence about how to identify a research mentor.

A panel of experienced undergraduate researchers will share details about their experiences in finding a research mentor and conducting undergraduate research.

BEFORE the session:

1. What general questions do you have for the panelists? Write at least two questions.

DURING the session:

1. Take notes on the experiences of each of the undergraduate researchers participating as a panelist. (Use the back of this sheet.)
2. What specific questions do you have for each panelist? Write at least two specific questions.

<u>Panelist</u> <u>Notes</u> <u>Questions</u>

Contributed by B. L. Montgomery with information from Branchaw, J. L., Pfund, C., and Rediske, R. (2010). *Entering Research: A Facilitator's Manual.* New York: W.H. Freeman & Co.

DIVERSITY IN STEM

Learning Objectives

Trainees will:

- Be more aware of disparities between majority and historically underrepresented groups in STEM careers.
- Consider different reasons for disparities in STEM career/degree attainment.
- Articulate strategies to reduce these disparities in STEM career/degree attainment.
- Learn about the prejudice encountered by researchers belonging to historically underrepresented groups.

Trainee level

undergraduate or graduate trainees
novice, intermediate, or advanced trainees

Areas of Trainee Development

- Equity and Inclusion Awareness and Skills
  - Advance equity and inclusion in the research environment.
  - Develop skills to deal with personal differences in the research environment.

Activity Components and Estimated Time for Completion

- Trainee Pre-Assignment 1 Time: 5 minutes
- In Session 1 Time: 20 minutes
- Trainee Pre-Assignment 2 Time: 15 minutes
- In Session 2 Time: 35 minutes

Total time: 1 hour, 15 minutes

When to Use This Activity

This activity is appropriate for use with undergraduate or graduate trainees at any stage whose understanding of race and race relations in the United States is at the novice stage. If the trainee population in the workshop is a mix of novice and more advanced trainees in their understanding of race and race relations in academia, this can be beneficial for the facilitator. Facilitators should encourage dialogue among students, but be aware of singling out individuals to be the spokesperson for their particular group or identity.

This activity is used for raising trainees' awareness that careers and career interest disparities exist in STEM fields between trainees from well-represented racial/ethnic backgrounds and trainees from groups historically underrepresented in STEM. This activity will also raise awareness in trainees that individuals who identify with racial or ethnic groups historically underrepresented in STEM frequently encounter prejudice, discrimination, and racism in their academic and professional journey as a STEM trainee and researcher.

Contributed by S. Keyl. (2018). *Diversity in STEM.*

When implementing this activity, it is important that facilitators make time to establish trust in the group. It may be useful for facilitators to work with trainees to establish some ground rules and display them on the whiteboard or flipchart throughout the session to remind participants of the guidelines that they co-constructed. Although this activity primarily focuses on racial, ethnic, and gender diversity in STEM, many of the concepts and ideas highlighted here can be applied to other groups underrepresented in STEM (LGBTQ, visible and invisible disabilities, socioeconomic status, etc.).

Reading/Viewing Material:

Neil deGrasse Tyson response to the question "What's up with chicks and science?" (YouTube; beginning at 3:37): https://www.youtube.com/watch?v=q5S7QD9dryI
Dr. Quiñones' story: http://www.doctorqmd.com/dr-q-s-story---a-doctor-without-borders/

Readings/Resources for Facilitators

- https://www.nsf.gov/statistics/2017/nsf17310/digest/introduction/
- https://phys.org/news/2017-01-women-minorities-persons-disabilities-science.html Women, Minorities, and Persons with Disabilities in Science and Engineering
- Ong, M., Wright, C., Espinosa, L., and Orfield, G. (2011). Inside the double bind: A synthesis of empirical research on undergraduate and graduate women of color in science, technology, engineering, and mathematics. *Harvard Educational Review*, *81*(2), 172–209.

Inclusion Considerations

Encourage all trainees to share their experiences with the group, but do not require it. Trainees from groups historically underrepresented in research should not be asked or expected to speak for their respective identity group. All shared experiences should be presented as individual experiences yet considered in light of the historical and social contexts the trainees learn about in the readings. If trainees from well-represented groups expect their peers from underrepresented groups to be the main contributors to the discussions, redirect questions that they ask to the entire group for consideration.

Implementation Guide

Trainee Pre-Assignment 1 (5 minutes)

► **Reflection Exercise**

- Prior to the session, distribute the handout provided in trainee materials and tell trainees to write about the ideas/images that come to mind for each of the words listed. Encourage trainees to focus on the first image that comes to mind.

Workshop Session 1 (20 minutes)

► **Introduction** (5 minutes)

- The goal of this guided reading activity is to:
  - Get trainees to think about their preconceived notions about historically underrepresented groups in STEM (immigrants, undocumented peoples, and African Americans).
  - Introduce trainees to empirical evidence that asserts career disparities exist between dominant male groups and underrepresented female groups.
  - Expose trainees to narratives of STEM researchers' lived experiences encountering prejudice and racism.
- Tell trainees that they will be doing a lot of reflection for the next hour on topics that they may or may not have a lot of experience with and that they will be discussing as a class issues related to identity and race.
- Establish ground rules. If this activity is used as part of a class or workshop series, then remind trainees of the ground rules they established at the beginning and add to them if necessary. A few ground rules that are important to this discussion are:
  - We respect the views of others.
  - No one talks unless they are holding the talking stick/wooden spoon/stress ball. When someone is done talking, they say done and toss the talking stick to the next person who wants to share.
  - Respect the confidentiality of what is shared in this room and do not talk about what is shared with individuals outside of this group.

► **Discussion of Reflection Exercise** (15 minutes)

- The purpose of this activity is to raise trainees' awareness of their assumptions about individuals and roles within STEM. Emphasize to trainees that we all carry around unconscious assumptions, and by raising our awareness about these assumptions, we can begin to challenge them.
- **Small-Group Discussion** (5 minutes)
  - In pairs or small groups, have trainees discuss their reflections.
- **Large-Group Discussion** (10 minutes)
  - What were some of the images and characteristics that came to mind?
  - Was there consensus in your small groups as to what these individuals looked like?
    - *Note:* If physical characteristics do not come up in discussion, facilitators may want to prompt trainees to consider the assumptions that they made about gender identity, racial/ethnic identity, socioeconomic status, age, or clothing, if such descriptions do not come up in your discussion. Facilitators may also want to acknowledge that we make assumptions about characteristics we cannot necessarily see, such as sexual orientation, political beliefs, religion, ability status, etc.
  - How were the pictures in your mind informed by what you have seen/heard in the news, what your friends, family members, or teachers have told you?

Contributed by S. Keyl. (2018). *Diversity in STEM.*

Trainee Pre-Assignment 2 (15 minutes)

► Have trainees review Dr. Quiñones' story prior to the next session: http://www.doctorqmd.com/dr-q-s-story---a-doctor-without-borders/

Workshop Session 2 (35 minutes)

► Video: Neil deGrasse Tyson (5 minutes)
- Share the video of Neil deGrasse Tyson responding to the question "What's up with chicks and science?" (beginning at 3:37): https://www.youtube.com/watch?v=q5S7QD9dryI

► **Reflection** (5 minutes)
- Invite trainees to reflect on how the story of Dr. Quinones, Dr. deGrasse Tyson, and the stories shared by trainees in this class compared to some of the characteristics that were identified in the reflection exercise in the previous session.

► **Large-Group Discussion** (20 minutes)
- How do the stories of Dr. Quinones and Dr. deGrasse Tyson match (or counter) your assumptions about the individuals that we reflected on in the first session (e.g., researcher, immigrant, minoritized group, PI, graduate trainee)?
- Were you aware of the discrimination that some minoritized groups encounter?
- Do you have friends or family members who encounter prejudice and discrimination?
- Which sections of the readings and videos were the most challenging to understand?
- How did the experiences of the scientists you read/viewed resonate with you? Did they reflect your own experience?

Note: Some trainees might have a difficult time with some aspects of this exercise. They may feel as if the facilitator is blaming dominant groups for the problems in diverse settings. Normalize this reaction (i.e., "many people in positions of power and privilege have this reaction at some point") and emphasize to trainees that we all play a part in reinforcing (or deconstructing) social norms and ideas.

► **Wrap-up** (5 minutes)
- Conclude the discussion by asking trainees to identify:
  - Questions they still have about the ideas presented in the reading and video
  - How the ideas described in the readings either directly or indirectly affect, in their opinion, STEM trainees and researchers

DIVERSITY IN STEM

Learning Objectives

Trainees will:

- Be more aware of disparities between majority and historically underrepresented groups in STEM careers.
- Consider different reasons for disparities in STEM career/degree attainment.
- Articulate strategies to reduce these disparities in STEM career/degree attainment.
- Learn about the prejudice encountered by researchers belonging to historically underrepresented groups.

Reflection Exercise: Using the space provided below, write about the ideas/images that come to mind when you hear each of the following words. Write about what you see/hear in the news, and what your friends, family members, and teachers have said:

Researcher:

Minoritized group:

Undergraduate trainee:

Professor:

Principal Investigator/PI:

Graduate trainee:

Immigrant:

Reflection exercise activity information from "Advancing Equity and Inclusion" in Pfund, Branchaw, & Handelsman (2014). *Entering Mentoring.* New York: W.H. Freeman & Co.
Contributed by S. Keyl. (2018). *Diversity in STEM.*

ELEVATOR SENTENCES

Learning Objectives

Trainees will:

- Engage general and expert audiences in conversations about their research.

Trainee Level

undergraduate or graduate trainees
novice, intermediate, or advanced trainees

Areas of Trainee Development

- Research Comprehension and Communication Skills
  - Develop research communication skills.
  - Develop logical/critical thinking skills.
- Researcher Identity
  - Develop identity as a researcher.

Activity Components and Estimated Time for Completion

- In Session Time: 45 minutes

Total time: 45 minutes

When to Use This Activity

This activity can be implemented at the beginning of a research experience to help trainees understand how their research can be communicated to both expert and general audiences. Trainees can share revised elevator sentences at the beginning of each session for the first few meetings as they deepen their understanding of their project. Other activities that can be used with this activity include:

- General Public Abstract
- Three-Minute Research Story

Inclusion Considerations

The background of trainees may influence how they discuss their research with different audiences. For example, first-generation college or graduate students might discuss their research with family members differently than they would the general public or another scientist. Acknowledge that their brief elevator sentence introduction may need to be accompanied by an explanation or justification of the research when presented to family members because the time and energy it takes to do research (i.e., working during the summer and getting a job right after college vs. participating in a summer research program and pursuing further training after graduation, etc.) may not align with familial notions of success.

Contributed by D. Wassarman with information from Branchaw, J. L., Pfund, C., and Rediske, R. (2010). *Entering Research Facilitator's Manual.* New York: W.H. Freeman & Co.

Implementation Guide

Workshop Session (45 minutes)

► **Introduction** (5 minutes)

The primary purpose of developing elevator sentences is to engage the audience in a conversation about the trainees' research. For example, an elevator sentence for a general audience might be "I am studying the effect of altitude on fruit fly size." This sentence might elicit a response such as "I didn't know that flies are different sizes." This response then provides the opportunity for the trainee to give more information about fly size, how it is measured, and why and how they are studying the effect of altitude. This elevator sentence had the right level of detail for the listener to further engage in the conversation.

On the other hand, an elevator sentence such as "I am studying the effect of hypoxia on thorax development in *Drosophila melanogaster*" says the same thing but is less likely to engage the audience in a conversation because it includes too much jargon for the general public. However, this elevator sentence would be appropriate to use with other scientists in the field. This activity is meant to give students practice refining their understanding of their research and communicating it in a way that is appropriate for the audience.

► **Activity: Elevator Sentences** (20 minutes)

- Distribute the Elevator Sentences worksheet to trainees to work on individually. They should generate one sentence to answer the question "What are you studying?" This activity could also be presented as an assignment for discussion at the next session.
- Trainees should work in pairs or trios and take turns asking each other this question as a nonexpert (or as an expert, if the trainees are all conducting similar research) and use their elevator sentences to start a conversation about research. Trainees should provide feedback to one another and work as a team to refine each of their elevator sentences. A peer-review assessment rubric is provided in the trainee materials.
- If time allows, mix trainee pairs and do the exercise again with the revised elevator sentence.

► **Large-Group Discussion** (15 minutes)

- As a large group, discuss the following questions:
  - How does your answer change when you are speaking to a nonexpert as opposed to an expert?
  - What types of questions did your sentence elicit from your partners/group?
  - How easy was it to talk about your research to individuals outside of your field?
  - Why might it be important to be able to communicate what you study to individuals outside of your field?

► **Wrap-up** (5 minutes)

- Invite trainees to reflect on the revisions they made during the session, share their elevator sentence with their mentor for feedback, and refine it once more before submitting it for review by the facilitator.

► ***Follow-up activity:*** If this activity is implemented as part of a semester course or summer seminar, facilitators may wish to revisit elevator sentences with trainees periodically. Encourage trainees to keep track of their revisions on the trainee materials worksheet and reflect on how their sentences change over the course of their research experience. Changes in sentences should reflect an increased understanding of their research and how it fits into the work of their mentor's research group.

Note: Optional activity: Video record students explaining their research in a short period of time (15 seconds) at the beginning, middle, and end of their research experience to document (and celebrate) their improvement over time.

Contributed by D. Wassarman with information from Branchaw, J. L., Pfund, C., and Rediske, R. (2010). *Entering Research Facilitator's Manual.* New York: W.H. Freeman & Co.

ELEVATOR SENTENCES

Learning Objectives

Trainees will:

- Engage general and expert audiences in conversations about research.

Use one sentence to answer the question "What are you studying?" if it were asked by **a nonexpert**, for example, your mailman.

Draft 1:

Draft 2:

Draft 3:

Use one sentence to answer the question "What are you studying?" if it were asked by **an expert**, for example, a graduate student in your lab.

Draft 1:

Draft 2:

Draft 3:

Contributed by D. Wassarman with information from Branchaw, J. L., Pfund, C., and Rediske, R. (2010). *Entering Research Facilitator's Manual.* New York: W.H. Freeman & Co.

ELEVATOR SENTENCES

Peer-Review Assessment Rubric

Name of the Presenter: ______________________________

Reviewer (circle one): expert/nonexpert in the trainee's area of research

Use the following scale to rate the presenter:

1 = strongly disagree, 2 = disagree, 3 = somewhat agree, 4 = agree, 5 = strongly agree

I was able to understand the subject of the presenter's research.	1	2	3	4	5
The presenter used terms and phrases that were appropriate to the audience's level of understanding.	1	2	3	4	5
I was able to come up with follow-up questions based on the presenter's elevator sentence.	1	2	3	4	5

Comments:

Contributed by D. Wassarman with information from Branchaw, J. L., Pfund, C., and Rediske, R. (2010). *Entering Research Facilitator's Manual.* New York: W.H. Freeman & Co.

ESTABLISHING YOUR IDEAL THESIS COMMITTEE

Learning Objectives

Trainees will:

- Identify the primary and secondary roles of a graduate thesis committee.
- Explore the roles that can be played by different members of a thesis committee.
- Begin identifying members of a thesis committee.

Trainee Level

graduate trainees
novice trainees

Areas of Trainee Development

- Professional and Career Development Skills
  - Develop confidence in pursuing a research career.
- Research Comprehension and Communication Skills
  - Develop effective interpersonal communication skills
- Researcher Confidence and Independence
  - Develop independence as a researcher.

Activity Components and Estimated Time for Completion

- In Session Time: 1 hour

Total time: 1 hour

When to Use This Activity

This activity is best suited for novice graduate trainees who have joined a research group and are ready to select a thesis committee. Other activities that may be used with this activity include:

- Fostering Your Own Research Self-Efficacy

Inclusion Considerations

Encourage trainees to consider what the benefits and challenges might be to having committee members whose backgrounds are either very similar to or very different from their own. What might these individuals contribute to the committee? Ask trainees to reflect on how well they should get to know prospective committee members before asking them to serve on their committee.

Contributed by G. Arrizabalaga. (2018). *Establishing Your Ideal Thesis Committee.*

Implementation Guide

Workshop Session (1 hour)

► **Introduction** (5 minutes)

- The goal of this activity is to help trainees understand the responsibilities of thesis committees in supporting trainee progress and development, including the specific roles different committee members can play. In addition, trainees will consider how academic and interpersonal politics can support or limit the effectiveness of the committee. Exploring potential committee members roles can help trainees identify people to invite to be members of their thesis committee.
- Discuss the general timeline and mechanics of how to set up a thesis committee. This introduction can use information specific to the particular doctoral program, such as the number of members and composition requirements.

► **Activity 1: What is the thesis committee for?** (25 minutes)

- Ask trainees to write down what they consider to be the main roles of a thesis committee.
- In pairs or groups of three have trainees compare and discuss their lists.
- Ask groups to report out on their lists and capture their ideas to create a comprehensive list on the board.

> Roles for a thesis committee should include, but are not limited to:
> - advisory (scientific, technical, conceptual),
> - mentoring,
> - professional and career development,
> - monitoring and facilitating progress toward graduation, and
> - advocacy.

- **Large-Group Discussion Questions**
  - How might the roles of a thesis committee change during different stages of your graduate school career?

    > - Before the qualifying exam, the committee might have an evaluative role, which may evolve into an advisory role as you become more independent.
    > - The committee's role in supporting the research project will evolve as the project progresses. It will provide guidance on the development, data collection, analysis, and publication of results as needed.
    > - The committee can support your transition to the next step in your career.

  - How might your committee help you make career decisions?

    > - Committee members will likely be reference letter writers.
    > - Committee members can serve as liaisons with those who can advise you about different career paths or different fields.

Contributed by G. Arrizabalaga. (2018). *Establishing Your Ideal Thesis Committee.*

- What can the role of the committee be in helping you navigate your relationship with your thesis mentor? How can you get your committee to advocate for you?

> - Committee members can give advice about how to approach your thesis mentor about expectations, graduation, and other issues.
> - Committee members can talk to your thesis mentor privately about expectations and goals.
> - Committee members who are familiar with your goals, background, and expectations can be advocates for you.
> - You can inform committee members about your goals, background, and expectations with regular meetings and one-on-one meetings.

- How can you ensure the committee understands what you expect of them? How and when do you convey those expectations?

> - Meeting face to face when inviting professors to be part of the committee provides an opportunity to explain your expectations.
> - To solicit feedback on training activities other than the research project (e.g., professional development), report on activities in which you've participated at committee meetings and link these activities to your career goals.

- How can you ensure you understand what your committee expects of you? How and when can you clarify those expectations?

> - After each committee meeting, write a summary of what was discussed and what was recommended, then ask all committee members to edit and approve it.
> - Meet one-on-one with your thesis mentor or with members of the committee to set an agenda for your committee meetings to clarify and reinforce what will be discussed.

► **Activity 2: The members of your thesis committee** (25 minutes)

- Ask trainees to write down what they consider pros and cons of the potential committee members listed on the Activity 2 worksheet.
- Ask trainees to share their answers and discuss how each of those might contribute to reaching the specific goals they identified in Activity 1.
- **Large-Group Discussion**
  - If your mentor is a senior professor, why would it be risky if all members of your committee are assistant or associate professors?
  - Is the racial, ethnic, and gender diversity of the committee important to you?
  - What other category of faculty not listed could be a good addition to your thesis committee?
  - What would you do if your mentor tells you not to add a particular faculty member who you were planning to invite to serve on your committee?
  - Think about your own thesis topic. What types of expertise might you need on your committee?

Contributed by G. Arrizabalaga. (2018). *Establishing Your Ideal Thesis Committee.*

► **Wrap-up** (5 minutes)

- Summarize the main ideas generated from the large group discussions.
- Reemphasize the multiple roles an effective committee plays in the success of the trainee. Remind trainees that the thesis committee's primary goal is to support them. Therefore, it is important the trainee actively participate in forming the thesis committees. Encourage the trainees to use what they learned in this session to start a list of possible committee members and to discuss this list with their thesis mentor.

Contributed by G. Arrizabalaga. (2018). *Establishing Your Ideal Thesis Committee.*

ESTABLISHING YOUR IDEAL THESIS COMMITTEE

Learning Objectives

Trainees will:

- Identify primary and secondary roles of a graduate committee.
- Explore the roles that can be played by different members of a thesis committee.
- Begin identifying members of a thesis committee.

Activity 1: What is the thesis committee for?

1. List the main roles that you anticipate your thesis committee will play during your graduate career.

2. How might the roles of a thesis committee change during different stages of your graduate school career?

3. How might your committee help you make career decisions?

4. What can the role of the committee be in helping you navigate your relationship with your thesis mentor? How can you get your committee to advocate for you?

5. How can you ensure the committee members understand what you expect of them? How and when do you convey those expectations?

6. How can you ensure you understand what your committee members expect of you? How and when can you clarify those expectations?

Contributed by G. Arrizabalaga. (2018). *Establishing Your Ideal Thesis Committee.*

Activity 2: The members of your thesis committee

What are the pros and cons of each of the potential committee members listed below?

1. Your research mentor's closest collaborator
   Pros:

   Cons:

2. A new untenured professor in your department
   Pros:

   Cons:

3. An expert in a particular technique you plan to utilize for your project
   Pros:

   Cons:

4. A senior professor who works on a field unrelated to your project and your lab
   Pros:

   Cons:

5. A professor at a different institution with expertise in your field of research
   Pros:

   Cons:

6. A professor with whom your mentor has scientific disagreements
   Pros:

   Cons:

Contributed by G. Arrizabalaga. (2018). *Establishing Your Ideal Thesis Committee.*

ETHICS CASE: DISCUSSION WITH MENTOR

Learning Objectives

Trainees will:

- Discuss with a mentor ethical issues associated with working in a research group.
- Learn strategies to deal with ethical situations associated with working in a research group.
- Learn about consequences of unethical behavior.

Trainee Level

undergraduate trainees
novice, intermediate, or advanced trainees

Areas of Trainee Development

- Research Ethics
  - Develop responsible and ethical research practices.
- Research Comprehension and Communication Skills
  - Develop effective interpersonal communication skills.

Activity Components and Estimated Time for Completion

- Trainee Pre-Assignment Time: 20 minutes
- In Session Time: 20 minutes

Total time: 40 minutes

When to Use This Activity

This activity can be used any time after a relationship has been established between a trainee and mentor. Ideally, it is used after the topics essential to forming the foundation of a research experience, such as documenting research, reading scientific literature, and defining expectations, have been addressed.

Inclusion Considerations

Discuss with trainees how understanding of ethical behavior may be different based on differences in culture, gender, or generational backgrounds. Ask trainees to consider how their background and their mentor's background may influence their interpretations of the case study. Emphasize that it can sometimes be as difficult to determine whether behavior is unethical as it is to decide how to deal with that behavior in a sensitive and respectful manner.

Contributed by C. Barta with information from Branchaw, J. L., Pfund, C., and Rediske, R. (2010). *Entering Research: A Facilitator's Manual.* New York: W.H. Freeman & Co.

Implementation Guide

Trainee Pre-Assignment (20 minutes)

Trainees should discuss the case study "Too Good to Be True?" with their research mentor. Encourage them to discuss this case study with the PI of the research group; if this is not possible, the trainee can discuss the case study with a graduate student, research scientist, lab manager, or post-doc who may also be acting as a research mentor. Trainees should record their mentors' responses to three follow-up questions to the case study.

Workshop Session (20 minutes)

- Trainees should share their mentors' response(s) to the case study "Too Good to Be True?" in either (1) small groups of three or four students or (2) a large group where three or four volunteers may be selected. (10 minutes)
- **Discussion Questions** (10 minutes)
  - Do you agree with your mentor's response to the scientific misconduct case? Why or why not?
  - How do the different responses from mentors reflect the different structures of research groups?
  - What can you do to establish a good relationship with your mentor(s) so that you are taken seriously when/if ethical issues arise?

> - Ask questions when you're unsure of what to do in potentially ethically difficult situations.
> - Be honest and forthcoming about mistakes you make in your research.
> - Be understanding and supportive when others on the research team admit to making mistakes.

  - With whom do you feel comfortable discussing ethical issues? Is this different from your classmates who are in different research groups?
  - Apart from fabrication of data, what other ethical issues exist in the scientific community?

> - Misuse of animals and resources.
> - Misuse of vulnerable populations for research (e.g., minors, pregnant women, people who cannot decide for themselves, low socioeconomic).
> - Designing research that is inclusive of gender, race, age, sex, etc.
> - Protection of sensitive data with human subjects.
> - Exploration of innovations or new ideas that can be used for harm.
> - Sexual harassment.
> - Appropriate spending of public funding.

  - Why is misconduct such an important issue in the scientific community?

> - Dishonesty can lead to corruption.
> - Small lies or fabrications can have large impacts on future work.
> - Wastes taxpayer money.
> - Defeats the purpose of doing research.

Contributed by C. Barta with information from Branchaw, J. L., Pfund, C., and Rediske, R. (2010). *Entering Research: A Facilitator's Manual.* New York: W.H. Freeman & Co.

- What measures are in place to prevent misconduct?

> - Code of ethics agreed upon by scientific society or community.
> - Peer review.
> - Trend toward open access data.

- What are the consequences of scientific misconduct?

> - Lose grant funding.
> - Lose the ability to apply for future grants.
> - Retract scientific articles.
> - Terminated from position.
> - Loss of your colleagues' respect.

Contributed by C. Barta with information from Branchaw, J. L., Pfund, C., and Rediske, R. (2010). *Entering Research: A Facilitator's Manual.* New York: W.H. Freeman & Co.

ETHICS CASE: DISCUSSION WITH MENTOR

Learning Objectives

Trainees will:

- Discuss with a mentor ethical issues associated with working in a research group.
- Learn strategies to deal with ethical situations associated with working in a research group.
- Learn about consequences of unethical behavior.

Discuss the case study with the person (Principle Investigator [PI], Professor) who leads your research group. If this person is not available, discuss the case study with another research mentor on your research team. Ask the mentor to consider the undergraduate researcher's perspective when reading the case study. Use the three questions to document your mentor's response to the case study.

Too Good to Be True?

Evelyn and John joined the lab at the same time as sophomores and have been doing research on related, yet separate projects for the past year. Evelyn, a quiet and very diligent worker, spends many hours in the lab working on her project. She has encountered several obstacles in her research, but is making slow, yet consistent progress. She sees John there infrequently and notices that he spends most of his time chatting with the other lab members. The PI of the lab travels a lot, but when he is there, John always seems to connect with him.

At lab meeting last week, John presented his research. The results he reported were exactly what the PI was looking for. The PI was ecstatic. Evelyn was stunned. She does not remember seeing John do any of the experiments he presented. She suspects that he is not being truthful, but has no proof. His research is linked to hers, so if the results are not valid, it will negatively impact her project, and the entire research group. Everyone really likes John, including the PI, and everyone knows that she has been dealing with a lot of setbacks in her research. She doesn't want to look like a jealous coworker by accusing John of fabricating data, but she truly suspects that he has.

Discussion Questions

1. What advice would you give Evelyn?

2. What are some of the potential repercussions if Evelyn does nothing at all?

3. Have you ever experienced a similar situation? If so, how was it handled?

Contributed by C. Barta with information from Branchaw, J. L., Pfund, C., and Rediske, R. (2010). *Entering Research: A Facilitator's Manual.* New York: W.H. Freeman & Co.

Trainee Materials

FINDING A RESEARCH MENTOR

Learning Objectives

Trainees will:

- ► Learn how to identify and contact potential research mentors.

Trainee Level

undergraduate trainees
novice trainees

Areas of Trainee Development

- ► Professional and Career Development Skills
  - Explore and pursue a research career.
- ► Research Comprehension and Communication Skills
  - Develop effective interpersonal communication skills.
  - Develop an understanding of the research environment.

Activity Components and Estimated Time for Completion

- ► Trainee Pre-Assignment Time: 1 hour

Total time: 1 hour

When to Use This Activity

Use this activity to guide aspiring undergraduate students who wish to engage in a research experience but need to identify and contact potential research mentors. For graduate trainees, facilitators should use the "Finding Potential Research Rotation Groups and Mentors" activity.

Note: For the purposes of the *Entering Research* activities, a "mentor" is defined as the person with whom the student is working most closely on a research project. This could be a professor, postdoctoral fellow, graduate student, laboratory technician, or a senior peer undergraduate researcher.

Inclusion Considerations

Trainees from diverse backgrounds may want to look for markers of attention to diversity and inclusion in the research teams they consider joining. Encourage all trainees, especially those from underrepresented groups, to ask questions of prospective mentors about management style and their thoughts around equity, inclusion, and diversity. They may also wish to ask about others from underrepresented groups who have been members of the research team.

Contributed by J. Branchaw with information from Branchaw, J. L., Pfund, C., and Rediske, R. (2010). *Entering Research: A Facilitator's Manual.* New York: W.H. Freeman & Co.

Implementation Guide

Trainee Pre-Assignment (1 hour)

Ideally, this activity is distributed before or on the first day of a workshop series or course. Once distributed, facilitators can communicate with their trainees frequently to monitor their progress in finding a mentor and to provide feedback, guidance, and encouragement along the way.

- If the activity is distributed as a resource to trainees to use on their own, then the name and contact information of a resource person to whom students can ask questions should be provided.
- When used as part of a workshop series or course, facilitators should contact the mentors directly, once they are identified, to describe what their research trainees will be expected to do as part of the workshop series or course. In particular, the facilitator should introduce assignments that will require trainees to meet with the mentor. Making mentors aware of assignments will help them prepare to work most effectively with their trainee. Below is a sample message from the facilitator to a mentor:

Dear <Mentor>,

I am <Facilitator's Name> and your undergraduate research trainee, <Student's Name>, will be participating in my *Entering Research* course. I'm writing to introduce myself and to tell you about the course.

The *Entering Research* course is designed for undergraduate students who are beginning research. The main goals are for students to establish a positive relationship with their mentor and to define and make progress on a research project. The syllabus is attached for your review.

In particular there are three assignments that require undergraduate trainees to meet with their mentors: (1) Your Research Group's Focus, (2) The Mentor Biography, and (3) a Mentor–Trainee Expectations Agreement. These assignments are attached for your reference. Please contact me if you have any questions about the workshop series or assignments.

Sincerely, <Facilitator's Name>

Attachments: Syllabus; Course Learning Objectives; Your Research Group's Focus; Mentor Biography; Mentor–Trainee Expectations Agreement

Contributed by J. Branchaw with information from Branchaw, J. L., Pfund, C., and Rediske, R. (2010). *Entering Research: A Facilitator's Manual.* New York: W.H. Freeman & Co.

FINDING A RESEARCH MENTOR

Learning Objectives

Trainees will:

- Learn how to identify and contact potential research mentors.

Use the attached pages to keep track of your progress in finding a mentor. Advice about how to identify and contact potential mentors is provided below.

Identifying Potential Research Mentors

1. Determine what most interests you in your discipline. In other words, define a research area (molecular biology, materials science, nanotechnology, plasma physics, analytical chemistry, computer architecture, etc.). What are you curious about?
2. Search websites to identify mentors working in your area of interest. Search through academic program listings, department websites, student job sites, and undergraduate research databases if they are available. Talk to friends who are already doing research to get their advice about potential mentors. If you're not sure what research area interests you, then start by doing a general review of faculty research in the academic department in which you are majoring. However, don't be afraid to think broadly and explore research outside of your academic department, too! Remember that the principles of research are the same and generally independent of the particular field of research. Gaining experience in one field will almost certainly translate to other fields.
3. Read the research descriptions and generate a ranked list of at least five potential mentors. Identify one or more aspects about each person's research that is interesting to you and that you would like to know more about. It may be necessary to contact several potential mentors before you find one with matching interests and space on their research team to add a new undergraduate research trainee. Do not get discouraged or give up!

Contacting Potential Research Mentors

Email is a good way to make initial contact with potential mentors. By sending an email you give the mentor a chance to review your materials before responding. It is like the first step in an interview, so be sure it reflects your best effort (no spelling or grammatical errors!). If you are comfortable, you may also ask to make an appointment to call or stop by a potential mentor's office to ask about a research experience.

Some things to consider when composing emails:

- Research mentors are very busy people, so keep it short and to the point (approximately 1 paragraph). However, do not treat the email like a text message. It should be a full, professional letter with proper grammar and spelling.
- Address the email using the mentor's official title (e.g., Professor, Dr.)
- Specifically refer to the mentor's research, and what you find interesting about it. Be sure to use your own words and not to copy text from the research description on their website. This is the most important part of the email to demonstrate your interest in research.
- Be clear that you are looking for a research experience (not a dishwashing job) and what your main goal will be (e.g., shadowing someone in the lab to get exposed to research; doing an honors thesis research project).
- Highlight what you have to offer; what distinguishes you from other students (e.g., hard worker, experience, eager to learn, willing to stay more than one semester, persistent, specific courses you've completed that are relevant to the research).

Contributed by J. Branchaw with information from Branchaw, J. L., Pfund, C., and Rediske, R. (2010). *Entering Research: A Facilitator's Manual.* New York: W.H. Freeman & Co.

- Show enthusiasm for learning how to do research!
- Request that if the mentor is not able to take an undergraduate research trainee at this time, that they could recommend a colleague who might be able to.
- Close the email as you would a letter, with a closing (e.g., "Sincerely," "Respectfully") and your full name, year of study, major and complete contact information (email, phone, mail).
- If requesting a meeting, give a large range of times that you are available or ask when the potential mentor has office hours.

Additional information you could include in an attached letter:

- Share that you are taking the *Entering Research* workshop/course and attach a copy of the syllabus.
- Give an estimate of the number of hours/credits you can be available to do research, and when you would like to begin, but leave room for negotiation.
- Give a *brief* overview of your academic credentials (e.g., GPA and relevant courses taken), or attach a transcript.

Interviewing with Potential Research Mentors

- Be on time, not early or late.
- Be yourself. Be enthusiastic and motivated. Smile!
- Be ready to discuss why you want to do research in general (What are your academic and career goals?), and why you want to do research with this mentor specifically (What is it about his/her research that is interesting to you? Is there a particular project on which you would like to work?).
- Read about the research BEFORE you go to the interview. There is usually a research overview on the researcher's website with references/links to the group's published papers. Try to read one or two of these papers and prepare some questions about them. Generally, mentors won't expect you to fully understand the research, but making an effort to learn about it on your own demonstrates independence and motivation.
- Ask about the expectations of undergraduate trainees in the group (time commitment, credits, type of work). In general, three to five hours of research per week is worth one academic credit. However, this varies, and you should ask how many hours the mentor expects per week per credit.
- Have a plan for the time you will be available to work in the lab during normal business hours. Be prepared to give the potential mentor the times that you are available (rather than the times you are unavailable). Think carefully about what is realistic for you.
- Ask about who would be your direct mentor in the group (professor, post-doc, graduate student, technician, senior peer undergraduate).
- Bring a copy of your transcript if you haven't already submitted one.
- If you are interviewing with multiple prospective mentors, tell the mentor exactly when you will follow up with him/her and do it. Mentors need to know if you will not be available if they are interviewing multiple candidates for a position.

Important

- It can be challenging to connect with research mentors, so ***be persistent, yet polite***. Ideally, give potential mentors a week to respond to your email before you follow up.
- Research groups have limited space, so it may be difficult to find a group that is looking for, or willing to take, another student. ***Do not take it personally if they decline your request***. You may go through all 10 (or more) potential mentors before you find a match. Stick with it! You *will* find someone.

Contributed by J. Branchaw with information from Branchaw, J. L., Pfund, C., and Rediske, R. (2010). *Entering Research: A Facilitator's Manual.* New York: W.H. Freeman & Co.

Potential Mentor #1:

Department: Initial Contact Date:

Email: Follow-up Contact Date:

Phone: Response:

Research Area: Interview Day & Time:

What I find interesting about the research:

Potential Mentor #2:

Department: Initial Contact Date:

Email: Follow-up Contact Date:

Phone: Response:

Research Area: Interview Day & Time:

What I find interesting about the research:

Potential Mentor #3:

Department: Initial Contact Date:

Email: Follow-up Contact Date:

Phone: Response:

Research Area: Interview Day & Time:

What I find interesting about the research:

Potential Mentor #4:

Department: Initial Contact Date:

Email: Follow-up Contact Date:

Phone: Response:

Research Area: Interview Day & Time:

What I find interesting about the research:

Contributed by J. Branchaw with information from Branchaw, J. L., Pfund, C., and Rediske, R. (2010). *Entering Research: A Facilitator's Manual.* New York: W.H. Freeman & Co.

Potential Mentor #5:

Department:

Email:

Phone:

Research Area:

What I find interesting about the research:

Initial Contact Date:

Follow-up Contact Date:

Response:

Interview Day & Time:

Potential Mentor #6:

Department:

Email:

Phone:

Research Area:

What I find interesting about the research:

Initial Contact Date:

Follow-up Contact Date:

Response:

Interview Day & Time:

Potential Mentor #7:

Department:

Email:

Phone:

Research Area:

What I find interesting about the research:

Initial Contact Date:

Follow-up Contact Date:

Response:

Interview Day & Time:

Potential Mentor #8:

Department:

Email:

Phone:

Research Area:

What I find interesting about the research:

Initial Contact Date:

Follow-up Contact Date:

Response:

Interview Day & Time:

Contributed by J. Branchaw with information from Branchaw, J. L., Pfund, C., and Rediske, R. (2010). *Entering Research: A Facilitator's Manual.* New York: W.H. Freeman & Co.

FINDING POTENTIAL RESEARCH ROTATION GROUPS AND MENTORS

Learning Objectives

Trainees will:

- ► Identify potential research groups for rotation.
- ► Learn about strategies for contacting potential research mentors.
- ► Learn about important considerations when choosing research groups for rotation.

Trainee Level

graduate trainees
novice or intermediate trainees

Areas of Trainee Development

- ► Professional and Career Development Skills
  - Explore and pursue a research career.
- ► Research Comprehension and Communication Skills
  - Develop effective interpersonal communication skills.
  - Develop an understanding of the research environment.
- ► Researcher Identity
  - Develop identity as a researcher.

Activity Components and Estimated Time for Completion

- ► Trainee Pre-Assignment Time: 1 hour
- ► In Session Time: 1 hour, 35 minutes

Total time: 2 hours, 35 minutes

When to Use This Activity

This activity should be used with novice to intermediate graduate trainees prior to and/or during the time when they are choosing research groups with which to rotate. This may be institution and program specific, but will likely occur as the trainee is entering graduate school. Facilitators may want to include "2nd year" trainees who have just completed their own rotations and chosen the research groups they will join.

Inclusion Considerations

Trainees from diverse backgrounds may want to look for markers of attention to diversity and inclusion in the research groups they consider joining. Encourage all trainees, especially those from underrepresented groups, to ask questions of prospective mentors about management style and their thoughts around equity, inclusion, and diversity. They may also wish to ask about others from underrepresented groups who have been members of the research team. Encourage trainees to meet with members of the research team (graduate students in particular) to discuss these issues.

Contributed by A. Sokac. (2018). *Finding Potential Research Rotation Groups and Mentors.*

Implementation Guide

Trainee Pre-Assignment (*optional*)

- ► This activity can be done entirely in session, if working time is incorporated into the session. Alternatively, activities can be distributed for completion before the session and brought to session for discussion. If doing the latter, then distribute these two activities for completion before the session:
  - "Identifying Research Rotation Groups"
  - "Contacting Potential Rotation Mentors"

Workshop Session (1 hour, 30 minutes–2 hours)

- ► **Introduction**
  - The goal of this activity is for trainees to identify their prioritized list of rotation mentors and develop good strategies for securing a rotation opportunity with those mentors.
- ► **Reflecting on the "Right Fit"** (40 minutes)
  - Give the trainees time to read over the activity and individually list characteristics they value in a research group and primary mentor. (5 minutes)
  - Ask trainees to pair up and discuss the characteristics they value in a research group and primary mentor. (10 minutes)
  - Come together as a group and generate a list of valued research group characteristics and a separate list of valued mentor characteristics. (10 minutes)
    - Suggested areas for discussion
      - Research
      - Research group/lab environment
      - Working relationships
      - Mentor's training philosophy
      - Training potential
      - Funding
  - Ask trainees to think of strategies for identifying these characteristics in specific mentors and research groups. (10 minutes)
  - Suppose a trainee fits well with a research group but not the mentor or vice versa. What considerations should a trainee make when in this difficult situation? (5 minutes)
    - For example, the size of the lab may help or hinder the misalignment of research mentor and research group:
      - If trainees are part of a small lab, they may interact with all of the research group members more frequently than if they are part of a larger lab. If there are a few members of the research group who are difficult to work with, it might OK to choose this lab if you do not need to work with them often.
      - If trainees are unsure, for example, that the primary faculty mentor will have enough time to mentor them in the way they need, they could ask others in the lab who could provide additional mentoring.
- ► **Identifying Research Rotation Groups** (5–20 minutes)
  - Either assign this activity to be completed before the session or give trainees 15 minutes to complete it in session.

Contributed by A. Sokac. (2018). *Finding Potential Research Rotation Groups and Mentors.*

- **Discussion** (5 minutes)
  - What were the challenges in identifying research groups you are interested in for a rotation?
  - What strategies did you use to find information about the research group?
  - *Note:* Some trainees may have difficulty identifying specific names for some of the questions on the shortlist worksheet. Reassure them that this list is just a place to start thinking about individuals who might fill these roles and reiterate that the names will likely change.

► **Contacting Potential Rotation Mentors** (25–40 minutes)
- Either assign this activity to be completed before the session or give trainees 15 minutes to complete it in session.
- In pairs, trainees share their email drafts and ask for feedback on both structure and content. (10 minutes)

► **Discussion** (15 minutes)
- Are you satisfied with your emails? Why or why not? Did you run into any challenges in writing them? (Volunteers can read from their drafts and ask for real-time feedback.)
- Do you have questions about what information should be included on your résumé/CV?
- When should emails be sent to mentors?
- How long should you wait before sending a mentor a reminder email?
- How should you prioritize between mentors?
  - Is the best strategy to rotate with your top choice mentor first? Last? Why or why not?
  - Be prepared to give trainees strategies based on the setup of your program. For example, if Term 1 includes significant coursework plus Rotation 1, then it is advisable to do the most desirable rotation in a later term because finding the balance between classes and lab can be challenging at first.

► Interviewing with Potential Rotation Mentors (20 minutes)
- Give the trainees time to read "Interviewing with Potential Rotation Mentors" individually. Ask them to circle the points that resonate with them or concern them. (3–5 minutes)
- Discussion (10 minutes)
  - Ask the trainees what resonated with them and what concerned them.
- Give the trainees time to generate their own list of important questions to ask at their interviews. (5 minutes)

► **Wrap-up** (5 minutes)
- Encourage trainees to revisit the worksheet after rotations and reflect on whether they might answer the questions differently. *Facilitators may also formally incorporate this follow-up as an activity in a later session.*

Alternative Implementations

► The "Three Professors" or "Roles for Your Research Mentor" activities can be used to supplement or used instead of the "Reflecting on the Right Fit" activity, particularly if trainees do not have significant prior research experience.

► If "2nd year" trainees are included, the session can end with a structured networking activity where the older trainees share with the group their single most important piece of advice about the rotation choice and process.

FINDING POTENTIAL RESEARCH ROTATION GROUPS AND MENTORS

Learning Objectives

Trainees will:

- Identify potential research groups for rotation.
- Learn about strategies for contacting potential research mentors.
- Learn about important considerations when choosing research groups for rotation.

Reflecting on the "Right Fit"

The goal of this activity is to explore what research group characteristics and mentoring styles will fit your personal and professional goals, and personality. This is particularly important for the rotations that you identify as being the most likely to translate into a home or significant collaboration for your future thesis project. Where can you be successful? Which mentor can help you achieve your career goals? Where will you be comfortable?

Individually, think about the characteristics you value in a research group and in a primary mentor. *Note:* These may be slightly different! Once you have generated a list, discuss your list with a neighbor.

Research Group	Primary Mentor

Contributed by A. Sokac. (2018). *Finding Potential Research Rotation Groups and Mentors.*

IDENTIFYING RESEARCH ROTATION GROUPS

The goal of this activity is to generate a prioritized list of rotation mentors and develop strategies for securing a rotation opportunity with those mentors. In considering which research groups to rotate in, complete the following:

I am interested in learning more about these scientific topics/methods:

- *topic 1:* ________________.
- *topic 2:* ________________.
- *method 1:* ________________.
- *others:* ________________.

These topics/methods are studied/used in the research groups of (search the departmental website and talk to trainee and faculty contacts for this information):

- *mentor 1:* ________________.
- *mentor 2:* ________________.
- *mentor 3:* ________________.
- *mentor 4:* ________________.

Use this list as a starting point to choose your rotations. Even if you do not rotate with a mentor or choose to join their research group, these individuals can still contribute to your graduate training and professional development. Consider inviting them to fill the roles listed below. Revisit this list throughout your graduate career because your answers may change as you acquire more experiences, get acquainted with the department, and take more classes.

I would like to invite ________________ to serve on my thesis committee.

I would like ________________ to be part of my professional network.

I would like to ask ________________ for a letter of recommendation in the future.

I really enjoyed my interaction with ________________ during recruitment, seminar, class, etc.

Contributed by A. Sokac. (2018). *Finding Potential Research Rotation Groups and Mentors.*

CONTACTING POTENTIAL ROTATION MENTORS

Email is a good way to make initial contact with potential rotation mentors. By sending an email you give the mentor a chance to review your materials before responding. It is like the first step in an interview, so be sure it reflects your best effort and be sure there are no spelling or grammatical errors.

Use the following tips to draft an email to ask one of the mentors you have identified if you can meet to discuss a rotation opportunity in their research group. Bring your draft to class. Also, update your résumé/CV as it should accompany your email to the mentor.

Here are some things to consider when composing emails:

- Research mentors are very busy people, so keep it short and to the point (approximately one paragraph).
- Make sure your email has a "subject." For example, you may want to reference the name of the lab or your interest in doing a research rotation.
- Address the email using the mentor's official title (e.g., Professor, Dr.)
- Specifically refer to the mentor's research, and what you find interesting about it. Be sure to use your own words and not to copy text from the research description on their website.
- If you are interested in a specific project in the mentor's research group, then clarify that.
- Show enthusiasm for the mentor's research!
- Be clear that you are looking for a research rotation and specify which term or semester you would like to do the rotation.
- Request that if the mentor is not able to take a rotation trainee for the term you asked for, that they recommend the terms where availability does exist.
- Highlight what you have to offer; what distinguishes you from other trainees (e.g., prior research experience, eager to learn, work you have published relevant to the mentor's research).
- Make sure it is clear which graduate program you are a part of (if your institution has multiple graduate training programs).
- Request a face-to-face meeting to discuss the rotation further.
- Include your current résumé/CV and PDFs of your research publications as an email attachment.

Information from: Branchaw, J. L., Pfund, C., and Rediske, R. (2010). *Entering Research: A Facilitator's Manual.* New York: W.H. Freeman & Company.
Contributed by A. Sokac. (2018). *Finding Potential Research Rotation Groups and Mentors.*

INTERVIEWING WITH POTENTIAL ROTATION MENTORS

The goal of this activity is to develop strategies to collect the information you need to make an informed choice when selecting research groups for rotation. Read the following tips for interviewing potential rotation mentors. Circle the points that resonate with you. It is also useful to ask current trainees in the research group some of these questions.

Tips for making a good impression with potential rotation mentors:

- Be on time.
- Be yourself. Be enthusiastic and motivated. Smile!
- Be ready to discuss why you want to do research in general and why you want to do research with this mentor specifically.
  - What are your academic and career goals?
  - What is it about their research that is interesting to you?
  - Is there a particular project on which you would like to work?
- Read about the research BEFORE you go to the interview. There is usually a research overview on the web with references/links to the group's published papers. Try to read one or two of these papers and prepare some questions about them. Generally, mentors won't expect you to fully understand the research, but making the effort to learn about it on your own shows independence and motivation. Carefully choose the papers that reflect the primary interests of the research groups, and not just collaborations (i.e., choose papers where the mentor is the senior author).
- Bring a hardcopy of your current résumé/CV and PDFs of your research publications if you have any.

Tips for getting the information you need to make an educated decision about whether to rotate with a mentor's research group:

- What are the expectations for rotation trainees in the group?
  - How much time should you spend in the research group each week?
  - How does the mentor view the balance between classes and benchwork?
  - What type of work will you do?
  - Will you be expected to give a research group meeting, final presentation, or write a paper on your rotation project?
- Who will serve as your direct mentor in the group for the rotation and beyond (professor, post-doc, senior graduate trainee, technician)?
- What is the big picture research direction of the research group and how will your rotation project contribute to it? Why is the rotation project important for the research group?
- How does the research group communicate and interact? (group meetings, weekly or monthly one-on-one meetings with the mentor, etc.).
- Does the mentor have a mentor–trainee compact or contract or a mentoring philosophy?
- If you have specific career goals that require specific activities (e.g., serving as a teaching assistant), then ask if the mentor will support you in that activity in the future.
- What is the funding situation of the research group? Are there funds available to support you if you join the group?
- What additional questions do you want to ask at your interviews?

Contributed by A. Sokac. (2018). *Finding Potential Research Rotation Groups and Mentors.*

FOSTERING YOUR OWN RESEARCH SELF-EFFICACY

Learning Objectives

Trainees will:

- Define self-efficacy and its sources.
- Clearly articulate steps to complete a research-related task.
- Articulate how the four sources of self-efficacy can be used to build confidence to complete research-related tasks.

Trainee Level

undergraduate or graduate trainees
novice, intermediate, or advanced trainees

Areas of Trainee Development

- Researcher Confidence and Independence
  - Develop confidence as a researcher.

Activity Components and Estimated Time for Completion

- Trainee Pre-Assignment Time: 20 minutes
- In Session Time: 55 minutes

Total time: 1 hour, 15 minutes

When to Use This Activity

Although this activity can be used at any stage of the research experience, it is most effective when a trainee has an understanding of the research process and has had at least some research experience with his or her mentor. Other activities that may be used with this activity include:

- The Power of Social Persuasion

Inclusion Considerations

Trainees who are members of underrepresented groups may be experiencing unconscious "threats" in the research environment that can undermine the development of their research self-efficacy. These threats must be acknowledged in order to mitigate them. Consider using the *Entering Research* resources on stereotype threat and imposter syndrome (available in "Stereotype Threat" and "Challenges Facing Diverse Teams") to help students identify and acknowledge these issues. Explain that overcoming or dealing with these types of challenges will contribute to building their research self-efficacy.

Contributed by A. R. Butz with information from *Promoting Mentee Research Self-Efficacy* (Byars-Winston, Leveritt, Branchaw, and Pfund, 2013, 2016).

Implementation Guide

Trainee Pre-Assignment (20 minutes)

- Trainees should complete the following prior to the session:
  - Read "Self-Efficacy" article (http://psychology.about.com/od/theoriesofpersonality/a/self_efficacy.htm?p=1)
  - Review Handout #1 in trainee materials

Workshop Session (55 minutes)

- **Introduction:** Define self-efficacy and its sources. (20 minutes)
  - Begin the session by providing a brief overview of self-efficacy and its sources. This can be done in a lecture format, but you may also want to use this as an opportunity for trainees to share what they learned from the pre-reading and their review of Handout #1. Below are some of the major points that should be discussed to ensure trainees have a good sense of what self-efficacy is and the sources of self-efficacy in the research experience.
  - **How would you define self-efficacy in your own words?**
    - Self-efficacy is a belief one has in his/her ability to successfully complete a given goal or task. In other words, it is situation-specific self-confidence. It answers the question "can I do this?"
  - **From the reading, what are the four sources of self-efficacy?**
    - Self-efficacy is informed by four sources: mastery experience, vicarious experience, social persuasion, and emotional/physiological state. Here are some examples:
      - **Mastery experience**—a past accomplishment or success: "I've done this before."
      - **Vicarious experience**—a model that has successfully completed the task: "I've seen others do this before."
      - **Social persuasion**—a social or verbal message reinforcing ability or effort: "Others have told me that I can do this."
      - **Emotional/physiological state**—an emotional, affective, or physiological response: "Doing research in the lab makes me happy," "I get excited when I'm doing field research," or "My heart starts racing when I begin to conduct an experiment."
  - **Conclude the discussion by reminding students of the following:**
    - Each of these four sources has the potential to both raise *or* lower one's self-efficacy for research. Although mastery experience tends to have the most powerful influence on a trainee's self-efficacy, another source may be a more powerful predictor of self-efficacy depending upon the situation and the individual. For example, trainees who have little research experience may rely more upon vicarious experience and social persuasion to gauge their self-efficacy at first, then gradually come to rely more upon mastery experience as they obtain more direct experience with research.
- **Activity: Significant Research/Learning Moment** (15 minutes)
  - Ask trainees to think of a significant moment in their own life when they felt more confident as a learner. This question can be projected onto a screen, written on the board, or prepared as a handout for trainees. (*Note:* For trainees with previous research experience, you may want to ask them to think of a significant moment when they felt more confident *as a researcher*.)
  - In pairs, have trainees discuss the questions below. (5 minutes)
    - How did that significant moment happen? What were the events, people, and experiences that contributed to the success?

Contributed by A. R. Butz with information from *Promoting Mentee Research Self-Efficacy* (Byars-Winston, Leveritt, Branchaw, and Pfund, 2013, 2016).

- What factors contributed to your sense of efficacy?
- Were some efficacy sources more common than others in your success story? If so, what are they?

- **Large-Group Discussion** (10 minutes)
  - Write each of the four sources on a flipchart or whiteboard. As trainees share their stories, encourage them to categorize the factors that contributed to their self-efficacy into one of the four sources of self-efficacy.
  - Questions for discussion:
    - What sources were common in your "success" stories?
    - Did some sources come up more than others? If so, what were they? What are examples of how the sources help to increase self-efficacy?
    - Do these same sources work well for you in other areas of your studies?
  - Use this as a time to talk about whether all sources work the same across domains. Some sources may be more effective domains than others. Some may work better for different ages/career stages than others. Also talk about differences. For example, some of the literature on self-efficacy suggests that individuals of different genders and cultural backgrounds may rely upon different sources or find certain sources of self-efficacy more salient than others. Encourage trainees to be mindful of the sources that work well for them, and to seek out opportunities in their research experience that will allow them to boost their self-efficacy for research.

► **Activity: Stair Steps** (Handout #2; 20 minutes)

- The goal of this activity is to have trainees clearly articulate steps to complete a research-related task and articulate how the four sources of self-efficacy can be used to support their confidence to complete research-related tasks.
- Have trainees write down something on their current research to-do list that they are not feeling confident in their ability to complete. Now have the trainees break the task down into at least two or three steps. If you are concerned that trainees will have difficulty identifying a task and/or breaking it down into steps, you can use one of the tasks listed in the alternate activity below. Alternatively, you could also have trainees think back to a task that they successfully completed and identify the steps that they took to complete that task.
- Ask trainees to consider the following questions:
  - What sources of self-efficacy could help you get to the first step? What about to get from the first step to the second?
  - How can you help build your own self-efficacy along the way?
- In pairs, have trainees discuss the ways that they could support their own self-efficacy at **one** of these steps using **at least three** of the sources of self-efficacy. Trainees can reference Handout # 1 to remind them of examples of each of the four sources of self-efficacy.
- Depending upon the size of your group, you may wish to have trainees summarize their discussions and talk about the different ways in which they incorporated the four sources to support them in a challenging task.
- Trainees can use these steps by themselves or in conversations with their mentor to help them recognize each successful step and celebrate it.

Contributed by A. R. Butz with information from *Promoting Mentee Research Self-Efficacy* (Byars-Winston, Leveritt, Branchaw, and Pfund, 2013, 2016).

Note: Facilitators can use one of the following scenarios for the Stair Steps Exercise. Ask participants as a group to help break the task down into manageable steps and identify ways to support self-efficacy at each step. These can also be adapted for different trainee levels.

"I need to learn a new technique in the lab."
Step 1: Ask someone in the lab to model the technique for you.
Step 2: Practice the technique a few times.
Step 3: Ask your PI/supervisor in the lab to provide you with feedback on your technique.

"I need to write an abstract for my research project, but I don't know where to begin."
Step 1: Read a journal article or conference abstracts for examples.
Step 2: Ask your mentor for examples of abstracts that they have written.
Step 3: Exchange abstracts with another student and give each other feedback.

"I need to do an analysis for my manuscript, but I don't feel confident in my ability to do it."
Step 1: Find online readings and tutorials that will walk you through the analysis process.
Step 2: Attend a workshop or training where you can learn how to do the analysis with others.
Step 3: Try the analysis with a different set of data with known results to check your process.

"I need to get started on my grant proposal, but I don't know where to start."
Step 1: Review the call for proposals and come up with a writing timeline by back-mapping your deadlines from the submission deadline.
Step 2: Ask your mentor for an example of a successful grant proposal.
Step 3: Invite colleagues to join a grant writing group with you.

References for Additional Reading

Chen, J. A., and Usher, E. L. (2013). Profiles of the sources of science self-efficacy. *Learning and Individual Differences, 24*: 11–21. http://doi.org/10.1016/j.lindif.2012.11.002

Hurtado, S., Cabrera, N. L., Lin, M. H., Arellano, L., and Espinosa, L. L. (2009). Diversifying science: Underrepresented student experiences in structured research programs. *Research in Higher Education, 50*(2): 189–214. http://doi.org/10.1007/s11162-008-9114-7

Byars-Winston, A., Rogers, J., Branchaw, J., Pribbenow, C., Hanke, R., and Pfund, C. (2016). New measures assessing predictors of academic persistence for historically underrepresented racial/ethnic undergraduates in science. *CBE-Life Sciences Education.*

Sawtelle, V., Brewe, E., and Kramer, L. H. (2012). Exploring the relationship between self-efficacy and retention in introductory physics. *Journal of Research in Science Teaching, 49*(9): 1096–1121. http://doi.org/10.1002/tea.21050

Usher, E. L., and Pajares, F. (2008). Sources of self-efficacy in school: A critical review of the literature and future directions. *Review of Educational Research, 78*: 751–796. doi:10.3102/0034654308321456

Zeldin, A. L., and Pajares, F. (2000). Against the odds: Self-efficacy beliefs of women in mathematical, scientific, and technological careers. *American Educational Research Journal, 37*(1): 215–246. http://doi.org/10.3102/00028312037001215

Zeldin, A. L., Britner, S. L., and Pajares, F. (2008). A comparative study of the self-efficacy beliefs of successful men and women in mathematics, science, and technology careers. *Journal of Research in Science Teaching, 45*(9): 1036–1058. http://doi.org/10.1002/tea.20195

Contributed by A. R. Butz with information from *Promoting Mentee Research Self-Efficacy* (Byars-Winston, Leveritt, Branchaw, and Pfund, 2013, 2016).

FOSTERING YOUR OWN RESEARCH SELF-EFFICACY

Learning Objectives

Trainees will:

- Define self-efficacy and its sources.
- Clearly articulate steps to complete a research-related task.
- Articulate how the four sources of self-efficacy can be used to build confidence to complete research-related tasks.

Handout #1

The Self-Efficacy Tool Box—What Can You Do to Be a More Confident Researcher?

From: http://psychology.about.com/od/theoriesofpersonality/a/self_efficacy.htm?p=1

Remember:

Self-efficacy: belief in one's ability to achieve a specific goal or task. Self-efficacy is situation-specific self-confidence. Simply put, *"Can I do this?"*

Strong self-efficacy beliefs create interest, persistence, actual college degree completion, and career pursuits in science and research fields.

Here are some efficacy-building strategies to try:

Mastery experience—a past accomplishment or success

- Think about your past success during the research experience in particular or academics in general ["If you did it before you can do it now"]. What contributed to that success? How can that be applied to your future research endeavors?
- Apply strategies and habits that have contributed to your past success in research to the task at hand.
- Recall the things you are doing right; devise strategies to improve your skill in areas that are challenging to you.

Vicarious experience—seeing or knowing of someone else who has successfully completed the task

- Do you know of others who have done similar research and have been successful? Talk to those researchers and use their actions as a model for your own. Consider your role models and what research skills (and attitudes) are being modeled by them.
- Be aware of what skills and behavior you are observing about coping with research challenges and setbacks; do they make you feel more or less confident about the work that you do? Ask others to share strategies for what they do when they hit a wall and how they cope with challenges/setbacks in research.
- Think of examples of others who struggled but made it (i.e., were successful in research).

Social persuasion—a social or verbal message reinforcing ability or effort

- Seek out individuals who provide encouragement and support to you in your research. Social persuasions relating to a specific effort or ability can be particularly influential.
- Ask for specific, constructive feedback from your mentors.

Emotional/Physiological state—an emotional, affective, or physiological response

- Be aware of positive (enjoyment) or negative moods (anxiety) that you have related to research/the lab.
- Attend to negative, anxiety-related feelings (e.g., negative self-talk that you are not as smart as other researchers).
- Acknowledge and normalize when things are difficult; "It's supposed to be hard, new things usually are."

Contributed by A. R. Butz with information from *Promoting Mentee Research Self-Efficacy* (Byars-Winston, Leveritt, Branchaw, and Pfund, 2013, 2016).

Handout #2

Stair Steps: Skill Development for Future Researchers

Instructions: Write your challenging task at the top of the stairs, then break it down into at least two or three steps. Place the steps in sequential order beginning with the skill you believe needs to come first on the bottom step.

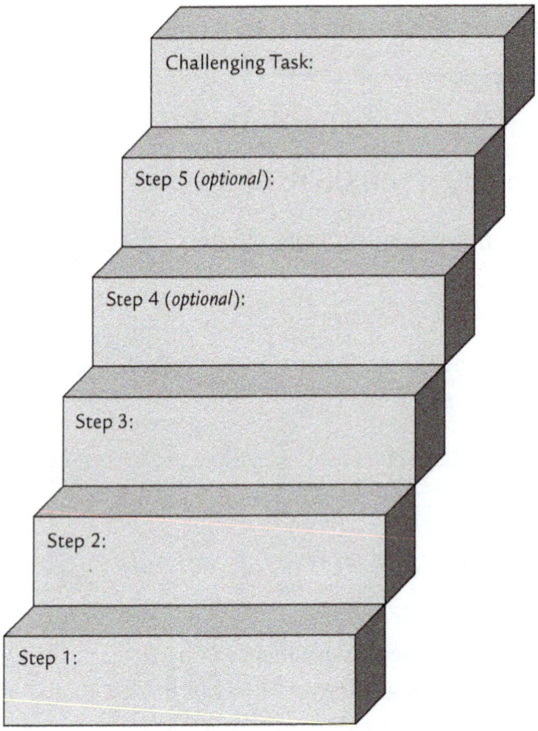

Challenging Task:

Step 5 (*optional*):

Step 4 (*optional*):

Step 3:

Step 2:

Step 1:

Contributed by A. R. Butz with information from *Promoting Mentee Research Self-Efficacy* (Byars-Winston, Leveritt, Branchaw, and Pfund, 2013, 2016).

CASE STUDY: FRUSTRATED

Learning Objectives

Trainees will:

- Practice strategies for communicating with mentors and research group members.

Trainee Level

undergraduate or graduate trainees
novice or intermediate trainees

Activity Components and Estimated Time for Completion

- In Session Time: 20 minutes

Total time: 20 minutes

When to Use This Activity

This activity can be used with undergraduate or graduate trainees at any career stage. Other activities that may be used with this activity include:

- Messages Sent and Received
- Barriers to Effective Communication
- Research Experience Reflections 1: Entering Research?
- Aligning Mentor and Trainee Expectations

Inclusion Considerations

Individuals from backgrounds historically underrepresented in STEM may struggle to find common ground with their mentors. Different mentor and trainee identities and backgrounds can create barriers to connecting and to the development of effective mentor–trainee relationships. Help trainees realize that even with different backgrounds and identities it is possible to develop positive and trusting relationships through open and honest communication. Reassure them that their mentors are there to support their training and development, but acknowledge that it may be difficult to speak up when things are not going well given the power and authority dynamics in these relationships. Establishing and aligning clear expectations with their mentor is important.

Areas of Trainee Development

- Research Comprehension and Communication Skills
  - Develop effective interpersonal communication skills.
  - Develop an understanding of the research environment.
- Researcher Confidence and Independence
  - Develop independence as a researcher.

Contributed by A. R. Butz with information from Branchaw, J. L., Pfund, C., and Rediske, R. (2010). *Entering Research: A Facilitator's Manual.* New York: W.H. Freeman & Co.

Implementation Guide

Workshop Session (20 minutes)

► **Case Study** (5 minutes)

- Distribute the case study and ask trainees to read and individually write down answers to the questions.

► **Discussion** (15 minutes)

- Bring the entire group together to discuss the case.
- If the group is large, put trainees in small groups to discuss the case, then ask each group to summarize their conversation for the large group at the end.

► **Discussion Questions**

1. To whom should Jamal go to discuss his frustration?
2. What strategies might he use to avoid appearing as though he is complaining?
3. How might establishing specific goals and expectations with his mentor help Jamal to avoid this situation?

Contributed by A. R. Butz with information from Branchaw, J. L., Pfund, C., and Rediske, R. (2010). *Entering Research: A Facilitator's Manual.* New York: W.H. Freeman & Co.

CASE STUDY: FRUSTRATED

Undergraduate

Learning Objectives

- Trainees will practice strategies for communicating with mentors and research group members.

Jamal has been in his research group for almost three weeks and is disappointed with his project so far. When he interviewed with Professor Stanley, she described a molecular biology project that he would work on. However, his graduate student mentor, Roxanne, has not given him any molecular biology experiments, but instead tasks such as making media and growing bacteria. Other undergraduate students in the lab seem to be doing things like cloning and sequencing genes. Jamal is getting frustrated, but doesn't want to complain or look ungrateful. What can he do?

1. To whom should Jamal go to discuss his frustration?

2. What strategies might he use to avoid appearing as though he is complaining?

3. How might establishing specific goals and expectations with his mentor help Jamal to avoid this situation?

Contributed by A. R. Butz with information from Branchaw, J. L., Pfund, C., and Rediske, R. (2010). *Entering Research: A Facilitator's Manual.* New York: W.H. Freeman & Co.

CASE STUDY: FRUSTRATED

Graduate

Learning Objectives

- Trainees will practice strategies for communicating with mentors and research group members.

Jamal is a first year Ph.D. student in molecular biology in Professor Stanley's lab. When he interviewed with Professor Stanley, she described a molecular biology project that he could work on as part of his dissertation research. He was excited to already have a potential dissertation topic identified! However, his first few months in the lab have been very frustrating. Professor Stanley has been out of town for much of the semester working on another grant-funded project, and Jamal has received little guidance from the post-docs, who are supervising the project in her absence. He feels like he is lagging behind some of the other students in his cohort, who are already working on manuscripts and conference presentations. Jamal wants to talk with Professor Stanley about this, but doesn't want to complain or look ungrateful; nor does he want the post-docs to feel like he is going over their head. What can he do?

Questions for Discussion

1. To whom should Jamal go to discuss his frustration?

2. What strategies might he use to avoid appearing as though he is complaining?

3. How might establishing specific goals and expectations with his mentor help Jamal to avoid this situation?

Contributed by A. R. Butz with information from Branchaw, J. L., Pfund, C., and Rediske, R. (2010). *Entering Research: A Facilitator's Manual.* New York: W.H. Freeman & Co.

FUNDING YOUR RESEARCH

Learning Objectives

Trainees will:

- Identify funding sources available to support research.

Trainee Level

undergraduate and graduate trainees
novice, intermediate, or advanced trainees

Areas of Trainee Development

- Professional and Career Development Skills
  - Explore and pursue a research career.
- Research Comprehension and Communication Skills
  - Develop research communication skills.

Activity Components and Estimated Time for Completion

- Trainee Pre-Assignment Time: 30 minutes
- In Session Time: 35 minutes

Total time: 1 hour, 5 minutes

When to Use This Activity

This activity is appropriate for trainees who are currently in a research group. Facilitators working with intermediate to advanced undergraduate and graduate trainees may want to implement the "Research Group Funding" activity in lieu of this activity. This activity can be implemented in several different ways to accommodate students at all levels of experience.

Inclusion Considerations

Include funding opportunities designed to support individuals from underrepresented groups (e.g., Ford Fellowships, HHMI Gilliam Fellowships) in pursuing research careers as examples.

Contributed by C. Barta with information from Branchaw, J. L., Pfund, C., and Rediske, R. (2010). *Entering Research: A Facilitator's Manual.* New York: W.H. Freeman & Co.

Implementation Guide

Trainee Pre-Assignment (30 minutes)

- Distribute the worksheet "Funding your Research" at least one week before the topic will be discussed. Ask trainees to complete the worksheet by either talking with their PI or a knowledgeable individual in their research group (graduate student, postdoctoral researcher, lab manager, program director, etc.).

Workshop Session

- **Introduction** (5 minutes)
  - This activity will inform trainees about sources of funding to support research in general and to identify specific funding sources available to support their research.
- **Discussion** (30 minutes)
  - **Small Groups** (10 minutes)
    - Trainees share the funding sources they identified.
  - **Large-Group Discussion Questions** (20 minutes)
    - What funding opportunities are currently available to you as an undergraduate or graduate student? (campus scholarships/research awards/fellowships, NSF GRFP, McNair Scholars, etc.)
    - What are the requirements for these funding opportunities? Deadlines?
    - What are the characteristics of a "strong" scholarship or fellowship application?
    - What are the characteristics of a "strong" research grant?
    - How is research funded on campus?
    - Who provides funding for researchers, such as graduate students, post-doc trainees, lab technicians, and PIs, on campus?
    - What skills do you need to write a successful application for scholarships, fellowships, or research grants?
    - What are some resources you can utilize to strengthen your application writing skills? (e.g., campus writing centers, courses in scientific writing, reviewing successfully funded research grants/scholarship applications, attending seminars/workshops, contributing to writing and/or submitting grants for their research group).

Optional Activities

- **Bring in examples of successful research grants, scholarship, and/or fellowship applications.** (30-minute discussion) Distribute the examples to small groups of three or four trainees to review. After 10–15 minutes, ask each group to report what they observed from reviewing the documents. Alternatively, distribute these examples the week before for trainees to review. During class, ask the trainees to reflect on what they liked, what they were surprised by, and what they plan to do in the future to be successful in writing research grants, proposals, scholarship, and/or fellowship applications.
- **Research funding at your institution.** (5- to 20-minute discussion) Discuss how grants are submitted, managed, and reviewed at your institution. Consider also discussing commonly used terms such as overhead or indirect costs, matching funds, 9-month vs. 12-month appointments, etc.
- **Invite a campus speaker to talk about funding opportunities for students.** (variable time limit) Invite the speaker to your course to talk about funding opportunities for students. Allow several minutes for questions.
- **Organize a panel of three to five trainees who have been successful in applying for funding.** (variable time limit) One variation to this panel is to invite three to five faculty members to talk about their success and failures in securing funding for their research and for their undergraduate, graduate students and/or post-doc trainees.

Contributed by C. Barta with information from Branchaw, J. L., Pfund, C., and Rediske, R. (2010). *Entering Research: A Facilitator's Manual.* New York: W.H. Freeman & Co.

FUNDING YOUR RESEARCH

Learning Objectives

Trainees will:

► Identify funding sources available to support research.

Arrange a meeting with either the Principal Investigator (PI) of your research group or another knowledgeable research group member (graduate student, postdoctoral researcher, lab manager, program director, etc.) and discuss the following questions regarding research funding.

1. What funding opportunities are available for trainees at my career stage? (i.e., scholarships, fellowships, research grants for student projects, etc.)

2. What are the requirements for these funding opportunities? Deadlines?

3. What makes for a "strong" fellowship, scholarship, or research grant application?

4. Where can trainees find additional funding opportunities for research?

5. What resources are available on campus to help trainees prepare a successful fellowship, scholarship, or research grant application?

6. What advice would you give a trainee who is applying for research funding for the first time?

Contributed by C. Barta with information from Branchaw, J. L., Pfund, C., and Rediske, R. (2010). *Entering Research: A Facilitator's Manual.* New York: W.H. Freeman & Co.

GENERAL PUBLIC ABSTRACT

Learning Objectives

Trainees will:

- ▶ Identify the attributes of a good public abstract.
- ▶ Learn to identify and eliminate or define research jargon.
- ▶ Give constructive feedback to peers on general public abstracts.

Trainee Level

undergraduate or graduate trainees
novice or intermediate trainees

Areas of Trainee Development

- ▶ Research Comprehension and Communication Skills
  - Develop research communication skills.
  - Develop logical/critical thinking skills.
- ▶ Research Ethics
  - Develop responsible and ethical research practices.

Activity Components and Estimated Time for Completion

- ▶ Session 1 Time: 1 hour, 10 minutes
- ▶ Pre-Assignment Time: 1 hour
- ▶ Session 2 Time: 1 hour, 10 minutes

Total time: 3 hours, 20 minutes

When to Use This Activity

It is best to implement this activity several months into the trainees' research experiences to ensure they understand their project well enough to construct an abstract and explain their project in terms the general public would understand. Other activities that may be used with this activity include:

- ▶ Communicating Research to the General Public
- ▶ Three-Minute Research Story

Inclusion Considerations

Trainees from non-research backgrounds may have a particularly valuable perspective to share in writing general public abstracts. Encourage trainees to share how they think their families will react to the general public abstract drafts. Point out that this exercise may be particularly helpful to them in finding ways to explain the value of their research and their motivation for doing it to their families.

Contributed by A. Bramson and L. Adams with information from Gilpin, A. A., and Patchet-Golubev, P. (2000). *A Guide to Writing in the Sciences.* Toronto: University of Toronto Press, and from Branchaw, J. L., Pfund, C., and Rediske, R. (2010). *Entering Research: A Facilitator's Manual.* New York: W.H. Freeman & Co.

Implementation Guide

Session One

► Introduction (30 minutes)

- **Overview of the elements of an abstract.** (20 minutes) If trainees have no experience writing abstracts, they will need to be introduced to them. The facilitator should bring examples of abstracts to class or ask trainees to bring examples of abstracts from their research groups. Discussion outline:
  - An abstract is:
    - A brief (often on the order of 100 words), yet comprehensive summary of a research project without added interpretation. It conveys the most significant information about the research, especially the results. Readers often use abstracts to decide whether or not they want to read the full article or attend a presentation.
  - An abstract should:
    - Be complete, concise, clear, and cohesive (the 4 C's).
    - Present a summary of the project in a way that is appropriate to the background knowledge of the intended reader.
    - Present the hypothesis or research question.
    - Outline the methodology used to test the hypothesis/ answer the research question.
    - Report the most important results, but not include figures or tables unless they are allowed.
    - Provide the main conclusions of the study.
  - Clear and concise phrases that are useful when writing an abstract:
    - "The goal of our research is to . . ."
    - "This study is important because . . ."
    - "To test our hypothesis, we used . . . [approach] . . . [rationale]"
    - "The results indicate that . . ."
    - "Our findings support the conclusion that . . ."
  - Including enough technical detail (and knowing how many/which details to include), while maintaining the readability of the abstract can be challenging. It is perhaps the hardest part about writing an abstract.
  - If implementing this session with trainees from very different fields of science, it may be useful to discuss how different disciplines can emphasize different aspects of a project (i.e., methods vs. conclusions vs. purpose). However, each abstract should still contain:
    - Purpose
    - Research problem
    - Methods used/data analyzed
    - Conclusions/preliminary results
    - Significance/usefulness of the results
- **Modifying an abstract for the general public.** (10 minutes) If trainees have written a scientific abstract and will be modifying it for the general public, have them reflect on the characteristics of a good abstract (regardless of the audience):
  - 4 C's: complete, concise, clear, and cohesive
  - err on the side of simpler vocabulary and sentence structures
  - eliminate extra words when possible; examples: "a large number of" = "many" and "as a general rule" = "generally"
  - avoid useless intensifiers; examples to avoid: *really, always, clearly, extremely*

Contributed by A. Bramson and L. Adams with information from Gilpin, A. A., and Patchet-Golubev, P. (2000). *A Guide to Writing in the Sciences.* Toronto: University of Toronto Press, and from Branchaw, J. L., Pfund, C., and Rediske, R. (2010). *Entering Research: A Facilitator's Manual.* New York: W.H. Freeman & Co.

- **"Talk Nerdy to me"** (video and discussion; 10 minutes)
  - http://www.ted.com/talks/melissa_marshall_talk_nerdy_to_me
  - Discussion: Review Melissa Marshall's suggestions to scientists to communicate to nonscientists. Come up with additional strategies as a group (suggestions from previous discussions below):
    - Relate to a personal experience
    - Use an analogy if possible, especially when trying to relate scale
    - Remove jargon
    - Tell stories
    - Tell why (relevance)

► **Example of a general public abstract** (20 minutes)

- Provide trainees with the sample general public abstract provided in the trainee materials.
- Ask trainees to identify the major components in the abstract individually: purpose, research problem, methods, conclusions, significance. Reconvene as a group and discuss answers.
- Ask trainees to discuss the potential ethical issues with the research described in the abstract. Tell trainees that when presenting research to the general public, they may be asked about ethical issues associated with their work. The general public may also ask about why the work is important to society.

► **Discussion Questions** (20 minutes)

- How should an abstract change for different audiences?
- Will the components of an abstract change?

> - In general, the components will not change.
> - How much different components are emphasized may change.

- Are there any components that will be most prominent in an abstract meant for other scientists in your field? What about for the public? How might abstracts change for different fields of science?

> - In a general public abstract, the purpose and significance will likely need to be the most prominent parts of the abstract.
> - For a conference focused on research methods, the abstract should focus on the data analysis techniques used.
> - For other scientists in the specific subfield, the implications of research are already well understood, so the abstract would focus on results.

- Will the language used when writing an abstract change?

> - Don't use jargon in an abstract meant for the public; they will not automatically know research or disciplinary terms.
> - Avoid acronyms in general public abstracts by spelling out terms.

Trainee Assignment (1 hour)

► Assign trainees to write an abstract for the general public and bring at least two copies to the next session. Remind them that this abstract should still contain all the major components discussed (purpose, research problem, methods, conclusions, significance), but it should explain the project in a way the general public would understand.

Contributed by A. Bramson and L. Adams with information from Gilpin, A. A., and Patchet-Golubev, P. (2000). *A Guide to Writing in the Sciences.* Toronto: University of Toronto Press, and from Branchaw, J. L., Pfund, C., and Rediske, R. (2010). *Entering Research: A Facilitator's Manual.* New York: W.H. Freeman & Co.

Session Two

► **Peer review of general public abstract** (50 minutes)

- Assign trainees in pairs to read each other's abstracts and provide feedback using the Abstract Review Form provided in the trainee materials. When ready, they should discuss their feedback.
- Ask trainees to:
  - identify the part of the abstract where the importance of the research to the general public is presented; and
  - identify words in the abstract that are research or discipline specific and likely not part of the vocabulary of a non-researcher.
- The focus of the discussion should be on how to clarify or better convey the importance of the research and how to replace the terminology with words or phrases better suited for a general audience.
- Time permitting, trainees can rotate partners during a session (20 minutes per pair).
- **Alternative activity: Jargon Police**
  - Trainees exchange "General Public Abstract" with a partner, read through their partner's abstract and highlight the jargon that requires further explanation or should be eliminated. (3–5 minutes)
  - Remind the trainees that if they are struggling to understand any aspect of the writing, it indicates that the writer needs to clarify and remove jargon.

► **Discussion** (20 minutes)

- The goal of this discussion is to compile the ideas for abstract improvement generated in small groups/pairs. Each group can share its lists of problems, solutions, and good features with the class.
- The facilitator captures ideas on the board to create a compiled list for trainees' future reference.
- **Discussion Questions**
  - What was the most difficult part about writing a general public abstract? The easiest?
  - What advice would you give trainees writing their first abstract?

> ▪ Ask friends, family, and members of the public to provide feedback. This also gives trainees the opportunity to explain their research to those individuals!

  - Why is describing the implications of the research results so important in a general public abstract?

> ▪ In academia the results are generally considered most important and implications are assumed. By contrast, the public usually needs the researcher to directly explain the implications to answer the question "Why should I care?"
>
> ✓ Example: "increased lift on airfoils" (language used in his science abstract) vs. "allows airplanes to go higher" (used in the general public version).

► **Wrap-up** (5 minutes)

- Revision of general public abstract. Assign trainees to use the feedback they received to write a final draft of their general public abstract.
- Revised general public abstracts can be used as a pre-assignment to the "3-Minute Research Story" activity.

Contributed by A. Bramson and L. Adams with information from Gilpin, A. A., and Patchet-Golubev, P. (2000). *A Guide to Writing in the Sciences.* Toronto: University of Toronto Press, and from Branchaw, J. L., Pfund, C., and Rediske, R. (2010). *Entering Research: A Facilitator's Manual.* New York: W.H. Freeman & Co.

GENERAL PUBLIC ABSTRACT

Learning Objectives

Trainees will:

- Identify the attributes of a good public abstract.
- Learn to identify and eliminate or define research jargon.
- Give constructive feedback to peers on general public abstracts.

General Public Abstract Example

by Jennifer Arens-Gubbels, University of Wisconsin–Madison

General Public Abstract

Epithelial ovarian cancer is the deadliest of gynecological cancers. Cells of this tumor type express an extremely large molecule called MUC16. This molecule is found covering the surface of the tumor cells, as well as floating in elevated amounts in the body fluids of ovarian cancer patients. We have shown previously that MUC16 in body fluid binds to immune cells and causes them to be non-functional. We, therefore, investigated the effects of cell surface MUC16 to determine if it also functions to inhibit immune cells. We incubated ovarian cancer cells that express MUC16 or ovarian cancer cells that lack MUC16 with immune cells to determine if the immune cells could create an activating immune synapse with the cancer cells. Activating immune synapses are indicators of target cell death, require cell-to-cell contact, and give detailed information about the functionality of immune cells. We found that cells that lack MUC16 form more activating synapses with immune cells compared to cells that have large amounts of MUC16 on their surface. Our results indicate that MUC16, because of its large size, may be physically inhibiting immune cells from forming functional activating immune synapses.

Ethical Issues

- Research Ethics: There could be bias in counting synapses if the cell type that the NK cells were incubated with is known. To address this, we used an experimental design in which the person counting the synapses was blind to the cell type. In addition, another member of the lab, who is unfamiliar with the project, counted the synapses to ensure accurate counts were taken.
- Bioethics: Cells used in this research are derived from human donors. Appropriate legal procedures were followed in harvesting and using the cells, which include full consent from the donor of the cells.

Contributed by A. Bramson and L. Adams with information from Gilpin, A. A., and Patchet-Golubev, P. (2000). *A Guide to Writing in the Sciences.* Toronto: University of Toronto Press, and from Branchaw, J. L., Pfund, C., and Rediske, R. (2010). *Entering Research: A Facilitator's Manual.* New York: W.H. Freeman & Co.

DRAFT OF GENERAL PUBLIC ABSTRACT

1. Write an abstract for the general public. (200 words or less) To help you think about how to write an abstract meant for a public audience, imagine you are sitting on an airplane and the person sitting next to you asks, "What do you do?" How would you explain your research to this person?

2. When presenting research to the general public, you may be asked about ethical issues associated with your work. The general public may also ask about why your work is important to society. Identify any possible ethical issues associated with your research, explain how you would address those issues to a general audience and how your work will impact the general public or why they should care.

Contributed by A. Bramson and L. Adams with information from Gilpin, A. A., and Patchet-Golubev, P. (2000). *A Guide to Writing in the Sciences.* Toronto: University of Toronto Press, and from Branchaw, J. L., Pfund, C., and Rediske, R. (2010). *Entering Research: A Facilitator's Manual.* New York: W.H. Freeman & Co.

FINAL VERSION OF GENERAL PUBLIC ABSTRACT

Revise your general public abstract based on feedback received from peer reviewers. Make sure to include a statement in your abstract explaining how your research will impact the general public and why they should care. If your research involves ethical issues that the public will be concerned about, you may also want to add a statement about how you are handling these issues.

Contributed by A. Bramson and L. Adams with information from Gilpin, A. A., and Patchet-Golubev, P. (2000). *A Guide to Writing in the Sciences.* Toronto: University of Toronto Press, and from Branchaw, J. L., Pfund, C., and Rediske, R. (2010). *Entering Research: A Facilitator's Manual.* New York: W.H. Freeman & Co.

GENERAL PUBLIC ABSTRACT

Peer-Review Form

Author: ______________________ Reviewer: ______________________

CONTENT

1. Title and Authors YES SORT OF NO

   Explain:

2. Purpose of the project, context, and/or relevance YES SORT OF NO

   Explain:

3. Hypothesis/research question YES SORT OF NO

   Explain:

4. Research methods/approach taken YES SORT OF NO

   Explain:

5. Results (or preliminary results) and conclusions YES SORT OF NO

   Explain:

6. Significance/importance/new understandings YES SORT OF NO

   Explain:

Contributed by A. Bramson and L. Adams with information from Gilpin, A. A., and Patchet-Golubev, P. (2000). *A Guide to Writing in the Sciences.* Toronto: University of Toronto Press, and from Branchaw, J. L., Pfund, C., and Rediske, R. (2010). *Entering Research: A Facilitator's Manual.* New York: W.H. Freeman & Co.

STYLE (The Four C's)

	YES	SORT OF	NO
1. Complete	YES	SORT OF	NO
Explain:			
2. Concise	YES	SORT OF	NO
Explain:			
3. Clear	YES	SORT OF	NO
Explain:			
4. Cohesive	YES	SORT OF	NO
Explain:			

Trainee Materials

LENGTH

Does the abstract fit the character/word limit?	YES	TOO LONG	TOO SHORT

ADDITIONAL COMMENTS:

Contributed by A. Bramson and L. Adams with information from Gilpin, A. A., and Patchet-Golubev, P. (2000). *A Guide to Writing in the Sciences.* Toronto: University of Toronto Press, and from Branchaw, J. L., Pfund, C., and Rediske, R. (2010). *Entering Research: A Facilitator's Manual.* New York: W.H. Freeman & Co.

IMPORTANCE OF READING IN GRADUATE SCHOOL

Learning Objectives

Trainees will:

- Learn how research literature can play an important role in research success.
- Develop strategies to read and comprehend appropriate and relevant research literature.

Trainee Level:

graduate trainees
intermediate or advanced trainees

Activity Components and Estimated Time for Completion

- Trainee Pre-Assignment Time: 20 minutes
- In Session Time: 45 minutes

Total time: 1 hour, 5 minutes

When to Use This Activity

This activity is recommended for beginning graduate students especially those in research rotations. Other activities that may be used with this activity include:

- Article Organization, Comprehension, and Recall
- Searching Online Databases
- Research Writing 4: Research Literature Review and Publishing Process

Inclusion Considerations

Learning and reading styles will vary among trainees. Invite trainees to share with you any learning accommodations they need or preferences they have and be flexible when setting reading and writing assignment deadlines. Encourage trainees to share with the group alternative ideas about how to approach reading research papers, organizing information, and constructing reviews.

Areas of Trainee Development

- Research Comprehension and Communication Skills
  - Develop disciplinary knowledge.
- Professional and Career Development Skills
  - Explore and pursue a research career.
- Researcher Confidence and Independence
  - Develop independence as a researcher.

Contributed by A. Sokac. (2018). *Importance of Reading in Graduate School.*

Implementation Guide

Trainee Pre-Assignment (20 minutes)

► Read Parker, R. (2012). Skill development in graduate education, *Molecular Cell, 46*: 377–381.

Workshop Session (45 minutes)

► **Activity: Building Skills by Reading Literature** (20 minutes)

- Reflecting on the article, "Skill development in graduate education," generate a list of the skills and competencies needed in graduate school and how reading literature can help one achieve those skills/competencies. (pairs, 10 minutes)
- Group discussion of skills and competencies needed in grad school. Document answers on a whiteboard or flipchart. (15 minutes)
    - What competencies do trainees believe they will need in order to successfully reach the following milestones in graduate school?
        - become a productive part of their rotation/permanent lab
        - pass their Qualifying Exam
        - complete their thesis work
        - become an expert in their field
    - How will reading the literature in the field impact each milestone?
    - Optional additions:
        - If an institutional or departmental rubric for the Qualifying Exam is used in the graduate training program, it could be distributed so students see the expectations for passing to degree status. The expectations listed on the rubric could substitute for the list of competencies generated in class.
        - After discussing the impact of reading the literature on professional development for the student, discuss what the scientific literature provides more broadly. For example, it reveals the remaining outstanding questions in a field, and serves as a "newspaper" for scientists. The literature also connects scientists across the world and centers their efforts.

► **Activity: When do I read what? Where do I start?** (20 minutes)

- Ask trainees to reflect on their research reading habits. (3 minutes)
- In pairs, discuss the following questions. (7 minutes)
    - How many papers have you read so far for your rotation lab/project?
    - Do you think that you are reading enough?
    - Are you finding the papers that you want?
    - What problems are you having in finding papers?
- Distribute the "When do I read what? Where do I start?" handout. Using the document as a source of ideas, ask the group to brainstorm strategies to find relevant papers to read. (10 minutes)

► **Wrap-up** (5 minutes)

- Summarize the main ideas generated from the large-group discussion.
    - Importance of reading in graduate school
    - Strategies to find relevant scientific literature

Trainee Post-Assignment: Ask trainees to interview their mentor or a colleague in their research group to complete the My Research Group's Core Journal List activity included in the trainee materials.

IMPORTANCE OF READING IN GRADUATE SCHOOL

Learning Objectives

Trainees will:

- Learn how research literature can play an important role in research success.
- Develop strategies to read and comprehend appropriate and relevant research literature.

Building Skills by Reading Literature

The goal of this activity is to see how reading the literature can promote your immediate and long-term success in graduate school and beyond.

Consider the following questions in order to generate a list of skills or competencies that you will need to be successful in graduate school. Then consider how reading the literature in your field will impact each skill or competency on the list. Use the table below to record your ideas.

What skills or competencies do you need to gain in order to become a productive part of:

- Your rotation lab?
- Your permanent research group?
- To pass your Qualifying Exam?
- To complete your thesis work?
- To become an expert in your eventual field of study?

Skill/Competency	Impact of Reading the Literature

Contributed by A. Sokac. (2018). *Importance of Reading in Graduate School.*

ACTIVITY: WHEN DO I READ WHAT? WHERE DO I START?

The goal of this activity is to identify strategies to filter through the innumerable papers that are available, so you can focus on those that are most useful to your purposes right now (e.g., your current rotation lab/project).

In pairs, reflect on your current rotation lab/project and answer the following:

- How many papers have you read so far for your rotation lab/project?

- Are you finding the papers that you want?

- Do you think that you are reading enough papers to excel in your rotation lab/project?

- What problems, if any, are you having in finding papers?

- What strategies could you use to read more relevant literature?

Contributed by A. Sokac. (2018). *Importance of Reading in Graduate School.*

WHEN DO I READ WHAT? WHERE DO I START?

When do I read what?

Reviews are a good start to get the big picture.

- On the hill with binoculars, overlooking the cornfield

An expert needs a deep understanding of the primary literature.

- Farmer in the cornfield identifying plant pathogens

Getting more familiar with methods, including their uses, strengths, and weaknesses, through methods articles can make the primary literature easier to understand and evaluate.

Where do I start?

Job 1: Acquire knowledge

- Start with a review. Dig deeper by looking up the papers cited in the most relevant parts of the review.
- Start with your mentor's paper, a paper from your mentor's research group, or a paper recommended by your mentor or other member of your research group. Dig deeper by looking up additional papers that are cited in the first paper.
- Start with older papers first, and read toward the newer papers to gain a historical perspective on the research subject.
- Brainstorm keywords for database searches with somebody knowledgeable in your research group.

Job 2: Stay current in the research field

- Sign up for eAlerts on journal websites (the table of contents for each issue of the journal will then be delivered to your email).
- Talk to people about new findings!
- Attend seminars and journal clubs.

Contributed by A. Sokac. (2018). *Importance of Reading in Graduate School.*

ACTIVITY: MY RESEARCH GROUP'S CORE JOURNAL LIST

Every research group will have a different set of core journals that they read and keep up with. You need to find out what that core set of journals is for the research group that you are rotating/working in. Knowing this can help you understand what literature the research group values, as well as how the research group sees itself fitting into the broader research community.

Broad interest journals (e.g., *Science* or *Nature*): Among these journals, the mentor or research team members look at every issue and get email notifications (eAlerts) for every issue.

Specialist journals (e.g., *Plant Cell, Human Brain Mapping*): Among these journals, the mentor or lab members look at every issue and get email notifications (eAlerts) for every issue.

Other journals with relevant articles, but of lesser importance: Time permitting, the mentor or research group members look at issues of these journals. They may or may not receive eAlerts for every issue.

Contributed by A. Sokac. (2018). *Importance of Reading in Graduate School.*

INTERVIEWING FOR GRADUATE SCHOOL

Learning Objectives

Trainees will:

- Learn what to expect when interviewing for graduate school.
- Articulate skills and areas of interest in research.
- Become empowered to succeed in graduate school interviews.

Trainee Level:

undergraduate trainees
intermediate or advanced trainees

Areas of Trainee Development

- Professional and Career Development Skills
  - Explore and pursue a research career.
- Research Comprehension and Communication Skills
  - Develop research communication skills.
  - Develop an understanding of the research environment.

Activity Components and Estimated Time for Completion

- Trainee Pre-Assignment Time: 1 hour
- In Session Time: 1 hour

Total time: 2 hours

When to Use This Activity

This activity can be used anytime, but is most relevant when trainees are ready to apply to graduate programs.

Inclusion Considerations

Explore with the group why it is important to consider how a trainees' background, interests, and motivations should factor into their selection of a graduate training program. One program will not be the right fit for every trainee. Critically reflecting on what they will need to be well, and to be a successful student and productive researcher, rather than what others are telling them is the "right choice," is an important step for trainees toward identifying the best fit.

Contributed by E. Frazier with information from Branchaw, J. L., Pfund, C., and Rediske, R. (2010). *Entering Research: A Facilitator's Manual.* New York: W.H. Freeman & Co.

Implementation Guide

Trainee Pre-Assignment (1 hour)

- Distribute the trainee materials and ask trainees to bring their prepared answers to use in mock interviews and discussion during the next session. Trainees should answer these questions with their top choice institution/program in mind.

Workshop Session (1 hour)

Introduction (5 minutes)

- The goal of this activity is for trainees to gain a better understanding of the different types of information that interviewers are trying to obtain in a graduate school interview.

Discussion: Types of Graduate Program Admission (10 minutes)

- Briefly discuss the different types of graduate programs. It is important that trainees know which kind of program they will be applying for so that they can better prepare themselves for the interview. Depending on trainee prior knowledge, facilitators may want to engage the group in a large discussion, or simply provide this information to the group.
  - **How does the application process work for a graduate program with rotations?**
    - Candidates apply to a graduate program first and once selected they do short rotations (~8 weeks) with two or three research groups before identifying which mentor they would like to work with.
    - When applying for a graduate program with rotations, candidates need to concentrate their interview preparation on gathering information about the specific graduate program and several potential mentors.
  - **How does the application process work for a graduate program with direct admissions?**
    - Some programs encourage prospective trainees to find a research mentor before they apply to the graduate program.
    - If trainees apply for a direct admission program, they would need to concentrate their interview preparation on gathering information about the specific research mentor they are interested in working with along with understanding the graduate program.
  - **What are the pros and cons of rotations and direct admission to graduate school programs?**
    - **Rotation Pros:**
      - Trainees choose a mentor after working with that person and gaining an understanding of the research environment.
      - Trainees get to experience the research before making a decision to engage in that research for graduate school.
      - Trainees have the opportunity to work with multiple mentors in the program, who may become part of their mentoring network, even if they do not become their primary research mentor.
      - Trainees have the opportunity to learn new techniques and to gain an understanding of the types of expertise within the graduate program and/or department.
    - **Rotation Cons:**
      - Trainees will spend much of the first year of graduate school identifying a research mentor and research group rather than immediately engaging in their research of interest.

Contributed by E. Frazier with information from Branchaw, J. L., Pfund, C., and Rediske, R. (2010). *Entering Research: A Facilitator's Manual.* New York: W.H. Freeman & Co.

- **Direct Admission Pros:**
  - Trainees will know which research mentor they will work with before agreeing to join the graduate program.
  - Trainees can immediately start working in their research area of choice.
- **Direct Admission Cons:**
  - Trainees may not have had a lot of time to interact with the primary research mentor and other research group members before accepting the invitation to join the group.
  - Trainees will need to cultivate relationships with other faculty and graduate students on their own, rather than through rotations.

► **Activity: Mock Interviews** (30 minutes)

- The goal of the mock interview is to give students an opportunity to practice talking about their interest in a particular graduate school program and their undergraduate (or other) research experience(s).
- To provide this practice, facilitators can meet one-on-one with each trainee, or facilitators can ask trainees to contact their mentors and ask them to conduct a mock interview. If interviews are conducted with mentors, this can be completed as an out-of-class assignment.
- The interview questions can be separated into two categories:
  - Questions assessing the candidate's knowledge of the program/research group/faculty mentor: Candidates should have a basic understanding of the programs they are applying to and why they chose that graduate program or research mentor.
  - Questions assessing the candidate's ability to talk about their research and to clearly articulate their research and career goals: Candidates should be prepared to talk about their research in depth, and show that they have clear career goals (even if these will change with time).
- Primary (numbered) interview questions and potential follow-up questions (lettered) are provided in the trainee materials. Emphasize to trainees that they should be prepared to answer the primary question and that the follow-up questions will likely be asked if they do not touch upon those topics in their response to the primary question.

► **Discussion** (10–15 minutes)

- How prepared were you to answer these questions? In which areas did you feel more/less prepared?
- What did you notice about the interviewer?
- What were some of the main points that interviewers tried to cover in their questions?
  - Candidate's prior research experience
  - Candidate's level of fit with the program/research group
  - Candidate's level of preparedness for interview/commitment to this program/research group
  - Candidate's critical thinking skills (by asking about previous research experience)
  - How well the candidate knows themselves

Contributed by E. Frazier with information from Branchaw, J. L., Pfund, C., and Rediske, R. (2010). *Entering Research: A Facilitator's Manual.* New York: W.H. Freeman & Co.

INTERVIEWING FOR GRADUATE SCHOOL

Learning Objectives

Trainees will:

- ► Learn what to expect when interviewing for graduate school.
- ► Articulate skills and areas of interest in research.
- ► Become empowered to succeed in graduate school interviews.

You have completed your undergraduate degree, have some research experience and will be applying for graduate programs. If your application is successful, you may be invited to interview for graduate school. It is very important to be prepared for an interview by reflecting on some of the topics that are commonly discussed in interviews. For this exercise, select your top choice of a graduate program and answer the questions below.

In some graduate programs, candidates apply to the program first and, once selected, are paired with several possible research mentors (rotations program). In other programs, candidates are encouraged to find a research mentor before they apply to the graduate program (direct admissions). It is important that you know which type program you will be applying for so that you can better prepare for the interview. For example, if applying for a rotation program, you will need to concentrate on gathering information about the specific graduate program and multiple possible mentors. If applying for a direct admissions program, you would need to gather information about the specific research mentor in addition to gathering information about the graduate program.

Questions assessing the candidate's knowledge of the program/research group/faculty mentor—Candidates should have a basic understanding of the programs they are applying to and why they chose that graduate program or research mentor.

1. What led you to apply to this particular research group or graduate training program?
   a. Why do you want to attend this institution?
   b. How will participation in this graduate program support your career goals?
   c. What research mentors/areas of research would you like to work with and why?
   d. What about this research group appeals to you?
   e. What about the type of research/work that is done in the research group or graduate program that you have applied for interests you?

Questions assessing the candidate's ability to talk about their research and to clearly articulate their research and career goals—Candidates should be prepared to talk about their research in-depth, and show they have clear career goals (even if these will change with time).

2. Describe your prior research experience.
   a. What were the major hypotheses or questions that you tested?
   b. What were your results?
   c. How do your results support or refute your hypotheses or questions?
   d. What was the significance of your research?
   e. What are the next steps in your research?
   f. How did your undergraduate research experience influence your decision to attend graduate school?

Contributed by E. Frazier with information from Branchaw, J. L., Pfund, C., and Rediske, R. (2010). *Entering Research: A Facilitator's Manual.* New York: W.H. Freeman & Co.

3. What skills have you gained from your undergraduate research experience that could be applied to your future position?
   a. research communications skills, written and oral
   b. obtaining funding
   c. networking skills
   d. leadership skills
   e. outreach—communicating research to the public
4. Tell us about yourself.
   a. What should we know about you?
   b. What are your strengths?
   c. What are your weaknesses?
   d. What are you doing to overcome your weaknesses?
   e. Can you give examples that show your leadership skills?
   f. Can you give examples of your communications skills?
   g. Can you give an example of your mentoring skills?
   h. Can you give an example of how you deal with conflict in a research group?
   i. Can you give examples of volunteer/outreach activities in which you have participated?
5. What are your career goals?
   a. How will this program help you meet your career goals?
   b. What additional training do you need for your career goals?
6. What are your research goals?
   a. Do you have any research questions of your own that you would like to pursue? What are they and why do they interest you?
   b. Do any research groups pursue similar research questions?
   c. What deficiencies in knowledge do you have in your research area? How do you plan to address them?
   d. What are the broader impacts of your research and what kind of broader impact activities do you expect to engage in?
7. Which courses do you anticipate taking to fulfill your training in this field?
8. What are your expectations of the program?
9. What contributions can you make to the program?
10. What are your expectations of your research mentor?
11. Do you have teaching experience?
12. Do you have any questions for us?

Contributed by E. Frazier with information from Branchaw, J. L., Pfund, C., and Rediske, R. (2010). *Entering Research: A Facilitator's Manual.* New York: W.H. Freeman & Co.

CASE STUDY: KEEPING THE DATA

Learning Objectives

Trainees will:

- Explain why it is important to accurately document research.
- Identify key elements in research documentation.
- Understand the ethical implications of documenting research.

Trainee Level:

undergraduate or graduate trainees
novice trainees

Areas of Trainee Development

- Research Ethics
  - Develop responsible and ethical research practices.

Activity Components and Estimated Time for Completion

- In Session Time: 15 minutes

Total time: 15 minutes

When to Use This Activity

This activity should be used with novice trainees, who are at the beginning of their research experience. The case introduces trainees to important procedures and considerations for documenting their research as they begin to perform experiments and collect and analyze data. Other activities that may be used with this activity include:

- Research Documentation Process
- Research Documentation: Can You Decipher This?

Inclusion Considerations

Consider learning styles, differences and disabilities when discussing effective documentation of research data. Ask whether trainees have concerns about the documentation protocols used in their research groups to empower them to talk about any of these barriers or concerns.

Contributed by J. Branchaw and K. Spencer with information from Branchaw, J. L., Pfund, C., and Rediske, R. (2010). *Entering Research: A Facilitator's Manual.* New York: W.H. Freeman & Co.

Implementation Guide

Workshop Session (15 minutes)

► **Case Study: Keeping the Data**

- Distribute the case study "Keeping the Data" (see trainee materials) and let participants read it individually (2–3 minutes).
- Ask trainees to write down their answers to each of the questions (5 minutes).
- Discuss the case within the large group. You may want to record the ideas on a whiteboard or flipchart (7 minutes). Some important points raised in previous discussions have included:

► **Discussion Questions**

1. Did May do anything wrong?
2. Is it important to keep hard-copy records of data?

- There is no one "right answer" to the questions posed in the case study; different research groups have different standards for keeping data and analysis records.
- If electronic records are backed up regularly and securely, then it may not be necessary to keep hard-copy records. However, regardless of whether a research group uses electronic or traditional hard-copy laboratory notebooks, real-time records of data collection with time and date information must be kept.
- The record of information must be complete enough to allow independent investigators to reproduce the experiments or analysis.

Contributed by J. Branchaw and K. Spencer with information from Branchaw, J. L., Pfund, C., and Rediske, R. (2010). *Entering Research: A Facilitator's Manual.* New York: W.H. Freeman & Co.

CASE STUDY: KEEPING THE DATA

Learning Objectives

Trainees will be able to:

- Explain why it is important to accurately document research.
- Identify key elements in research documentation.
- Understand the ethical implications of documenting research.

May, who has been doing research with Professor Gonzalez for two years, is preparing to present her research results at the campus-wide Research Symposium. Because some of her findings are quite novel and contradict reports of similar experiments in the literature, Professor Gonzalez asks to review the raw data before signing off on her presentation. When he reviews May's notebook, however, there are no hard-copy records of the data. Instead he finds the data on May's computer.

1. Did May do anything wrong? Why or why not?

2. Is it important to keep hard-copy records of data? Why or why not?

Contributed by J. Branchaw and K. Spencer with information from Branchaw, J. L., Pfund, C., and Rediske, R. (2010). *Entering Research: A Facilitator's Manual.* New York: W.H. Freeman & Co.

LETTER OF RECOMMENDATION

Learning Objectives

Trainees will:

- Identify the characteristics and attributes desired in a letter of recommendation.
- Learn how to request a letter of recommendation from a mentor.

Trainee Level:

undergraduate or graduate trainees
novice, intermediate, or advanced trainees

Areas of Trainee Development

- Professional and Career Development Skills
  - Explore and pursue a research career.
- Research Comprehension and Communication Skills
  - Develop effective interpersonal communication skills.
- Researcher Identity
  - Develop identity as a researcher.

Activity Components and Estimated Time for Completion

- In Session Time: 25–45 minutes

Total time: 25–45 minutes

When to Use This Activity

This activity should be implemented after a trainee has chosen a mentor and has discussed goals and expectations. Other activities that may be used with this activity include:

- Research Experience Reflections 1: Entering Research?
- Aligning Mentor and Trainee Expectations

Alternatively, this activity could be used as a professional development opportunity to help intermediate and advanced trainees request letters of recommendation from their mentor in preparation to apply to graduate school, postdoctoral fellowships, and jobs.

Inclusion Considerations

Consider how power and authority dynamics might be at play for trainees when asking mentors for a letter of recommendation, especially trainees who are members of groups underrepresented in research careers. Some trainees may come from backgrounds where humility is valued, and they may be uncomfortable or consider it disrespectful to request a letter of recommendation. Discuss strategies for approaching mentors with these requests and reassure trainees that mentors expect their trainees to request letters. Encourage trainees to establish clear expectations to which they, their mentor, and others can agree.

Contributed by A. R. Butz and J. Gleason with information from Branchaw, J. L., Pfund, C., and Rediske, R. (2010). *Entering Research: A Facilitator's Manual.* New York: W.H. Freeman & Co.

Implementation Guide for Novice Undergraduate and Graduate Trainees

Workshop Session (45 minutes)

► **Introduction** (5 minutes)

- Explain to trainees that one of the benefits of doing research is that they will get to know their mentor well and he or she will be able to write a detailed letter of recommendation on their behalf when they move to the next stage of their academic or professional career.
- When mentors write letters of recommendation, they are often asked to comment on key attributes or characteristics that will make the trainee successful in the next phase of their career.
- It is beneficial for trainees to begin thinking about what types of attributes (e.g., skills, knowledge, experiences) will help them succeed in the future as a graduate student or employee, and to establish goals and expectations in the mentoring relationship that will help them to develop those attributes.

► **Activity: Letter of Recommendation Worksheet** (15 minutes)

- Have trainees complete the Letter of Recommendation Worksheet on their own first. This can be done as outside work prior to the next session or in class (10 minutes).
- In pairs or trios, have trainees talk about the attributes that they selected (5 minutes). Bring the large group together to discuss the attributes.

► **Discussion Questions** (20 minutes)

- How well did the attributes you selected align with the goals and expectations you discussed with your mentor?
- If the skills and characteristics you identified are not easily linked to the experiences you will have with your mentor or your research experience, what can you do?
- What can you do if you want your mentor to focus on a certain attribute or skill in your letter, but he or she wants to focus on other skills or aspects of your research experience?

► **Wrap-up** (5 minutes)

- If trainees are comfortable with their mentors, they may take these letters to them to initiate a discussion about whether meeting the goals and expectations they agreed on will allow the mentor to address these things in a letter.

Contributed by A. R. Butz and J. Gleason with information from Branchaw, J. L., Pfund, C., and Rediske, R. (2010). *Entering Research: A Facilitator's Manual.* New York: W.H. Freeman & Co.

Implementation Guide for Intermediate and Advanced Undergraduate and Graduate Trainees

Workshop Session (25 minutes)

► **Introduction** (5 minutes)
- Distribute the materials for advanced trainees. Have trainees complete the Letter of Recommendation Worksheet on their own, either in class (10 minutes) or as homework prior to the session where you plan to discuss letters of recommendation.

► **Discussion Questions** (20 minutes)
- As a group, discuss different strategies for approaching one's mentor about writing a letter of recommendation. Have trainees review the "Requesting a Letter of Recommendation: Questions to Ask Your Mentor" handout that immediately follows the worksheet.
  - Have you already discussed the process for requesting letters of recommendation from your mentor?
  - If you have requested letters from other individuals in the past, how did you go about making that request?
  - What can you do if you want your mentor to focus on a certain attribute or skill in your letter, but they want to focus on other skills or aspects of your research experience?
  - What can you do to ensure that your mentor is able to write you a strong letter of recommendation?

Contributed by A. R. Butz and J. Gleason with information from Branchaw, J. L., Pfund, C., and Rediske, R. (2010). *Entering Research: A Facilitator's Manual.* New York: W.H. Freeman & Co.

LETTER OF RECOMMENDATION

Novice Trainees

Learning Objectives

Trainees will:

- Identify the characteristics and attributes desired in a letter of recommendation.
- Learn how to request a letter of recommendation from a mentor.

One of the benefits of doing research is that you will get to know your mentor well and he or she will be able to write a detailed letter of recommendation on your behalf when you move to the next stage of your academic or professional career. As you begin the research experience, consider what you would like your mentor to be able to say about you at the end of the research experience and complete this draft letter of recommendation.

What are some of the characteristics that make a good researcher or scientist? Some examples are listed below. Circle three characteristics or qualities that you believe you can demonstrate through your research experience, or use the spaces provided below to write down additional characteristics.

Ingenuity	Innovation	Initiative
Work ethic	Responsibility	Reliability
Team player	Problem-solving skills	Tenacity
Professionalism	Knowledge of field	Analytical Thinking
Oral/written communication skills	Receptive to constructive feedback	Independence as a researcher
Motivation	Maturity	Ability as a researcher
Other:	Other:	Other:

Trainee Materials

Contributed by A. R. Butz and J. Gleason with information from Branchaw, J. L., Pfund, C., and Rediske, R. (2010). *Entering Research: A Facilitator's Manual.* New York: W.H. Freeman & Co.

For each characteristic you selected, describe *how* you will be able to demonstrate that quality in the context of your research experience, and a goal or expectation that you established with your mentor that aligns with that characteristic.

Characteristic	**I will demonstrate [characteristic] by:**	**Goal/Expectation that provides the opportunity to demonstrate or improve your skills in this area:**
1.		
2.		
3.		

Will the goals and expectations you outlined with your mentor allow you to demonstrate these characteristics? If not, how can you adjust your goals and expectations so that you will have the opportunity to engage in activities that allow your mentor to comment positively on these attributes? Use this worksheet to discuss with your mentor how you can further develop your desired characteristics through your research experience.

Contributed by A. R. Butz and J. Gleason with information from Branchaw, J. L., Pfund, C., and Rediske, R. (2010). *Entering Research: A Facilitator's Manual.* New York: W.H. Freeman & Co.

LETTER OF RECOMMENDATION

Intermediate/Advanced Trainees

Learning Objectives

- ▶ Identify the characteristics and attributes desired in a letter of recommendation.
- ▶ Learn how to request a letter of recommendation from a mentor.

One of the benefits of doing research is that you know your mentor well and he or she is able to write a detailed letter of recommendation on your behalf when you prepare for the next stage of your academic or professional career. As you begin to think about your next steps, whether that be continuing your education or your career, consider what you would like your mentor to be able to say about you and use this worksheet as a starting point for a discussion with your mentor about letters of recommendation. Some additional points to discuss with your mentor are listed below (see resource: Requesting a Letter of Recommendation: Questions to Ask Your Mentor).

What are some of the characteristics that make a good researcher or scientist? Some examples are listed below. Circle three characteristics or qualities that you believe you can demonstrate through your research experience, or use the spaces provided below to write down additional characteristics.

Ingenuity	Innovation	Initiative
Work ethic	Responsibility	Reliability
Team player	Problem-solving skills	Tenacity
Professionalism	Knowledge of field	Analytical thinking
Oral/written communication skills	Receptive to constructive feedback	Independence as a researcher
Motivation	Maturity	Ability as a researcher
Other:	Other:	Other:

Contributed by A. R. Butz and J. Gleason with information from Branchaw, J. L., Pfund, C., and Rediske, R. (2010). *Entering Research: A Facilitator's Manual.* New York: W.H. Freeman & Co.

For each characteristic you selected, describe *how* you have demonstrated that quality in the context of your research experience, and a goal or expectation that you established with your mentor that aligns with that characteristic.

Characteristic	**I will demonstrate [characteristic] by:**	**Goal/Expectation that provides the opportunity to demonstrate or improve your skills in this area:**
1.		
2.		
3.		

Contributed by A. R. Butz and J. Gleason with information from Branchaw, J. L., Pfund, C., and Rediske, R. (2010). *Entering Research: A Facilitator's Manual.* New York: W.H. Freeman & Co.

LETTER OF RECOMMENDATION REQUEST: QUESTIONS TO ASK YOUR MENTOR

Below is a list of questions you may want to ask your mentor when you are starting to think about requesting letters of recommendation. You can use this guide as a starting point for your conversation with your mentor.

1. Would you be willing to write a strong/positive letter of recommendation for me?

2. How much lead time do you need to write letters of recommendation?

3. What type of supporting information/materials would be helpful in writing my letter? For example, your mentor may wish to see some or all of the following:
   - Your resume/CV
   - The job posting or application requirements for your desired graduate program or job
   - Key areas of your relationship/research experience that you would like them to discuss in your letter
   - A spreadsheet with names, addresses, and deadlines for each place that you would like your mentor to send a letter
   - Pre-addressed, stamped envelopes or details about how to submit letters online
   - Personal statement or other essays required by the application

4. Would you prefer that I remind you about this letter as the deadline nears? If so, would you prefer that I remind you in person? by email? another method?

Contributed by A. R. Butz and J. Gleason with information from Branchaw, J. L., Pfund, C., and Rediske, R. (2010). *Entering Research: A Facilitator's Manual.* New York: W.H. Freeman & Co.

MENTOR BIOGRAPHY

Learning Objectives

Trainees will:

- ► Begin to establish a positive relationship with a research mentor by getting to know them as a researcher and a person.
- ► Learn about the diversity of experiences in research careers by comparing a mentor's experiences to their own.

Trainee Level:

undergraduate or graduate trainees
novice or intermediate trainees

Areas of Trainee Development

- ► Researcher Identity
  - Develop identity as a researcher.
- ► Research Comprehension and Communication Skills
  - Develop effective interpersonal communication skills.
- ► Equity and Inclusion Awareness and Skills
  - Develop skills to deal with personal differences in the research environment.
- ► Professional and Career Development Skills
  - Develop confidence in pursuing a research career.

Activity Components and Estimated Time for Completion

- ► Session 1 Time: 10–15 minutes
- ► Trainee Pre-Assignment for Session 2 Time: 15–20 minutes
- ► Session 2 Time: 10–15 minutes

Total time: 35–50 minutes

When to Use This Activity

This activity is most relevant at the beginning of a new mentor–trainee relationship. Facilitators may wish to contact the mentors of the trainees in advance of the assignment to let them know that trainees are required to meet with them in order to successfully complete this activity.

Inclusion Considerations

Encourage trainees to celebrate the differences and similarities they have with their mentor. Point out that they should consider what they learn about their mentor when working to develop a positive relationship with that person and they should take the opportunity to share their own background and motivations.

Contributed by A. R. Butz with information from Branchaw, J. L., Pfund, C., and Rediske, R. (2010). *Entering Research: A Facilitator's Manual.* New York: W.H. Freeman & Company.

Implementation Guide

Session 1 (10–15 minutes)

- Introduce the activity by distributing copies of the Trainee Materials and ask trainees to review the questions and brainstorm (either in small groups or as a class) additional questions they might want to ask their mentor. Write ideas for questions on the whiteboard or flipchart.
- Have trainees identify five to seven questions that they intend to ask their mentor during their interview.

Trainee Pre-Assignment for Session Two

- Assign trainees to schedule a 15- to 20-minute interview with their mentor.

Session 2 (10–15 minutes)

- **Large-Group Discussion** (10–15 minutes)
  - Invite trainees to share what they learned about their mentors during the interview. Discussion questions:
    - What is the most interesting thing you learned about your mentor from this interview?
    - Does your mentor do more than research as part of their job (e.g., teaching and service responsibilities in the academy; management and supervision responsibilities in industry)? How do they manage the different types of demands of their job?
    - How did your mentor end up in their field? Are there any patterns that emerge from the various mentors' responses? Was it always an interest? A random circumstance? Something someone did or said?
    - Was there anything about your mentor's experiences that made you more confident in your ability to be a scientist?
    - What interesting similarities/differences did you discover between your mentor and yourself as a result of this interview? How do differences between you and your mentor benefit your relationship and the research?
    - What did you learn about your mentor as a person? Was there anything that surprised you? Research mentors are people too, with families and obligations/interests outside of work.
    - How will this interview impact your relationship with your mentor?
- Graduate trainees should be encouraged to ask questions about their mentor's graduate training experience and reflect upon how that experience compares to their experience or expectations.

Alternative Activity
In lieu of a large-group discussion, trainees can write a paragraph about their mentor's background and a paragraph about what they learned about their mentor's life before coming to the next session. In pairs or small groups, trainees can share what they have learned about their mentors and identify similarities and differences. The discussion questions listed above can be used to help guide discussion in small groups.

Contributed by A. R. Butz with information from Branchaw, J. L., Pfund, C., and Rediske, R. (2010). *Entering Research: A Facilitator's Manual.* New York: W.H. Freeman & Company.

MENTOR BIOGRAPHY

Learning Objectives

Trainees will:

- Begin to establish a positive relationship with a research mentor by getting to know them as a researcher and a person.
- Learn about the diversity of experiences in research careers by comparing a mentor's experiences to their own.

Interview your mentor using the questions below and at least three of your own questions. Take notes and write a two or three paragraph biography about your mentor that summarizes the information you learned in your interview.

Questions to Ask your Mentor

1. Where did you grow up and what was it like there?

2. Why did you decide to become a researcher?

3. What challenges or obstacles did you have to overcome in order to be successful in your field?

4. What keeps you motivated to do research?

5. Where did you do your training and why did you decide to attend those institutions? (undergraduate degree, graduate degree, etc.)

6. Why did you decide on your disciplinary/research area? Have you done research in any other areas? If so, which?

7. What classes do you currently teach, or have you taught? Which was your favorite and why?

8. Outside of your research responsibilities, what else do you do as part of your job?

9. What do you do when you are not at work?

Contributed by A. R. Butz with information from Branchaw, J. L., Pfund, C., and Rediske, R. (2010). *Entering Research: A Facilitator's Manual.* New York: W.H. Freeman & Company.

Your Questions:

1.

2.

3.

Write a two or three paragraph biography about your mentor summarizing the information obtained in your interview.

Contributed by A. R. Butz with information from Branchaw, J. L., Pfund, C., and Rediske, R. (2010). *Entering Research: A Facilitator's Manual.* New York: W.H. Freeman & Company.

MENTOR INTERVIEW ABOUT MAKING RESEARCH POSTERS

Learning Objectives

Trainees will:

- Learn how to create a research poster.
- Learn disciplinary norms for graph or image construction.
- Set deadline(s) for poster completion.

Trainee Level:

undergraduate or graduate trainees
novice or intermediate trainees

Areas of Trainee Development

- Research Comprehension and Communication Skills
  - Develop research communication skills.
  - Develop disciplinary knowledge.

Activity Components and Estimated Time for Completion:

- In Session Time: 10 minutes
- Assignment: 1 hour

Total time: 1 hour, 10 minutes

When to Use This Activity

This assignment should be used before trainees make a research poster. It provides the mentor and trainee with important information about deadlines and due dates. It also provides an opportunity for trainees to learn research group and discipline specific norms for poster construction.

The assignment can be used with both undergraduate and graduate students. For undergraduate students it is useful to scaffold deadlines for the poster. For example, several drafts of different sections of the poster may be required before a final poster is due. This may not be necessary for graduate students, depending on their previous experience with making research posters.

The assignment can be used early in a course or a few weeks before the poster is due. Other activities that may be used with this activity include:

- Communicating Research Findings 1: Poster Presentations
- Communicating Research Findings 3: Developing Your Presentation

Inclusion Considerations

This will likely be the first time that novice trainees, especially those who have traveled nontraditional academic pathways, will make a research poster. Reassure them that their research poster-making skills will improve with practice and encourage them to ask their mentor for help and to harness the collective experience of the members of their research groups. If trainees have challenges with writing (e.g., disabilities), talk about these challenges and encourage them to seek help from professionals and to share this information with their research mentor, so that they are aware and can help.

Contributed by K. Eskine. (2018). *Mentor Interview about Making Research Posters.*

Implementation Guide

Workshop Session (10 minutes)

- **Introduction** (5 minutes)
  - The goal of this assignment is for trainees to obtain guidance from their mentor on how to make a research poster. The interview is a formal way for students to talk about the process of making a poster, to learn lab norms for creating graphs or images, and to set reasonable deadlines.
- **Assignment: Interviewing Mentor about Poster** (5 minutes)
  - Assign trainees to interview their mentors about the poster requirements and norms in their research group. Trainees should use the interview guide included in the trainee materials.
  - Emphasize to trainees that they should have this conversation with their mentors prior to preparing their research posters.

Contributed by K. Eskine. (2018). *Mentor Interview about Making Research Posters.*

MENTOR INTERVIEW ABOUT MAKING RESEARCH POSTERS

Learning Objectives

Trainees will:

- Learn how to create a research poster.
- Learn disciplinary norms for graph or image construction.
- Set deadline(s) for poster completion.

Before you begin creating your research poster, interview your mentor to learn about the policy and formatting standards that you should use as a member of the research group.

1. How independent would you like me to be in making this poster (do you want to see the intro, then the methods, etc., or are you happy to see a completed poster for edits)?
2. Is there a template our research group uses that I should use? Is there an example that I can refer to?
3. What software program does the research group use to make posters? Are there particular parameters that I need to set, like height and width?
4. What software program does the research group use to make graphs? Do you have any tips or are there any formatting conventions that I should use to make the graphs?
5. What software program does the research group use to manipulate images? Do you have any tips or are there any formatting conventions that I should use to create images?
6. I will present the poster on [insert date]. By what date would you like to see the poster so that you have enough time to review it and provide feedback for revision?
7. What research funding sources should I cite on the poster? Are there any specific funding agencies or grant numbers that I should include?

Notes:

Deadlines:

Contributed by K. Eskine. (2018). *Mentor Interview about Making Research Posters.*

MESSAGES SENT AND RECEIVED

Learning Objectives

Trainees will:

- ► Identify the intent behind statements and questions.
- ► Practice effective ways to communicate with their research mentor.

Trainee Level:

undergraduate or graduate trainees
novice or intermediate trainees

Areas of Trainee Development

- ► Researcher Confidence and Independence
  - Develop confidence as a researcher.
- ► Professional and Career Development Skills
  - Develop confidence in pursuing a research career.
- ► Research Comprehension and Communication Skills
  - Develop effective interpersonal communication skills.

Activity Components and Estimated Time for Completion

- ► In Session Time: 25 minutes

Total time: 25 minutes

When to Use This Activity

Other activities that may be used with this activity include:

- ► The Power of Social Persuasion
- ► Case Study: Responding to Feedback

Inclusion Considerations

Consider that trainees who are members of underrepresented groups in research may already feel marginalized in the research environment. Given this, how might the statements in this exercise impact those trainees, compared to trainees from majority groups who feel more comfortable and welcome in the research environment? The facilitator could invite the trainees to consider how they might support one another in their research groups in situations like this.

Contributed by C. Pfund and A. R. Butz with information from Handelsman, J., Pfund, C., Miller Lauffer, S., and Pribbenow, C. M. 2005. *Entering Mentoring: A Seminar to Train a New Generation of Scientists.* Madison, WI: University of Wisconsin Press.

Implementation Guide

Workshop Session (25 minutes)

- Trainees select and complete the rows for two or three statements in the table in the trainee materials. (5 minutes)
- Trainees discuss their responses in pairs or trios. (5 minutes)
- **Large-Group Discussion Questions** (10 minutes)
  - How do you determine that you are receiving the message that your mentor intended to send?
  - How might your interpretation differ if you and your mentor's communication styles don't align?
  - How do status and power play into how you and your mentor communicate with one another?
  - How do you deal with feedback that makes you feel stressed or less confident in your abilities as a researcher?
- **Wrap-up** (5 minutes)
  - Emphasize that trainees should give their mentors the benefit of the doubt when it comes to feedback. However, if communication is a problem, the trainee should work with their mentor to make sure that expectations are clear and to identify a method of communication that works for each of them.

Contributed by C. Pfund and A. R. Butz with information from Handelsman, J., Pfund, C., Miller Lauffer, S., and Pribbenow, C. M. 2005. ***Entering Mentoring:*** *A Seminar to Train a New Generation of Scientists.* Madison, WI: University of Wisconsin Press.

MESSAGES SENT AND RECEIVED

Learning Objectives

Trainees will:

- Identify the intent behind statements and questions.
- Practice effective ways to communicate with their research mentor.

Instructions: Fill in the blank columns for two or three statements below.

Statement or Question	What is the likely intention of this statement?	How might the statement be heard?	How could you respond to this statement in a constructive manner?
"Be on time to our group meetings from now on."			
"How much longer do you think it will take you to finish that project?"			
"You will never get anywhere in this field if you don't dig in and stick with problems until you solve them."			
"If you think you are busy now, wait until you're a faculty member."			

Contributed by C. Pfund and A. R. Butz with information from Handelsman, J., Pfund, C., Miller Lauffer, S., and Pribbenow, C. M. 2005. *Entering Mentoring: A Seminar to Train a New Generation of Scientists.* Madison, WI: University of Wisconsin Press.

Statement or Question	**What is the likely intention of this statement?**	**How might the statement be heard?**	**How could you respond to this statement in a constructive manner?**
"Clean up your work area."			
"I haven't seen you around the lab much. Are you taking time off?"			
"I'm not sure you have your priorities in order."			
"What's it like to be a minority in this program, anyway?"			
"It seems you might be better suited for an 'alternative' career."			

Contributed by C. Pfund and A. R. Butz with information from Handelsman, J., Pfund, C., Miller Lauffer, S., and Pribbenow, C. M. 2005. *Entering Mentoring: A Seminar to Train a New Generation of Scientists.* Madison, WI: University of Wisconsin Press.

MINI-CASE STUDIES: STICKY SITUATIONS

Learning Objectives

Trainees will:

- Develop strategies to deal with difficult situations that may arise during the course of their research experience.

Trainee Level:

undergraduate or graduate trainees
novice, intermediate, or advanced trainees

Areas of Trainee Development

- Research Comprehension and Communication Skills
  - Develop effective interpersonal communication skills.
  - Develop an understanding of the research environment.
- Equity and Inclusion Awareness and Skills
  - Develop skills to deal with personal differences in the research environment.
- Research Ethics
  - Develop responsible and ethical research practices.
- Professional and Career Development Skills
  - Develop confidence in pursuing a research career.

Activity Components and Estimated Time for Completion

- In Session Time: 35 minutes

Total time: 35 minutes

When to Use This Activity

This activity is appropriate for undergraduate and graduate trainees who already have a mentor or are considering joining a research group. It can be implemented in one session or facilitators can use individual sticky situations as "conversation starters" at the start of any session. Facilitators may want to have trainees obtain a sense of their research group's structure prior to completing this activity. Other activities that may be used with this activity include:

- Research Group Diagram

Inclusion Considerations

Discuss how different backgrounds can play a role in how difficult (or sticky) situations are perceived and handled. Trainees may have different ideas about the best strategies for dealing with difficult situations depending on their backgrounds. As a facilitator, use the differences in trainee responses as an opportunity to reflect upon assumptions that may have been made in interpreting the situation. Encourage them to seek to understand what others are saying and what they intend.

Contributed by C. Barta and C. Pfund with information from Branchaw, J. L., Pfund, C., and Rediske, R. (2010). *Entering Research: A Facilitator's Manual.* New York: W.H. Freeman & Co.

Implementation Guide

Workshop Session (35 minutes)

► **Introduction** (5 minutes)

- The goal of this activity is to explore difficult situations that trainees may encounter in their research experience and devise concrete strategies to address each challenge.
- Tell trainees that these scenarios may remind them of sticky situations that they wish to share from their own experience. Remind them to respect the confidentiality of their peers by not sharing identifying information and by not sharing the discussion outside of this session. If your group established ground rules in an earlier session, you may wish to revisit them at this time.

► **Activity** (10 minutes)

- Distribute the list of "sticky situations" included in the trainee materials and ask trainees to individually write down what they would do in each situation. *Separate sticky situations appropriate for undergraduate and graduate-level trainees are included in the trainee materials.*
- Alternatively, trainees may be randomly assigned to one or more of the cases and discuss those situations in small groups. Facilitators can also encourage trainees to come up with and share their own sticky situations.

► **Discussion** (20 minutes)

- As a large group, discuss each situation and ask trainees to reflect on the structure of their own research groups. *If trainees have completed a research group diagram, it may be useful for them to refer to it during this discussion.*
- **Discussion Questions**
  - How do the lines of communication that have been established in your research group help to define the best course of action in each situation?
  - How does the background or identity of the mentor and trainee impact how you might react to these different scenarios? For example, does the situation change:
    - If the mentor and trainee have the same gender identity?
    - If the mentor and trainee are from different racial/ethnic backgrounds?
    - If the trainee has an invisible disability?
    - If the mentor was born in another country?
    - If the trainee is the first in their family to go to college?
    - If the mentor is from a big urban city and the trainee is from a small rural community?
  - In general, how might differences between mentors and their trainees change these scenarios and the assumptions that we might make about each person?

Contributed by C. Barta and C. Pfund with information from Branchaw, J. L., Pfund, C., and Rediske, R. (2010). *Entering Research: A Facilitator's Manual.* New York: W.H. Freeman & Co.

MINI-CASE STUDIES: STICKY SITUATIONS

Undergraduate

Learning Objectives

Trainees will:

► Develop strategies to deal with difficult situations that may arise during the course of their research experience.

---

1. A trainee's mentor wants an experiment done this week, but the trainee does not have the time to do it because of upcoming exams. What should the trainee do? How would you handle this situation if it were your mentor? Do you believe that your mentor values research over coursework? How do you know (or how would you go about finding out)?
2. A trainee has been working in a research group for a couple of months and found that he is not interested in the research project. He wants to change to a different project, but does not know how to approach his mentor? How would you advise this student? Would you handle this situation differently if this student was you? If yes, how so?
3. Someone in your friend's research group gives your friend a new protocol that they say is better than the one that was given to her by her mentor. Which protocol do you advise your friend to use? What would you do if you were in a similar situation? What are the unintended consequences (positive or negative) of using a protocol from somebody other than your direct mentor?
4. Your mentor asks you to write an abstract for an upcoming conference. You've been in the research group for several months now, but still do not understand the project very well. You are worried that if you attempt to write the abstract, your mentor will find out how little you understand. What do you do? How could this have been prevented?
5. Your friend has been working in her research group for almost a semester, but still has not been given a research project. She feels that she is ready to do some of her own research, but instead is starting to feel like this research opportunity is a waste of her and her mentor's time. What would you advise your friend to do? If this were happening to you, who in your research group would you approach to resolve it?
6. A student is extremely excited to have found a research position on campus. He arrived full of energy the first day and learned that he was "assigned" to a graduate student mentor by the PI of the research group, against the graduate student's wishes to serve as a mentor. The graduate student acts distant, disinterested, and visibly annoyed every time the student comes to lab. What should the student do? Have you or anyone else in your research group experienced a similar situation? If so, how was it handled?
7. A classmate has an extremely busy semester juggling four courses (many with lab components), a part-time job, participation in two student organizations, and undergraduate research. She is devoting 10 hours per week to research and thought that this was a fair commitment. During her last meeting with her research mentor, however, the mentor was upset that she had not made more progress on her project. In fact, the mentor mentioned that he would not be able to write her a strong letter of recommendation for graduate school unless she started acting "more like a graduate student." What should this classmate do? If this were you, what would you do?

Contributed by C. Barta and C. Pfund with information from Branchaw, J. L., Pfund, C., and Rediske, R. (2010). *Entering Research: A Facilitator's Manual.* New York: W.H. Freeman & Co.

8. Your friend has been assigned a mentor who does not share the same first language and is from a very different culture. He is having a hard time understanding what she is saying. He also feels that she has different expectations of him compared to his friend's research mentors' expectations. He mentions to you that he feels a little uneasy always asking her to repeat herself and also thinks that it is unfair that he is being held to a different standard than you and the rest of your friends. What would you advise him to do in this situation?

9. A trainee has been working in a research group for 3 months and has noticed that many of the other researchers in the lab do not use proper safety protocols (not wearing gloves when handling certain chemicals, not wearing lab coats or safety glasses when working in the cell culture room, not properly cleaning up chemical spills, etc.). She does not want to make a bad impression in the research group, but she is truly worried about the others' personal safety. What would you advise this trainee to do? How would you approach your mentor to resolve your concerns if this was occurring in your research group?

Contributed by C. Barta and C. Pfund with information from Branchaw, J. L., Pfund, C., and Rediske, R. (2010). *Entering Research: A Facilitator's Manual.* New York: W.H. Freeman & Co.

MINI-CASE STUDIES: STICKY SITUATIONS

Graduate

Learning Objectives

Trainees will:

- ► Devise concrete strategies to deal with difficult situations that may arise during the course of their research experience.

---

1. Your mentor wants an experiment done this week, but you do not have the time because of an upcoming exam in your graduate course. In general, you feel the mentor does not appreciate how much time it takes to succeed in graduate courses. What do you do?
2. You have been working in the research group for several months and have found that you are not interested in the project you have started. You want to focus on a different research topic, one that is linked to a project you previously worked on. You mention this to your mentor, but she says that they do not want to pursue the new research because the group does not have the proper equipment or funding to support it. How should you respond?
3. Someone in your research group gives you a new protocol that they say is better than the one given to you by your mentor. Which protocol do you use? What are the unintended consequences (positive or negative) of using a protocol from somebody other than your direct mentor?
4. You want to go home for 10 days over the holiday break to visit your family, but your mentor is stringent on time spent away from the research group. You don't recall anyone ever being away for more than a long weekend. How can you approach your mentor with this request?
5. You learn that your thesis advisor has assigned you an undergraduate student to mentor without discussing it with you. You are very busy, with multiple experiments that are critical to your thesis research to run in the next couple of months. You do not trust an inexperienced undergraduate trainee to do the experiments because they are very complicated. Your undergraduate student arrives on the first day, full of energy and eager to learn. You hope that your mentor has a research plan in mind for the student, but she acts distant and tells you that the student is "your responsibility." What do you do?

Contributed by C. Barta and C. Pfund with information from Branchaw, J. L., Pfund, C., and Rediske, R. (2010). *Entering Research: A Facilitator's Manual.* New York: W.H. Freeman & Co.

MINI-GRANT PROPOSAL

Learning Objectives

Trainees will:

- Learn how to find and apply for grants.
- Develop a logical progression of ideas.
- Develop disciplinary written communication skills.

Trainee Level:

undergraduate or graduate trainees
novice or intermediate trainees

Areas of Trainee Development

- Research Comprehension and Communication Skills
  - Develop research communication skills.
  - Develop disciplinary knowledge.
  - Develop logical/critical thinking skills.
- Research Confidence and Independence
  - Develop independence as a researcher.

Activity Components and Estimated Time for Completion

- Trainee Pre-Assignment Time: 30 minutes
- Session 1 Time: 30 minutes
- Session 2 Time: 1 hour

Total time: 2 hours

When to Use This Activity

This activity is appropriate for novice to intermediate undergraduate and graduate trainees. To successfully implement this activity, trainees must be doing research, have a well-defined research question and methods, and have conducted a background literature search. This activity can be adapted to follow guidelines for specific grants or funding opportunities including fellowships, scholarships, travel grants, and internships. Facilitators working with trainees with no prior experience writing research proposals should implement the activities listed below prior to this activity or ensure that trainees have had the opportunity to draft these components of a research proposal prior to implementing this activity.

- Research Writing 1: Background Information and Hypothesis or Research Question
- Research Writing 2: Research Project Outline and Abstract
- Research Writing 3: Project Design
- Research Writing 4: Research Literature Review and Publishing Process
- Research Writing 5: The Peer-Review Process
- Tips for Technical Writers
- Research Group Funding
- Funding Your Research
- Communicating Research Findings 1, 2, or 3

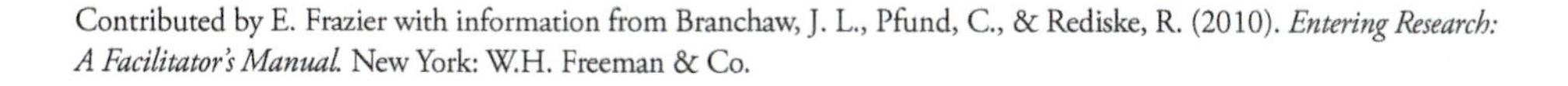
Contributed by E. Frazier with information from Branchaw, J. L., Pfund, C., & Rediske, R. (2010). *Entering Research: A Facilitator's Manual.* New York: W.H. Freeman & Co.

Inclusion Considerations

Learning and reading styles will vary among trainees. Invite them to share with you any learning accommodations they need or preferences they have. Be flexible when setting reading and writing assignment deadlines. Encourage trainees to share with the group alternative ideas about how to approach writing and encourage them to talk about any challenges or concerns they have about writing with their mentors.

Contributed by E. Frazier with information from Branchaw, J. L., Pfund, C., & Rediske, R. (2010). *Entering Research: A Facilitator's Manual.* New York: W.H. Freeman & Co.

Implementation Guide

Trainee Pre-Assignment (*optional;* 30 minutes)

- Ask trainees to search for a funding opportunity that is relevant to their field of study and to bring the call for applications or proposals to the next session.

Workshop Session #1—Introduction to Mini-Grant Assignment (30 minutes)

- **Introduction** (5 minutes)
  - The objective of this activity is for trainees to write a mini-grant proposal. Grant proposals can be used to apply for research funding and are usually reviewed by peers (peer reviewers). In this activity, trainees work with their mentors to integrate the pieces they have written thus far, plan the next step of their project, and create a proposal draft that could be revised and submitted for funding.
- **Components of a Mini-Grant Proposal** (15 minutes)
  - Review the components of a proposal. This will vary by funding agency, so it may be useful to show trainees examples from specific funding agencies (NSF, NIH, NASA, etc.). Facilitators may wish to provide copies of complete proposals in the appropriate format for trainees to review. Tips on applying for grants and examples of proposal guidelines are provided in the trainee materials.
  - The template provided in the trainee materials covers the major components of proposals for most funding agencies, but the order/formatting of these components will vary depending upon the type of funding for which they wish to apply. If working with intermediate trainees who are drafting proposals for specific grants, facilitators may alter the template or encourage trainees to develop a proposal that is aligned with the requirements of their funding agencies.
  - If facilitators asked trainees to research their own funding opportunities as part of the pre-assignment, they can engage the group in a discussion about the guidelines. Using a whiteboard or flipchart, ask trainees to write down the different components that are required for the various proposals. Discuss similarities and differences among the different funding agencies.
- **Assignment: Proposal Draft and Peer Review** (10 minutes)
  - Introduce the mini-grant proposal and peer-review assignments.
    - Trainees should integrate drafts of their abstract, research question, introduction, literature review, project design, preliminary results, proposed experiments with timeline, and reference list into a complete proposal. If these components have not been drafted as part of prior activities (see "When to Use This Activity"), then facilitators should allow time either in or outside of the class session for trainees to prepare these components of the mini-grant proposal. Encourage trainees to schedule time with their mentors to review their proposal prior to sharing it with their peers.
    - Assign trainees to review draft proposals of two of their peers.
  - Trainees should share first drafts of their proposals with their assigned peer reviewers at least 48 hours before Session 2. The proposal may either be a written document (we recommend three to five pages), a scientific poster, or an oral presentation. For poster and oral presentation assignments, see "Communicating Research Findings 1 or 2."
  - In addition to preparing their mini-grant proposals, trainees should complete reviews of their assigned proposals prior to session #2 using the worksheet provided in the trainee materials. Peer reviewers should identify the strengths and weakness of the proposal, and be prepared to discuss:
    - The best part of the mini-grant proposal; and
    - At least one specific suggestion for how to improve the mini-grant proposal.

Contributed by E. Frazier with information from Branchaw, J. L., Pfund, C., & Rediske, R. (2010). *Entering Research: A Facilitator's Manual.* New York: W.H. Freeman & Co.

Workshop Session # 2: Peer Review of Mini-Grant Proposal (1 hour)

► **Introduction** (5 minutes)

- Debrief the proposal writing experience with trainees.
  - What part of the proposal was most challenging for them to write? Why?
  - How easy was it to review their peers' work? Did they know what to look for when reviewing?

► **Activity: Peer Review** (40 minutes)

- Invite trainees to meet with their peer reviewers to go over the review comments in general.
- Reviews can take place in one session or, if time permits, you may wish to offer a few opportunities for trainees to receive feedback on iterative drafts of their proposals as they work toward a final draft.
- Encourage trainees to ask questions and discuss specific strategies for how to improve the draft. Trainees should leave with a list of items from the review to discuss with their mentor.

► **Discussion** (5–10 minutes)

- What are the major issues you need to address in your proposal?
- How will you address these issues in your final draft?
- To whom, if anyone, will you go for help in revising your mini-grant proposal?

► **Wrap-up** (5 minutes)

- Outline the next steps in the revision process with your trainees. Indicate when trainees should be prepared to turn in a final draft of their mini-grant proposal. Depending upon trainees' progress with their proposal, you may wish to dedicate additional sessions to peer review.

► **Assessment Rubric**

- The assessment rubric is provided for facilitators who wish to evaluate trainees' mini-grant proposals as part of an assignment or to give formative feedback on the proposal.

Contributed by E. Frazier with information from Branchaw, J. L., Pfund, C., & Rediske, R. (2010). *Entering Research: A Facilitator's Manual.* New York: W.H. Freeman & Co.

MINI-GRANT PROPOSAL

Assessment Rubric

	0	1	2	3
Abstract	Absent	Does not summarize the research and preliminary results. Provides only basic background information.	Summarizes research and preliminary results, but provides too much information, is not concise, or uses jargon without definitions.	Concise descriptive summary of research and preliminary results; provides only relevant information with limited jargon.
Introduction	Absent	The background information presented lacks the content and support needed to understand the scientific basis of the hypothesis or research question and the gap of knowledge the research seeks to address is not apparent.	The relevant background information is presented, but poorly supported by peer-reviewed literature and/or organized such that the hypothesis or research question does not follow logically from the background. The gap of knowledge the research seeks to address may be vague and not explicitly stated.	The relevant background information is presented, supported by primary and secondary research papers, and is organized such that the hypothesis or research question follows logically from it. The gap of knowledge the research seeks to address is readily apparent to the reader.
Hypothesis/ Research Question and Specific Aims	Absent	The hypothesis/ research questions or project aims are not clearly defined or articulated.	The hypothesis/research questions, or project aims are somewhat clearly defined and loosely connected to the literature cited in the introduction.	The hypothesis/research questions, or project aims, are clearly defined and well connected to the literature cited in the introduction.
Preliminary Results	Absent	Preliminary results are stated, though disorganized or poorly recorded.	Preliminary data are summarized and presented in a mostly organized and logical format. Connections of results to proposed project are present, but not logically organized. Figure tags are absent or are not used appropriately.	Preliminary data are summarized in a logical format and organized so it is easy for the reader to see the relevance to the proposed project. Figures are referenced in the text using figure tags (e.g., Figure 1) when appropriate.

Contributed by E. Frazier with information from Branchaw, J. L., Pfund, C., & Rediske, R. (2010). *Entering Research: A Facilitator's Manual.* New York: W.H. Freeman & Co.

	0	1	2	3
Proposed Experiments or Methods of Data Collection with Timeline	Absent	The description of the materials and methods used for experiments or data collection lack detail. The procedures are not outlined and/or are difficult to follow. Timeline is vague or missing.	Most of the materials and the setup used in the experiment or data collection procedures are accurately described. Some ambiguity remains OR too much unnecessary detail is included. Timeline is generally outlined.	All materials and methods used for experiments or data collection presented are described. All steps are written in a precise and concise manner such that the methods can be repeated by another research group. Timeline is clearly outlined and reasonable given the scope of the project.
Appendix: Budget	Absent	Budget is not detailed, well organized, or realistic given the scope of the project.	Budget is well organized, but lacks detail or is unrealistic given the scope of the project.	Budget is well organized and details both supply costs and personnel time needed to complete the proposed project. Budget is realistic given the scope of the project.

Contributed by E. Frazier with information from Branchaw, J. L., Pfund, C., & Rediske, R. (2010). *Entering Research: A Facilitator's Manual.* New York: W.H. Freeman & Co.

MINI-GRANT PROPOSAL

Learning Objectives

Trainees will:

- Learn how to find and apply for grants.
- Develop a logical progression of ideas.
- Develop disciplinary written communication skills.

The Next Step in Your Research Project

The goal of your mini-grant proposal is to convince fellow researchers (peer reviewers) that your research is worth funding. It is meant to be a continuation of the research you have been doing (the next step), not a new project. Writing it is an opportunity to **work with your mentor** to pull together the pieces you've written so far, plan the next step of your project, and create a proposal draft that could be revised and submitted for funding.

The proposal should be **three to five pages**, including the five sections outlined below, and an appendix with a budget.

1. Abstract
A brief summary of the proposed research (200 words or less).

2. Introduction: Background and Relevance
A brief description of the background and importance/significance/gap in knowledge of the proposed research.

3. Hypothesis/Research Question OR Specific Aims
EITHER a statement of the hypothesis/research question OR a list of one to three specific objectives.

4. Preliminary Results
A brief presentation of preliminary data that is relevant to the proposed research project. This could be data collected by you or others in the research group, or found in the published literature.

5. Proposed Experiments with Timeline
A detailed description of the proposed experiments, including methods and a timeline for completion.

Appendix: Budget
An estimate of the supply costs (in dollars) and personnel time (in hours) needed to complete the proposed research project. Personnel time should include both trainee and mentor time. It is important to work with your mentor on the budget to ensure your estimates are realistic.

Contributed by E. Frazier with information from Branchaw, J. L., Pfund, C., & Rediske, R. (2010). *Entering Research: A Facilitator's Manual.* New York: W.H. Freeman & Co.

MINI-GRANT PROPOSAL

Peer Review

Author: ______________________ Reviewer: ______________________

Strengths:

Weaknesses:

Specific Suggestions for Improvement:

Overall Score (circle one):

1 **Excellent**—Outstanding proposal in all respects, deserves highest priority for support

2 **Very Good**—High-quality proposal in nearly all respects, should be supported if at all possible

3 **Good**—A quality proposal, worthy of support

4 **Fair**—Proposal lacking in one or more critical aspects, key issues need to be addressed

5 **Poor**—Proposal has serious deficiencies

Contributed by E. Frazier with information from Branchaw, J. L., Pfund, C., & Rediske, R. (2010). *Entering Research: A Facilitator's Manual.* New York: W.H. Freeman & Co.

TIPS ON APPLYING FOR GRANTS, SCHOLARSHIPS, AND INTERNSHIPS

Funding opportunities usually come with detailed guidelines that applicants must follow for their submissions to be accepted by the funding agency. Make sure that you read the guidelines carefully and submit ALL supporting documentation requested. Failure to follow procedures will result in your application not being reviewed.

Grant Guideline Examples

Undergraduate Funding Opportunities

- **Undergraduate Research Grants Program (Florida Atlantic University):** http://www.fau.edu/ouri/undergraduate_grants.php
- **Barry Goldwater Scholarship:** https://goldwater.scholarsapply.org/steps-in-process/
- **Council for Undergraduate Research Conference grants:** http://www.cur.org/conferences_and_events/cur_conference_grants/

Graduate Funding Opportunities

- **National Science Foundation Graduate Research Fellowship Program:** http://www.nsf.gov/publications/pub_summ.jsp?WT.z_pims_id=6201&ods_key=nsf12599
- **Ford Foundation Predoctoral Fellowship:** http://sites.nationalacademies.org/PGA/FordFellowships/PGA_047958
- **American Association of University Women Dissertation Fellowships:** http://www.aauw.org/what-we-do/educational-funding-and-awards/american-fellowships/

Contributed by E. Frazier with information from Branchaw, J. L., Pfund, C., & Rediske, R. (2010). *Entering Research: A Facilitator's Manual.* New York: W.H. Freeman & Co.

MY MENTORING AND SUPPORT NETWORK

Learning Objectives

Trainees will:

- ► Define current support network.
- ► Explore how to establish professional relationships.
- ► Discuss how trainees can engage a mentoring network to advance their research career.
- ► Articulate the role(s) that each mentor in a network plays and develop strategies to fill any missing roles.

Trainee Level

undergraduate or graduate trainees
novice, intermediate, or advanced trainees

Areas of Trainee Development

- ► Professional and Career Development Skills
  - Explore and pursue a research career.
- ► Research Comprehension and Communication Skills
  - Develop effective interpersonal communication skills.

Activity Components and Estimated Time for Completion

- ► Trainee Pre-Assignment Time: 20 minutes
- ► In Session Time: 45 minutes

Total time: 1 hour, 5 minutes

When to Use This Activity

This activity can be used with trainees at any career stage and at any time during their training. Once created, trainees should revisit their mentoring and support map regularly and reflect on whether they are effectively engaging their network and whether the people in it are able to meet their emerging needs as they advance in their training. Other activities that may be used with this activity include:

- ► Research Group Diagram
- ► Prioritizing Research Mentor Roles
- ► The Next Step in Your Career
- ► Networking 1: Introduction to Networking

Inclusion Considerations

It may be particularly important for trainees from underrepresented groups in research to identify mentors who have similar backgrounds to include in their networks. Encourage them to consider multiple dimensions when looking for mentors and reassure them that mentors who support their personal well-being and growth are as important as mentors who support their career development and growth. Challenge them to consider all aspects of their identity when building their mentoring network.

Contributed by J. Branchaw. (2018). *My Mentoring and Support Network.*

Implementation Guide

Trainee Pre-Assignment (20 minutes)

- Distribute the pre-assignment provided in the trainee materials at least 24 hours, but ideally 1 week before meeting. The pre-assignment requires trainees to identify the people in their life who already do or could possibly provide mentoring. For existing mentors, students should briefly describe their relationships. For students who do not yet have mentors, they should identify the types of mentors they would like to have and people currently in their lives who might fill these roles.

Workshop Session (45 minutes)

- **Introduction** (5 minutes)
  - The goal of this activity is for trainees to identify their mentoring network or determine the type of network they should develop and to learn strategies for how to develop such a network.
- **Activity: My Mentoring and Support Network** (20 minutes)
  - Distribute and discuss the sheet describing different types of mentors. Briefly review the list as a large group and ask trainees if they know of any other types of mentors that should be added to the list.
  - Direct trainees to individually look at the list of mentors they generated for the pre-assignment and match their mentors to different types on the list of descriptions. What kinds of mentoring are they already getting? What types are they missing?
  - Pair up trainees to share their lists. Direct them to consider the challenges they face in identifying and establishing relationships with potential mentors and to brainstorm strategies they could use to overcome these challenges and fill the gaps in their mentoring networks.
- **Discussion** (10 minutes)
  - Bring trainees back together as a large group and invite them to share the challenges and strategies they came up with.
- **Action Plan to Build Mentoring Network** (5 minutes)
  - Direct trainees to make an action plan to build their mentoring network. Each trainee should identify at least two types of mentors they are missing in their network and list strategies that they will use to fill those gaps.
- **Wrap-up** (5 minutes)
  - Summarize the main ideas generated from the discussion and (as time permits) invite trainees to share their action plans.

Optional Follow-Up Assignment:
Facilitators could assign trainees to follow through on their action plans and then report back to the group at some set time in the future.

Contributed by J. Branchaw. (2018). *My Mentoring and Support Network.*

MY MENTORING AND SUPPORT NETWORK

Learning Objectives

Trainees will:

- Define current mentoring and support network.
- Explore how to establish professional relationships.
- Discuss how trainees can engage a mentoring network to advance their research career.
- Articulate the role(s) that each mentor in a network plays and develop strategies to fill any missing roles.

Pre-Assignment

Make a list of the people in your life who provide mentoring and/or support to you. Next to each person's name, briefly describe your relationship with that person and how they support you. Space is provided to list 10 individuals; however, you may list fewer or more.

People **How They Support You**

1. ____________________
2. ____________________
3. ____________________
4. ____________________
5. ____________________
6. ____________________
7. ____________________
8. ____________________
9. ____________________
10. ____________________

Contributed by J. Branchaw. (2018). *My Mentoring and Support Network.*

MY MENTORING AND SUPPORT NETWORK

Learning Objectives

Trainees will:

- ▶ Define current mentoring and support network.
- ▶ Explore how to establish professional relationships.
- ▶ Discuss how trainees can engage a mentoring network to advance their research career.
- ▶ Articulate the role(s) that each mentor in a network plays and develop strategies to fill any missing roles.

Different types of mentors provide distinct types of support and guidance to trainees as they navigate their research learning experiences and careers. Rarely is one person able to provide all of the support and guidance that a trainee needs. Therefore, it is a good idea to build a network of mentors. The types of guidance and support that mentors can provide include:

Mentor Roles	My Mentors
Intellectual Feedback People who provide critical feedback that helps you to improve your work.	
Intellectual Community People with whom you can brainstorm ideas and discuss your work.	
Sponsorship People who use their power and influence on your behalf to shape the story about who you are and the importance of your work.	
Access to Opportunities People who provide access to opportunities that will advance your work and career.	
Accountability People who check to make sure that you are making quality and timely progress in your work.	
Role Models People who exhibit the values, attitudes, and behaviors that you aspire to emulate as a person and a professional.	
Professional Development People (or organizations) who provide professional development training to advance your work and career.	

Contributed by J. Branchaw. (2018). *My Mentoring and Support Network.*

Mentor Roles	My Mentors
Emotional Support People who provide support to help you maintain personal well-being (psychological and physical) as you navigate the ups and downs of your work and career.	
Safe Space People with whom you can talk about anything and everything who will not judge or share the information.	
Financial Support People who you could go to in hard situations for financial support.	
Other:	

Action Plan: Building My Mentoring Network

1. Identify at least two types of mentors that are missing from your mentoring network.

2. What strategies will you use to identify and invite people to fill these mentoring roles?

Contributed by J. Branchaw. (2018). *My Mentoring and Support Network.*

NETWORKING 1: INTRODUCTION TO NETWORKING

Learning Objectives

Trainees will:

- ► Learn the positive career impact of building local, regional, and national networks.
- ► Begin to learn strategies for effective networking.

Trainee Level

undergraduate or graduate trainees
novice or intermediate trainees

Areas of Trainee Development

- ► Professional and Career Development Skills
  - Explore and pursue a research career.
- ► Research Comprehension and Communication Skills
  - Develop effective interpersonal communication skills.
  - Develop research communication skills.
- ► Researcher Confidence and Independence
  - Develop confidence as a researcher.
- ► Researcher Identity
  - Develop an identity as a researcher.

Activity Components and Estimated Time for Completion

- ► In Session Time: 1 hour, 10 minutes

Total time: 1 hour, 10 minutes

When to Use This Activity

For undergraduates, this activity can be implemented at the start of a semester or summer program as a way for trainees to get to know one another. For graduate students, this activity can be used prior to and/or during research rotations as the students are beginning graduate school.

Given that networking is a career-long endeavor, and that it pertains to all fields (even those where it is uncommon to do research rotations), this activity can be adapted for many career stages and student populations. It may be particularly useful for those times in a trainee's career when they will have a significant networking opportunity. For example, the activity could be used to prepare undergraduates for their interviews and visits to graduate school programs. It could also be used for more advanced graduate students who are preparing to participate in local or national meetings, or postdoctoral interviews.

It can be implemented as a stand-alone activity or as part of a series of activities focused on networking. If implemented as part of a series, the following order is recommended:

1. Networking 1: Introduction to Networking
2. Networking 2: What Should Your Network Look Like?
3. Networking 3: Your Brand
4. Networking 4: Planning for Networking Opportunities and Engaging in Purposeful Interactions

Other activities that may be used with this series include:

- ► My Mentoring and Support Network

Contributed by A. Sokac. (2018). *Networking 1: Introduction to Networking.*

Optional Reading:

- *Networking for Nerds: Find, Access and Land Hidden Game-Changing Career Opportunities Everywhere.* Alaina G. Levine. (2015). Wiley Blackwell. Hoboken, NJ.
  - This book served as a key resource for the development of this suite of activities and has proven useful to trainees. Therefore, departments or programs may consider buying a copy for each incoming trainee. Alternatively, this book could be given as a prize for Networking Challenges, etc., that fit appropriate institutional events.

Inclusion Considerations

Demystify networking and empower trainees to be themselves by discussing how networking is simply talking or having a conversation. This can be especially important for women and others historically underrepresented in research careers, who may feel intimidated or isolated when attending research events in their discipline or interacting with others in their department or future profession. Encourage them to come up with answers to the conversation starter questions in this activity that are genuine and will give them confidence to continue conversations.

Implementation Guide

Facilitator Pre-session Preparation

- Invite advanced students, postdocs, and/or faculty to join the session to talk about how networking has benefited them professionally. Invite "guests" based on their aptitude for networking, and/or career-building experiences with networking. Ask guests to be prepared to share a brief networking success story that precipitated or marked a significant event in their career trajectory. The director of the Career Development Office, or equivalent could also be invited to the session.

Workshop Session (1 hour, 10 minutes)

- Introduction
  - The goal of this activity is to familiarize trainees with the concept of networking and to provide a low-stakes environment for practicing networking.
- ***Activity 1: Introductions*** (20 minutes)
  - Have each person pair with someone they do not know and spend a few minutes getting to know one another (2–3 minutes). Information that each person could share about themselves:
    - What type of research are you doing or are you interested in doing?
    - What is your long-term career goal? What type of position would you like to have after your training/education is complete?
    - Where did you grow up?
    - Who is in your family?
  - Have each person introduce their partner to the group (17–18 minutes).
- ***Activity 2: Share a Networking Success*** (40 minutes)
  - Introduction (1 minute): Networking is a critical component of developing and promoting one's career. By letting others know what your value is, and by recognizing their value too, interactions can be initiated that lead to mutual success (e.g., collaborations and jobs).
  - Lead a discussion about what networking is, using the following questions. (10 minutes)
    - In your own words, what is networking?
    - What is the relationship between networking and self-promotion?
    - How can networking bring mutual benefit?
    - What are some obstacles to networking?
  - Introduce the guests and ask each to share a specific networking story and how it benefited their career. Allow time for questions after each story. (20 minutes)
  - Ask the entire group if anyone else would like to share a networking success story. (10 minutes)

Contributed by A. Sokac. (2018). *Networking 1: Introduction to Networking.*

NETWORKING 1: INTRODUCTION TO NETWORKING

Learning Objectives

Trainees will:

- Learn the positive career impact of building local, regional, and national networks.
- Begin to learn strategies for effective networking.

Activity 1: Introductions

Pair up with someone you do not know, spend two to three minutes learning about your partner. Switch so your partner can learn about you. When we come back to the large group, you will introduce your partner. Some questions that you might ask your partner:

- What type of research are you doing or are interested in doing?
- What is your long-term career goal? What type of position would you like to have after your training/education is complete?
- Where did you grow up?
- Who is in your family?

Activity 2: Share a Networking Success

Networking can have a positive impact and is a critical component of developing and promoting research careers. By letting others know what your value is and by recognizing their value too, interactions can be initiated that lead to mutual success (e.g., collaborations and jobs).

Consider the following questions:

- In your own words, what is networking?

- What is the relationship between networking and self-promotion?

- How can networking bring mutual benefit?

- What are some obstacles to networking?

Contributed by A. Sokac. (2018). *Networking 1: Introduction to Networking.*

NETWORKING 2: WHAT SHOULD YOUR NETWORK LOOK LIKE?

Learning Objectives

Trainees will:

- ► Think strategically about the individuals and types of individuals who should be in a professional network.

Trainee Level

undergraduate or graduate trainees
novice or intermediate trainees

Areas of Trainee Development

- ► Professional and Career Development Skills
  - Explore and pursue a research career.
- ► Research Comprehension and Communication Skills
  - Develop effective interpersonal communication skills.
  - Develop research communication skills.
- ► Researcher Confidence and Independence
  - Develop confidence as a researcher.
- ► Researcher Identity
  - Develop identity as a researcher.

Activity Components and Estimated Time for Completion

- ► In Session Time: 25 minutes

Total time: 25 minutes

When to Use This Activity

For undergraduate trainees, this activity is best implemented once a trainee has been paired with a research mentor and has a general idea of the career path that might be of most interest to them. In most cases, this activity may be more appropriate for intermediate undergraduate trainees, but can also be implemented with novice trainees.

For graduate students, this activity can be used prior to and/or during the time when they are doing research rotations. This may be institution and program specific, but will likely be just as the student is entering graduate school.

Given that networking is a career-long endeavor, and that it pertains to all fields (even those where it is uncommon to do research rotations), this activity can be adapted for many career stages and student populations. It may be particularly useful in proximity to those times in a trainee's career when they will have a significant networking opportunity. For example, the activity could be used to prepare undergraduates for their interviews and visits to graduate school programs. It could also be used for more advanced graduate students who are preparing to participate in local or national meetings, or postdoctoral interviews.

This activity can be implemented as a stand-alone activity or as part of a series of activities focused on networking. If implemented as part of a series, the following order is recommended:

1. Networking 1: Introduction to Networking
2. Networking 2: What Should Your Network Look Like?
3. Networking 3: Your Brand
4. Networking 4: Planning for Networking Opportunities and Engaging in Purposeful Interactions

Contributed by A. Sokac. (2018). *Networking 2: What Should Your Network Look Like?*

Other activities that may be used with this series include:

- My Mentoring and Support Network

Optional Reading:

- *Networking for Nerds: Find, Access and Land Hidden Game-Changing Career Opportunities Everywhere*. Alaina G. Levine. (2015). Hoboken, NJ: Wiley Blackwell.
    - This book served as a key resource for the development of these activities. This book is so useful to students that departments or programs may consider buying a copy for each incoming student. Alternatively, consider giving this book as a prize for Networking Challenges, etc., that fit appropriate institutional events.

Inclusion Considerations

Demystify networking and empower trainees to be themselves by discussing how networking is simply talking or having a conversation. This can be especially important for women and others historically underrepresented in research careers, who may feel intimidated or isolated when attending research events in their discipline or interacting with others in their department or future profession. Encourage them to think about how their unique background will contribute to (not be a barrier to) pursuing their future profession.

Implementation Guide

Workshop Session (25 minutes)

► **Introduction** (5 minutes)

- Building your network is a career-long task. It will grow throughout every career stage and it is never too early or too late to build a network! While every professional (and casual) conversation can lead to a professional opportunity, trainees should also have a thoughtful plan for what their own network should look like. This plan can take into account both short-term and/or long-term career goals.

► **Small-Group Discussion** (10 minutes)

- Give trainees time to consider the following questions at their tables:
    - Who should be in your network if you want a career in industry? In academia? In publishing? In law? In public policy? In consulting?

      Alternatively, this question can be posed as an activity. Post career options around the room (either on a whiteboard or flipchart) and have students write their responses under each career path. Discuss with the group the similarities and differences in networks across career pathways.
    - If you are settled on a career path, when should you start building your network?
    - What if you are not settled on a career path yet? Is networking still worthwhile? Why or why not?
    - If you have not established your research reputation yet, will anyone important want to network with you? Why or why not?

► **Wrap-up** (10 minutes)

- Bring everyone together as a large group and discuss responses to the questions above.

Contributed by A. Sokac. (2018). *Networking 2: What Should Your Network Look Like?*

NETWORKING 2: WHAT SHOULD YOUR NETWORK LOOK LIKE?

Learning Objectives

Trainees will:

- Think strategically about the individuals and types of individuals who should be in a professional network.

The goal of this activity is to start thinking about the composition of the network that you need in order to promote your future success.

Building your network is a career-long task. It will grow throughout every career stage and it is never too early or too late to build your network! While every professional (and casual) conversation can lead to a professional opportunity, you should also have a thoughtful plan for what your own network should look like. This plan can take into account both short-term and/or long-term career goals.

Discuss the following questions at your tables:

- Who should be in your network if you want a career in:
  - Industry?
  - Academia?
  - Publishing?
  - Law?
  - Public policy?
  - Consulting?
- If you are settled on a career path, when should you start building your network?
- What if you are not settled on a career path yet? Is networking still worthwhile? Why or why not?
- If you have not established your scientific reputation yet, will anyone important want to network with you? Why or why not?

Contributed by A. Sokac. (2018). *Networking 2: What Should Your Network Look Like?*

NETWORKING 3: YOUR BRAND

Learning Objectives

Trainees will:

- ► Learn the positive career impact of building local, regional, and national networks.
- ► Identify what a "brand" is and the network that is required to promote a trainee's future success.

Trainee Level

undergraduate or graduate trainees
novice or intermediate trainees

Areas of Trainee Development

- ► Researcher Identity
  - Develop identity as a researcher.
- ► Research Comprehension and Communication Skills
  - Develop effective interpersonal communication skills.
  - Develop research communication skills.
- ► Professional and Career Development Skills
  - Explore and pursue a research career.
- ► Researcher Confidence and Independence
  - Develop confidence as a researcher.

Activity Components and Estimated Time for Completion

- ► In Session Time: 45 minutes

Total time: 45 minutes

When to Use This Activity

For undergraduate students, this activity is best implemented after a trainee has been paired with a research mentor and has a general idea of the career paths that might be of most interest to them. In most cases, this activity may be more appropriate for intermediate undergraduate trainees, but can also be implemented with novice trainees.

For graduate students, this activity can be used prior to and/or during the time when graduate students are doing their research rotations. This may be institution and program specific, but will likely be just as the student is entering graduate school.

Given that networking is a career-long endeavor, and that it pertains to all fields (even those where it is uncommon to do research rotations), this activity can be adapted for many career stages and student populations. It may be particularly useful in proximity to those times in a trainee's career when they will have a significant networking opportunity. For example, the activity could be used to prepare undergraduates for their interviews and visits to graduate school programs. It could also be used for more advanced graduate students who are preparing to participate in local or national meetings, or postdoctoral interviews.

This activity can be implemented as a stand-alone activity or as part of a series of activities focused on networking. If implemented as part of a series, the following order is recommended:

1. Networking 1: Introduction to Networking
2. Networking 2: What Should Your Network Look Like?

Contributed by A. Sokac. (2018). *Networking 3: Your Brand.*

3. Networking 3: Your Brand
4. Networking 4: Planning for Networking Opportunities and Engaging in Purposeful Interactions

Other activities that may be used with this series include:

- My Mentoring and Support Network
- Professional Development Plans

Optional Reading:

- *Networking for Nerds: Find, Access and Land Hidden Game-Changing Career Opportunities Everywhere*. Alaina G. Levine. (2015). Hoboken, NJ: Wiley Blackwell.
    - This book served as a key resource for the development of these activities. This book is so useful to students that departments or programs may consider buying a copy for each incoming student. Alternatively, consider giving this book as a prize for Networking Challenges, etc., that fit appropriate institutional events.

Inclusion Considerations

Demystify networking and empower trainees to be themselves by discussing how networking is simply talking or having a conversation. This can be especially important for women and others historically underrepresented in research careers, who may feel intimidated or isolated when attending research events in their discipline or interacting with others in their department or future profession. Encourage them to consider how their unique background will intersect and contribute to their brand. Socioeconomic status, race, class, gender, age, sexual orientation, etc., do not exist separately from one another but are interwoven.

Contributed by A. Sokac. (2018). *Networking 3: Your Brand.*

Implementation Guide

Workshop Session (45 minutes)

► **Introduction** (5 minutes)

- Networking is about conveying to others that trainees can bring them some special benefit. What is their value? Why should people want to work with them? The answer to these questions represents one's "brand." The goal of this activity is to identify a trainee's "brand" and the network that is required to promote their future success.
- Distribute the "Developing Your Brand" handouts included in the trainee materials at the start of the session.

► **Activity: Networking Strategies: Your Brand** (30 minutes)

Trainees should spend about 15 minutes working independently on the "Developing Your Brand" table in the trainee materials. Invite them to share thoughts about their "brand." Specifically, what strengths, skills, and attributes did they include in their tables? You may want to make a list on the board. If the group is large, this discussion can be done in pairs or small groups (15 minutes).

► **Wrap-up** (5 minutes)

- Encourage students to complete a more detailed skills or strengths inventory. For example, they may use (1) the IDP workflow on the myIDP page at science.careers; or (2) the *Strengths Finder 2.0* book.
    - http://myidp.sciencecareers.org/
    - *Strengths Finder 2.0*. Tom Rath. (2007). New York, NY: Gallup Press.

Contributed by A. Sokac. (2018). *Networking 3: Your Brand.*

NETWORKING 3: YOUR BRAND

Learning Objectives

Trainees will:

- ► Learn the positive career impact of building local, regional, and national networks.
- ► Identify what a "brand" is and the network that is required to promote a trainee's future success.

Networking is about conveying to others that you can bring them some special benefit. What is your value? Why should people want to work with you? The answer to these questions represents your "brand."

The goal of this activity is to begin to develop the message that you want to convey to others about the skills, strengths, goals, etc., that give you value. This will be your brand (i.e., the professional identity that will make others want to work with you). Complete the "Developing Your Brand" table to begin to define your strengths, interests, areas of expertise, and areas for growth.

More detailed skills or strengths inventories are also available, as in (1) the IDP workflow on the myIDP page at science.careers; or (2) the *Strengths Finder 2.0* book.

- ► http://myidp.sciencecareers.org/
- ► *Strengths Finder 2.0.* Tom Rath. (2007). New York, NY: Gallup Press.

Contributed by A. Sokac. (2018). *Networking 3: Your Brand.*

DEVELOPING YOUR BRAND

Knowing your strengths, skills, moral code, work ethic, and passions will help you think about your brand (and how to develop your brand).

My technical skills include:	
My best personality traits are:	
My best workplace attributes are:	
My career goals are:	
I have already done research (published) in the following area(s):	
I want to get better at:	
When people think of me, I want them to think that I am:	
When people think of me, I want them to think of my work on:	
I have specialized scientific knowledge in:	
My special experiences include:	
My scientific and/or professional passions include:	

For an in-depth discussion on developing your brand, see the following book: *Networking for Nerds: Find, Access and Land Hidden Game-Changing Career Opportunities Everywhere*. Alaina G. Levine. (2015). Hoboken, NJ: Wiley Blackwell.

Contributed by A. Sokac. (2018). *Networking 3: Your Brand.*

NETWORKING 4: PLANNING FOR NETWORKING OPPORTUNITIES AND ENGAGING IN PURPOSEFUL INTERACTIONS

Learning Objectives

Trainees will:

- Learn the positive career impact of building local, regional, and national networks.
- Discuss and employ strategies for effective networking.

Trainee Level

undergraduate or graduate trainees
novice or intermediate trainees

Areas of Trainee Development

- Researcher Confidence and Independence
  - Develop confidence as a researcher.
- Professional and Career Development Skills
  - Explore and pursue a research career.
- Research Comprehension and Communication Skills
  - Develop effective interpersonal communication skills.
  - Develop research communication skills.
- Researcher Identity
  - Develop identity as a researcher.

Activity Components and Estimated Time for Completion

- Session 1 Time: 40 minutes
- Session 2 Time: 40 minutes
- Session 3 Time: 55 minutes

Total time: 2 hour, 15 minutes

When to Use This Activity

For undergraduate students, this activity is best implemented once a trainee has been paired with a research mentor and has a general idea of the career paths that might be of most interest to them. In most cases, this activity may be more appropriate for intermediate undergraduate trainees, but can also be implemented with novice trainees.

For graduate students, this activity can be used prior to and/or during the time when graduate students are doing their research rotations. This may be institution and program specific, but will likely be just as the student is entering graduate school.

Given that networking is a career-long endeavor, and that it pertains to all fields (even those where it is uncommon to do research rotations), this activity can be adapted for many career stages and student populations. It may be particularly useful in proximity to those times in a trainee's career when they will have a significant networking opportunity. For example, the activity could be used to prepare undergraduates for their interviews and visits to graduate school programs. It could also be used for more advanced graduate students who are preparing to participate in local or national meetings, or postdoctoral interviews.

Contributed by A. Sokac. (2018). *Networking 4: Planning for networking opportunities and engaging in purposeful interactions.*

This activity can be implemented as a stand-alone activity or as part of a series of activities focused on networking. To successfully implement this module, trainees should feel comfortable briefly summarizing their research project to individuals outside of their research group. Facilitators working with trainees who have no prior networking experience should implement the following activities prior to implementing this activity:

1. Networking 1: Introduction to Networking
2. Networking 2: What Should Your Network Look Like?
3. Networking 3: Your Brand
4. Networking 4: Planning for Networking Opportunities and Engaging in Purposeful Interactions

Other activities that may be used with this series include:

- My Mentoring and Support Network
- Three-Minute Research Story
- Elevator Sentences

This activity is most useful when students can immediately try out some networking strategies in a low-risk environment. Facilitators can create a low-risk networking environment (e.g., a poster session with older grad students and postdocs from the department), or implement the activity at or in proximity to a departmental retreat, institutional graduate symposium, etc. If everyone knows that one goal of this activity is to practice or try networking strategies, then there is a comfort level that allows even the most reserved individuals to reach out to the other trainees.

Optional Reading:

- *Networking for Nerds: Find, Access and Land Hidden Game-Changing Career Opportunities Everywhere*. Alaina G. Levine. (2015). Hoboken, NJ: Wiley Blackwell.
  - This book served as a key resource for the development of these activities. This book is so useful to students that departments or programs may consider buying a copy for each incoming student. Alternatively, consider giving this book as a prize for Networking Challenges, etc., that fit appropriate institutional events.

Inclusion Considerations

Demystify networking and empower trainees to be themselves by discussing how networking is simply talking or having a conversation. This can be especially important for women and others historically underrepresented in research careers, who may feel intimidated or isolated when attending research events in their discipline or interacting with others in their department or future profession. After trainees have participated in the networking opportunity, invite them to discuss how they felt and how others reacted when aspects about their identity or unique background came up in conversation.

Implementation Guide

Workshop Session 1: Planning for Networking Practice (40 minutes)

- **Introduction** (5 minutes)
  - Thoughtful preparation for networking opportunities can go a long way to increasing the benefit trainees get from the experience, with each interaction serving as a purposeful career development maneuver or advancement of interpersonal skills. While it may seem tedious at first, actively practicing and engaging in networking strategies and reflecting on experiences afterward can help successful networking techniques become professional habits.

- Give trainees time to discuss the following questions at their tables. (15 minutes)
  - What are some venues for face-to-face networking? For electronic networking?
  - How do you overcome your fears about networking?
  - What specific questions or ideas can you use to start a conversation with someone in your field? Outside your field?
  - If you are going to a conference or event, what can you do ahead of time to prepare for networking opportunities? What can you do to prepare for networking on a local level or within your department?
  - When and how might you bring up "your value" in a conversation with a new or preexisting contact in your network?
  - How long should an initial networking conversation last?
  - What strategies can you implement after a networking experience to remember your interaction and further your relationship with this contact in the future?
  - How can you take advantage of your current network of contacts to build a bigger network?

- **Large-Group Discussion** (20 minutes)
  - As a large group, summarize the key points that were raised in small groups. You may want to make lists on the board.

- **Wrap-up** (3 minutes)
  - Explain the idea of "Elevator Sentences," and propose as a conversation starter. *For more on this topic, see the* "Elevator Sentences" and *"3-Minute Research Story" activities.*

Workshop Session 2: Networking Practice Preparation (40 minutes)

- **Facilitator Preparation**
  - This activity should occur in a low-risk environment where students can practice networking and begin building a local network with other students, staff, and faculty. Consider hosting this event after a seminar or other common meeting. You may want to provide coffee and cookies to create a more social environment and encourage more networking.
  - Arrange for a cadre of advanced students, postdocs, and faculty to join the group. These "guests" should be invited to the activity based on their aptitude for networking, and/or career-building experiences with networking. It may also be a good idea to invite the director of your Career Development Office, or equivalent. The facilitator should ask guests to be as positive and supportive during this practice as possible, but also provide constructive feedback if needed.
  - If the facilitator feels that this practice activity works better for their group by assigning pairs of guests and students, then that should be planned ahead of time (an equal number of guests to students is required), and a pairing mechanism should be developed (give each guest a number, give each student a number, and ask guests and students with the same number to network with each other).

► **Trainee Preparation:**

- Tell trainees to collect their thoughts or write down some notes about what networking strategies they want to try. Recommend that trainees write down notes after each interaction as well. If planning for a big networking opportunity such as a conference, it can often be useful to keep a notebook or electronic file on a phone to record notes on each networking experience (e.g., name of person you talked to, that person's affiliation, where you met them, what you talked about) to be able to remember details for any follow-up contact with them, or to jog your memory for future encounters with that person.

Workshop Session 3: Networking Practice and Discussion (55 minutes)

► **Activity: Networking Practice** (30 minutes)

- Invite trainees and guests to eat some cookies, drink some coffee, and start a conversation with somebody in the room who they do not know. Encourage trainees to reflect on how their approach went. Recalibrate and repeat. This is a low-risk and confidential environment. The goal is to practice and feel more comfortable when networking.

Discussion (20 minutes)

- Ask trainees and guests to regroup after the networking and take time to reflect on effective strategies and ask questions.
  - What strategies did you try during your networking practice?
  - Were there times that a strategy worked well with one person, but not as well with another person? Why might that be?
  - What strategies did the guests use? Were there any that you would like to try in the future?
  - What strategies or content (e.g., elevator pitch) would you like to work on before your next networking opportunity?

NETWORKING 4: PLANNING FOR NETWORKING OPPORTUNITIES AND ENGAGING IN PURPOSEFUL INTERACTIONS

Learning Objectives

Trainees will:

- Learn the positive career impact of building local, regional, and national networks.
- Discuss and employ strategies for effective networking.

The goal of this activity is to develop well-thought-out strategies for networking that will bring you solid career outcomes, such as finding new mentors, unearthing new opportunities, starting collaborations, etc.

Brainstorm answers to the following questions at your table:

- What are some venues for face-to-face networking? For electronic networking?
- What fears do you have about networking and how can you overcome them?
- What specific questions or ideas can you use to start a conversation with someone in your field? Outside your field?
- If you are going to a conference or event, what can you do ahead of time to prepare for networking opportunities? What can you do to prepare for networking on a local level or within your department?
- When and how might you bring up "your value" in a conversation with a new or preexisting contact in your network?
- How long should an initial networking conversation last?
- What strategies can you implement after a networking experience to remember your interaction and further your relationship with this contact in the future?
- How can you take advantage of your current network of contacts to build a bigger network?
- How can you enlist your research mentor's help in building your network?

Contributed by A. Sokac. (2018). *Networking 4: Planning for networking opportunities and engaging in purposeful interactions.*

CASE STUDY: OVERWHELMED

Learning Objectives

Trainees will:
- Practice strategies to ask for help from mentor(s).
- Identify additional resources for help.

Trainee Level

undergraduate or graduate trainees
novice or intermediate trainees

Activity Components and Estimated Time for Completion

- In Session Time: 20 minutes

Total time: 20 minutes

When to Use This Activity

This activity is most appropriate for undergraduate and graduate trainees in their first semester of research.

Inclusion Considerations

Discuss how power and authority dynamics might be at play for trainees, especially those underrepresented in research careers, and why trainees might resist asking for help from a mentor. Encourage trainees to establish clear expectations to which they, their mentor, and others can agree. Remind trainees of their past accomplishments to help them build the confidence to ask for what they need.

Areas of Trainee Development

- Researcher Confidence and Independence
  - Develop confidence as a researcher.
- Research Comprehension and Communication Skills
  - Develop effective interpersonal communication skills.
- Professional and Career Development Skills
  - Develop confidence in pursuing a research career.

Contributed by A. R. Butz and C. Pfund with information from Branchaw, J. L., Pfund, C., and Rediske, R. (2010). *Entering Research: A Facilitator's Manual.* New York: W.H. Freeman & Co.

Implementation Guide

- **Case Study** (5 minutes)
  - Pass out the appropriate case study included in the trainee materials, there are undergraduate and graduate student versions.
  - Ask trainees to read and individually write down answers to the questions.
- **Discussion** (15 minutes)
  - Bring the group together to discuss the case using the questions provided in the trainee materials.
  - Alternatively, ask trainees to discuss the case study in small groups and then summarize their conversation for the large group.
- **Discussion Questions:**

Undergraduate Case

1. Is there a way for Ashley to approach her mentor to ask questions that respects his busy schedule?
2. Who else beside her mentor could Ashley turn to for help?
3. What resources might she use to help herself better understand the research on her own?

Graduate Case

1. Should Sam talk to his mentor? Why or why not?
2. Who else beside his mentor could Sam turn to for help?
3. What resources might he use to help him run his team more efficiently?

Contributed by A. R. Butz and C. Pfund with information from Branchaw, J. L., Pfund, C., and Rediske, R. (2010). *Entering Research: A Facilitator's Manual.* New York: W.H. Freeman & Co.

CASE STUDY: OVERWHELMED

Undergraduate

Learning Objectives

Trainees will:

- Practice strategies to ask for help from mentor(s).
- Identify additional resources for help.

Ashley, a sophomore majoring in chemistry, has found an undergraduate research position at the Center for Nanotechnology. She started a couple of weeks ago and is excited about her research project, which involves working on the development of an automatic gene synthesizer, but she doesn't really understand it. She is shy and was completely overwhelmed at the first research group meeting. It was like nothing she had ever experienced. She understood little of what was discussed, and she won't take introductory biology until next year. At the meeting, she just nodded whenever they asked if she understood, because she didn't want to look stupid. Now she is terrified to talk to the scientists for fear that they will realize how little she really understands. Her mentor, Sam, a biomedical engineering graduate student, is really nice, but also very busy. He told her to ask questions when she didn't understand something, but he is always engrossed in his work and she doesn't want to interrupt him. She has to write a 1-page summary of her research project for the undergraduate research seminar class by the end of next week and has no idea where to begin. What should she do?

1. Is there a way for Ashley to approach her mentor to ask questions that respects his busy schedule?

Trainee Materials

2. Who else beside her mentor could Ashley turn to for help?

3. What resources might she use to help herself better understand the research on her own?

Contributed by A. R. Butz and C. Pfund with information from Branchaw, J. L., Pfund, C., and Rediske, R. (2010). *Entering Research: A Facilitator's Manual.* New York: W.H. Freeman & Co.

CASE STUDY: OVERWHELMED

Graduate

Learning Objectives

Trainees will:

- Practice strategies to ask for help from mentor(s).
- Identify additional resources for help.

Sam, a biomedical engineering graduate student, has a research assistantship through the Center for Nanotechnology. This is his first year in the graduate program and he is excited about leading a research project for his mentor on the development of an automatic gene synthesizer and mentoring two undergraduate students on the project. Sam knows what needs to be done, but is not sure how to mentor the undergraduates and delegate tasks to get things done. As a result, he ends up doing a lot of the work himself, leading to long hours in the lab. The extended lab hours are compromising his ability to keep up with his coursework. When Sam's mentor asks him how things are going, he always replies "great!" and gives her a brief update on the project. Things are coming along, but with all of the work that he is doing, he has little time to mentor the undergraduate students on his project and the students are hesitant to ask him questions. Sam knows that he should be providing more direction to his undergraduate trainees and allow them to help out more with the project, but is not certain what changes need to be made to make his team function more efficiently. He thought about asking his mentor, but is worried that he will look like he is not ready for graduate school if he asks for help. What should he do?

1. Should Sam talk to his mentor? Why or why not?

2. Who else beside his mentor could Sam turn to for help?

3. What resources might he use to help him run his team more efficiently?

Contributed by A. R. Butz and C. Pfund with information from Branchaw, J. L., Pfund, C., and Rediske, R. (2010). *Entering Research: A Facilitator's Manual.* New York: W.H. Freeman & Co.

PERSONAL STATEMENT

Learning Objectives

Trainees will:

- Articulate career goals and reflect on research experiences.
- Draft a personal statement.
- Review peers' personal statements.

Trainee Level:

undergraduate or graduate trainees
intermediate or advanced trainees

Areas of Trainee Development

- Researcher Identity
  - Develop identity as a researcher.
- Professional and Career Development Skills
  - Explore and pursue a research career.

Activity Components and Estimated Time for Completion:

- In Session 1 Time: 15 minutes
- Pre-Assignment Time: 1 hour
- In Session 2 Time: 45 minutes
- *This activity is presented as a 2-workshop series in which the draft of a personal statement is assigned between sessions. Alternatively, facilitators can implement the entire activity as a 2-hour workshop in which trainees are given time to work on their personal statements.*

Total time: 2 hours

When to Use This Activity

This activity can be used at the end of a research experience with undergraduate or graduate trainees, when they are able to discuss their research in the context of applying for graduate school or other future positions. Graduate trainees should be familiar with personal statements, since most graduate school applications require them, so this activity provides an opportunity for them to update their personal statements. Other activities that may be used with this activity include:

- Developing a Curriculum Vitae

Inclusion Considerations

Encourage trainees to convey their unique backgrounds and experiences in their personal statements. Those from nontraditional backgrounds may have motivations for pursuing a research position or career that are different from those who are continuing generation college students and have traveled traditional academic pathways. Acknowledge and affirm that all pathways and motivations are valid and discuss how personal statements provide opportunities to highlight the unique contributions and perspectives one has to offer. Trainees who come from backgrounds where humility is valued may have been acculturated not to speak of their achievements freely. Let trainees know that it is important to speak about their research positively and with confidence in their personal statement. Reassure trainees that they can do this while remaining true to their ideals around humility.

Contributed by Bramson, A. (2018). *Personal Statement.*

Implementation Guide

Workshop Session 1 (15 minutes)

► **Discussion** (15 minutes)

- What is a personal statement and what is it used for?
  - A personal statement is a brief statement (one or two pages typically) describing your aspirations and experiences.
  - Writing a personal statement is an opportunity to convey your hopes, ambitions, life experiences, and inspirations. What makes you special, unique, distinctive, and/or impressive?
  - Graduate school admissions committees, internship application reviewers, and employers use personal statements to evaluate candidates.
- What topics should be addressed in a personal statement? What are the features of a strong personal statement?
  - Describe what you aspire to become as a professional and why. What are your career goals?
  - Emphasize what you have learned from your research experience and how this experience has contributed to your growth (as a trainee, as a researcher, as a person, etc.).
  - Specifically connect your aspirations to the position to which you are applying (summer research program, graduate school, postdoctoral position, etc.). How will this opportunity help you to clarify and/or achieve your goals?
  - Specifically describe how the organization or program will benefit from accepting you. What do you bring to them and why should they choose you for this position?
  - Use details when describing your previous experiences, skills, and motivations to set yourself apart.

Trainee Assignment (1 hour)

► **Personal Statement Assignment** (1 hour)

- Distribute the trainee materials. This assignment can be completed between workshop sessions if part of a series, or during the workshop if a single event.
  - Trainees will select a research position to which they would like to apply (e.g., summer program, graduate school, fellowship). They use the guiding questions on the handout to identify the points they would like to make and then draft their statement.

Workshop Session 2 (45 minutes)

► **Peer Review of Personal Statements** (45 minutes)

- **Peer Review** (30 minutes)
  - Trainees pair with a classmate to review one another's statement drafts using the peer-review rubric in the trainee materials. (15 minutes)
  - Exchange and discuss reviews. (15 minutes to ~7 minutes for each partner)
- **Large-Group Discussion** (15 minutes)
  - Bring the large group together to discuss their writing and reviewing experiences.
  - Which aspects of the personal statement were the easiest to write? Which were the most difficult?
  - What writing strategies did you find to be particularly useful?
  - What changes would you make to your statement based on the peer feedback that you received?

Contributed by Bramson, A. (2018). *Personal Statement.*

- What changes would you make based on what you learned from reviewing your partner's statement? What did they do that you would like to emulate in your statement?
- What parts of your statement would you change if applying for a different position? What parts would remain the same regardless of the position?

Optional Activity: Personal Statement Examples (35 minutes)

- This activity can be used in conjunction with the pre-assignment discussion to expose trainees to examples of personal statements before they attempt to write their own. It can also be used in conjunction with the peer-review activity to demonstrate to trainees how to provide constructive feedback using example statements that could use some improvement.
- Collect example statements from online sources or from members of your community who are willing to share them. If using the statements to teach trainees how to give constructive feedback, then make sure the statements could use some improvement. For example, first draft statements from a previous workshop could be used as examples.
- Read/review statements (15 minutes)
  - Distribute example personal statements and allow trainees time to read them. If you are asking trainees to complete reviews, then provide extra time.
- **Discussion Questions** (20 minutes)
  - What similarities or differences did you notice across statements?
  - What made these personal statements unique?
  - What was the most impactful from the statements that you read? Why?
  - What parts of a personal statement do you think are most effective for a diverse set of readers (for example, those sitting on an admissions committee)?

Note: Facilitators wishing to provide formative feedback on trainees' personal statements may use the peer-review form as a rubric.

Contributed by Bramson, A. (2018). *Personal Statement.*

PERSONAL STATEMENT

Learning Objectives

Trainees will:

- ► Articulate career goals and reflect on research experiences.
- ► Draft a personal statement.
- ► Review peers' personal statements.

Graduate school admissions committees, internship application reviewers, and employers usually ask for personal statements as part of their application. Since these statements are used to evaluate candidates, learning to write them well is an important skill. Use the questions below to define what you'd like to convey in a statement for a specific position. Use the answers to the questions to draft your personal statement (one or two pages).

1. What is your long-term career or professional goal? What do you aspire to be?

2. Why do you have this goal? What have you learned from your previous experiences that have contributed to your decision to pursue this goal? What is most interesting or motivating to you about working toward this goal?

3. Identify a position (summer research program, graduate school, a job, etc.) to which you are interested in applying. Briefly describe that opportunity or position.

4. What will you learn or do in this position that is interesting to you and will help you achieve your goal?

5. What have you learned from your previous experiences that you will be able to contribute to this position?

6. What sets you apart from others applying to this position? What unique assets or skills do you possess?

Contributed by Bramson, A. (2018). *Personal Statement.*

PERSONAL STATEMENTS—PEER REVIEW

Assessment Rubric

Evaluate three features of the personal statement—voice, organization, and language—using the scale provided. Explain your rating and offer suggestions for improvement.

Writer's Voice

Does the writer use details to make themselves stand out? Does the personal statement speak to the writer's hopes, ambitions, life experiences and inspirations? Does the personal statement come across with enthusiasm and honesty?

Circle: NEEDS IMPROVEMENT OK GOOD GREAT

Explanation of Rating:

Suggestions for Improvement:

Organization

Does the structure of the personal statement make sense? Does the content flow throughout the document and follow a logical progression?

Circle: NEEDS IMPROVEMENT OK GOOD GREAT

Explanation of Rating:

Suggestions for Improvement:

Language

Does the author use correct grammar, spelling, and punctuation?

Circle: NEEDS IMPROVEMENT OK GOOD GREAT

Explanation of Rating:

Suggestions for Improvement:

Contributed by Bramson, A. (2018). *Personal Statement.*

PRIORITIZING RESEARCH MENTOR ROLES

Learning Objectives

Trainees will:

- Identify the different roles that research mentors can play and prioritize those roles based on needs.
- Establish and align expectations with a mentor(s).

Trainee Level:

undergraduate or graduate trainees
novice, intermediate, or advanced trainees

Areas of Trainee Development

- Professional and Career Development Skills
  - Explore and pursue a research career
- Research Comprehension and Communication Skills
  - Develop effective interpersonal communication skills.
  - Develop an understanding of the research environment.
- Researcher Identity
  - Develop identity as a researcher.

Activity Components and Estimated Time for Completion

- In Session Time: 30–35 minutes

Total time: 35 minutes

When to Use This Activity

This activity can be used when trainees have been working with a mentor long enough to be able to use their experiences to think about the different roles that mentors can play. Alternatively, this activity can be used with advanced undergraduate or graduate trainees as part of a discussion about choosing a mentor to help them reflect on which mentoring roles they most value. This activity can help trainees to understand that one mentor may not fulfill all of their needs and that establishing a network of mentors can be valuable.

Other activities that may be used with this activity include:

- Research Group Diagram
- My Mentoring and Support Network
- The Next Step in Your Career
- Networking 1: Introduction to Networking

Inclusion Considerations

Consider that students from nonacademic backgrounds, especially first-generation college and graduate school trainees, may prioritize the roles of their research mentors differently. Discuss the differences and similarities between high school and college teachers/professors and their relationships with students/research trainees. Discuss the formal and informal interactions in mentor–trainee relationships. Encourage trainees to find a diverse group of mentors to support their training.

Contributed by A. Bramson, C. Pfund, and A. R. Butz with information from Branchaw, J. L., Pfund, C., and Rediske, R. (2010). *Entering Research: A Facilitator's Manual.* New York: W.H. Freeman & Co.

Implementation Guide for Undergraduate Trainees

Workshop Session (30 minutes)

- Cut out the boxes of different roles of research mentors. Have trainees move the items around to rank the roles based on their priorities. (10 minutes)
- Ask trainees to explain their highest and lowest priorities and any additional roles they added to their list. This can be done as a large group or in small groups of two or three followed by sharing with the large group. (20 minutes)
- **Discussion Questions**
  - How did your priorities differ from those of other trainees?
  - What roles were NOT important to you? Why?
  - What roles were missing from the list that are important to you?

Implementation Guide for Graduate Trainees

Workshop Session (35 minutes)

- Using the "Roles for your Research Mentor Worksheet" provided in the graduate trainee materials, have trainees complete the following (10 minutes):
  - Rank the roles they expect their mentor to play.
  - For the top 5 roles, have trainees consider whether their primary research mentor effectively addresses each priority. *Skip this question if using with trainees who are in the process of choosing a research mentor.*
  - Ask trainees to list another individual who could serve each of the top 5 roles during their graduate school training.
  - Lead a group discussion to emphasize that each person needs and expects different things from a mentor to have a good experience and achieve their goals from the experience.

Discussion Questions

- Do all these roles need to be fulfilled by one mentor?
- Who else in your group or beyond can be a mentor for you?
- Do you expect the rankings you gave these roles to stay the same as you continue in your research experience? In your career?

Contributed by A. Bramson, C. Pfund, and A. R. Butz with information from Branchaw, J. L., Pfund, C., and Rediske, R. (2010). *Entering Research: A Facilitator's Manual.* New York: W.H. Freeman & Co.

PRIORITIZING RESEARCH MENTOR ROLES

Undergraduate

Learning Objectives

Trainees will:

- Identify the different roles that research mentors can play and prioritize those roles based on needs.
- Establish and align expectations with a mentor(s).

Consider the different roles of research mentors listed below. Add additional roles that may be missing from the list. Cut out the boxes and rank these roles according to your priorities and expectations of a research mentor.

Teach by example
Train you in disciplinary research
Improve your writing and communication skills
Provide growth experiences
Help build your self-confidence as a researcher
Model and promote professional behavior
Inspire
Offer encouragement
Assist with advancement of your career
Facilitate networking with colleagues
Help build the bridge between research and application (i.e., industry, clinical work, etc.)
Provide guidance into future career options
Other:
Other:
Other:

Contributed by A. Bramson, C. Pfund, and A. R. Butz with information from Branchaw, J. L., Pfund, C., and Rediske, R. (2010). *Entering Research: A Facilitator's Manual.* New York: W.H. Freeman & Co.

PRIORITIZING RESEARCH MENTOR ROLES

Graduate

Learning Objectives

Trainees will:

- ► Identify the different roles that research mentors can play and prioritize those roles based on needs.
- ► Establish and align expectations with a mentor(s).

Prioritize the roles you expect your mentor to perform with #1 being the most important role. Consider whether your primary research mentor effectively addresses each of your top 5 priorities. List others who could serve these roles in during your graduate school training.

Role	Priority	Does your primary research mentor do this effectively?	Who else could serve this role in your training during graduate school?
Teach by example			
Train you in disciplinary research			
Improve your writing and communication skills			
Provide funding			
Provide growth experiences			
Help build your self-confidence as a researcher			
Promote professional behavior			
Inspire			
Offer encouragement			
Assist with advancement of career			
Facilitate networking with colleagues			
Other:			
Other:			

Contributed by A. Bramson, C. Pfund, and A. R. Butz with information from Branchaw, J. L., Pfund, C., and Rediske, R. (2010). *Entering Research: A Facilitator's Manual.* New York: W.H. Freeman & Co.

PRIVILEGE AND WHITE FRAGILITY

Learning Objectives

Trainees will:

- Learn about privileges often not afforded to members of groups historically underrepresented in STEM.
- Learn about the myth of meritocracy.
- Learn about the concept of "White fragility" (DiAngelo, 2011).

Trainee Level:

undergraduate or graduate trainees
novice, intermediate, or advanced trainees

Areas of Trainee Development

- Equity and Inclusion Awareness and Skills
  - Advance equity and inclusion in the research environment.
  - Develop skills to deal with personal differences in the research environment.

Activity Components and Estimated Time for Completion

- Trainee Pre-Assignment Time: 30 minutes
- In Session Time:1 hour, 25 minutes

Total time: 1 hour, 55 minutes

When to Use This Activity

This activity may be used with individuals who possess a beginner-level understanding of race and race relations in the United States. It should be used only after the group has established trust and has practiced ground rules for discussing challenging topics. The ground rules should be displayed on a whiteboard or flipchart throughout the session. *If the facilitator is not comfortable leading this activity, a guest with expertise in this area may be invited to co-facilitate.*

Facilitators can assess trainees' understanding of race and race relations by using a journaling exercise in which they ask trainees to reflect on a time when they have experienced racism, prejudice, or bias. If the group includes a mix of members with beginner and advanced levels of understanding of race and race relations, the facilitator can engage the more advanced trainees to lay the groundwork for the less-experienced trainees. In well-established groups, trainees may be more apt to believe their peers than the facilitator (an authority figure).

Other activities that may be used with this activity include:

- Counter-Storytelling

Facilitator Preparation/Additional Resources

Facilitators should review the following to prepare for this session:

- What Is Privilege? https://www.youtube.com/watch?v=hD5f8GuNuGQ
- DiAngelo, R. (2011). White fragility. *The International Journal of Critical Pedagogy*, *3*(3): 54–70.

Contributed by S. Keyl. (2018). *Privilege & White Fragility.*

Inclusion Considerations

Encourage all trainees to share their experiences with the group, but do not require it. Trainees from groups historically underrepresented in research should not be asked or expected to speak for their respective identity group. All shared experiences should be presented as individual experiences, yet considered in light of the historical and social contexts the trainees learn about in the readings. Challenge all trainees to share in identifying challenges and barriers to creating more inclusive teams and diversifying the research workforce.

Implementation Guide

Trainee Pre-Assignment (30 minutes)

- Trainees should review the following articles and resources prior to the workshop session:
  - DiAngelo, R. (2011). White fragility. *The International Journal of Critical Pedagogy, 3*(3): 54–70.

Workshop Session (1 hour, 25 minutes)

- **Introduction** (5 minutes)
  - The goal of the guided reading activity and Privilege Walk is to introduce trainees to the idea of privilege in research and to help them understand manifestations of privilege within both the research environment and in daily life.
- **Discussion** (10 minutes)
  - Discuss the reading:
    - In what context have you heard the term White fragility? How did the reading enhance or broaden your understanding of the topic?
    - As you read about White fragility and considered the assertions found in the DiAngelo article, what were your immediate reactions?
- **Privilege Walk** (10 minutes)
  - Explain to trainees that they will be participating in an exercise that will help them examine the privileges that they have in the context of their research experience.
  - Distribute copies of the trainee materials.
  - Have trainees form one line across the classroom.
  - Read each statement aloud or have trainees take turns reading each statement aloud. Trainees should take one step forward for each statement that accurately reflects their experiences as a student at this institution.
- **Privilege Walk Discussion Questions** (20 minutes)
  - For those who were at the front of the classroom (i.e., had the most privilege), how did that make you feel? How did those with less privilege feel?
  - What questions were not asked?
  - What privileges do you have outside of the research experience?
  - Do individuals have different amounts of privilege in different contexts?
  - Are some privileges universal?
- **Reflection Exercise: Privilege and White Fragility in STEM/Research** (10 minutes)
  - When you picture an Organic Chemistry class of 250 students, what do you see? Who are the people in the room? White? Black? Hispanic? Male? Female? On a piece of paper, write down what immediately comes to mind. (2 minutes)
- **Small-Group Discussion Questions** (5 minutes)
  - What do your responses say about who is represented in STEM?
  - Do the majority of students in your STEM courses look like you or not?
    - If not, what is that like for you?
  - How comfortable or uncomfortable are you discussing race and race relations in the United States, especially as they pertain to STEM?

- Do you identify with the discomfort that may come when discussing race and race relations? Why or why not?
- If you have, in the past, had classroom discussions about White privilege and White fragility, how did those discussions go? Was there noticeable tension? When you think about it in retrospect, do you think about it differently than you did in the moment?

► **Large-Group Discussion Questions** (20 minutes)

- What are the common ideas/themes captured in the paired/small-group discussions?
- Based on the video you watched and your participation in the privilege walk, does everyone have an equal chance of being successful in STEM? Why or not?
- Do some groups have unique privileges or advantages in STEM over others?
- Why might we say that meritocracy is a myth?
- *These discussion questions highlight what is known as "the myth of meritocracy," which questions the assumption that progress and achievement are based solely on ability and not on privilege or class.*

► **Wrap-up** (5 minutes)

- Ask trainees:
  - What questions do you still have about White fragility?
  - How do you feel about White fragility?

Contributed by S. Keyl. (2018). *Privilege & White Fragility.*

PRIVILEGE AND WHITE FRAGILITY

Learning Objectives

Trainees will:

- Learn about frequently taken for granted privileges not afforded to members of groups historically underrepresented in the sciences.
- Learn about the myth of meritocracy.
- Learn about the concept of "White fragility" (DiAngelo, 2011).

Privilege Walk Instructions. Take one step forward for each statement that accurately reflects your experience as a student at this institution.

- At least one of my parents completed a 4-year college degree (bachelor's).
- At least one of my parents completed a master's or Ph.D. degree.
- I have family members or close friends who are employed or who have been employed in the career field that I want to pursue.
- I have role models in my field who look like me.
- My primary language is the one that is spoken in my research group.
- People do not assume that I was admitted to this university/program because of any aspect of my identity (e.g., racial/ethnic group, gender, or ability status).
- I have family members or close friends who have previously completed research experiences.
- I frequently see people who look like me as I walk around campus.
- People do not avoid me in my research group.
- I am never asked to speak for all of the people of my racial/ethnic group.
- I do not have to worry about how I will pay for my college education.
- I grew up with more than 50 books in my house.

References for Additional Reading

Applebaum, B. (2017). Comforting discomfort as complicity: White fragility and the pursuit of invulnerability. *Hypatia, 32*(4): 862–875.

As/Is. (2015, July 4). What is privilege? [Video file]. Retrieved from https://www.youtube.com/watch?v=hD5f8GuNuGQ

Deconstructing White privilege with Dr. Robin Diangelo. Retrieved from http://www.gcorr.org/video/vital-conversations-racism-dr-robin-diangelo/

DiAngelo, R. (2011). White fragility. *The International Journal of Critical Pedagogy*, 3(3): 54–70.

Lambert, Diana, and Chabria, Anita. High school science fair project questioning African American intelligence sparks outrage. Retrieved from http://www.sacbee.com/news/local/education/article199440204.html#2

Contributed by S. Keyl. (2018). *Privilege & White Fragility.*

PROFESSIONAL DEVELOPMENT PLANS

Learning Objectives

Trainees will:

- ▶ Develop a plan to guide career and professional development.

Trainee Level:

undergraduate or graduate trainees
novice, intermediate, or advanced trainees

Area of Trainee Development

- ▶ Professional and Career Development Skills
  - Explore and pursue a research career.
- ▶ Researcher Confidence and Independence
  - Develop independence as a researcher.
- ▶ Researcher Identity
  - Develop identity as a researcher.

Activity Components and Estimated Time for Completion

- ▶ Session 1 Time: 35 minutes
- ▶ Trainee Pre-Assignment 2 Time: 30 minutes–1 hour
- ▶ Session 2 Time: 25 minutes

Total time: 1 hour–1 hour, 35 minutes

When to Use This Activity

The ideal time to implement this activity is during or immediately following the mentor selection process, but material from this activity can be discussed at any point in the mentoring relationship. This activity may be better suited for trainees who have worked in their mentor's research group for at least one month. Other activities that may be used with this activity include:

- ▶ Research Experience Reflections 1: Entering Research?
- ▶ Aligning Mentor and Trainee Expectations
- ▶ My Mentoring and Support Network

Inclusion Considerations

Consider the importance of cultural, ethnic, and family concerns that may influence trainees' career choices, decisions, and planning. Encourage trainees to incorporate these issues as they draft their Professional Development Plans. If possible, invite professionals who have taken nontraditional career paths to talk about their journeys with the trainees.

Contributed by A. R. Butz and C. Pfund with information from Pfund, Brace, Branchaw, Handelsman, Masters, and Nanney. (2012). *Mentor Training for Biomedical Researchers.* New York: W.H. Freeman & Co and from Pfund, C., Branchaw, J. L., and Handelsman, J. (2014), *Entering Mentoring: A Seminar to Train a New Generation of Scientists,* 2nd ed. New York: W.H. Freeman & Company.

Implementation Guide

Workshop Session #1: Introduction of Professional Development Plans (35 minutes)

► **Introduction of Individual Development Plan (IDP) assignment** (5 minutes)

- Explain to trainees that it is never too early to start thinking about how their research experience can prepare them for additional training, including graduate study or a career. Share examples from your own experience.
- Distribute the "Developing a Professional Development Plan" worksheet included in the trainee materials and read through the instructions as a class.
- Tell trainees that this activity is designed to guide them through the development of an Individual Development Plan and to introduce them to the concept of SMART goals.
- SMART[1] goals can be used to develop individual professional development goals.
    - **S**pecific—The goal is clear and focused.
    - **M**easurable—You can easily determine whether or not you have achieved it.
    - **A**ction-oriented—The action plan for achieving your goal is clear and logical.
    - **R**ealistic—The goal is attainable given the difficulty of the task and the timeframe in which you have to complete it.
    - **T**ime-bound—You have specified a deadline.

► **Reviewing and Developing Professional Development Plans Activity** (15 minutes)

- Ask trainees to individually review the example development plan provided in the trainee materials and make notes indicating which aspects of the plan they would like to adopt for use in their own professional development plan.
- Ask trainees to come up with at least one immediate, one short-term, and one long-term SMART goal.
- Additional examples are available at http://cimerproject.org. Mentors may also wish to refer their trainees to http://myidp.sciencecareers.org where they can develop an IDP through a guided, online process.
- Allow trainees time to review their goals, or do peer review in pairs, to see if the goals meet the SMART goal criteria. Ask trainees to modify goals as appropriate. (5 minutes)

► **Large-Group Discussion** (10 minutes)

- What elements in the IDP example would you like to add to your own IDP?
- How difficult was it to create goals that meet the SMART goal criteria? Which criteria were most challenging to meet?
- How can discussing your IDP with prospective mentors help you assess fit when choosing a mentor, or help you keep on track with progress toward your goals if you already have a mentor?
- What are some of the concerns or fears you have about discussing career plans with your mentor?

► **Wrap-up** (5 minutes)

- Ask trainees to complete the "Trainee Pre-Assignment for Workshop Session 2."

[1] Note: There are many variations on the SMART acronym for goal setting. We took our example from Fuhrman, C. N., Hobin, J. A., Clifford, P. S., and Lindstaedt, B. (2013). Goal setting strategies for scientific and career success. *Science Careers.* doi: 10.1126/science.caredit.a1300263

Contributed by A. R. Butz and C. Pfund with information from Pfund, Brace, Branchaw, Handelsman, Masters, and Nanney. (2012). *Mentor Training for Biomedical Researchers.* New York: W.H. Freeman & Co and from Pfund, C., Branchaw, J. L., and Handelsman, J. (2014), *Entering Mentoring: A Seminar to Train a New Generation of Scientists,* 2nd ed. New York: W.H. Freeman & Company.

► **Trainee Assignment**

- Ask trainees to complete a professional development plan using the example provided in the trainee materials, or another template, and bring it to the second session.
- Additional resources: Guided online IDP tool: http://myidp.sciencecareers.org

Workshop Session 2: Discussion of Professional Development Plans (25 minutes)

► **Small-Group Discussion** (15 minutes)

- In pairs or groups of three, have trainees review each other's IDPs. Encourage them to discuss whether the goals listed are SMART goals and whether the plan to achieve the goals is realistic. They should work together to revise and improve their goals and plans.

► **Large-Group Discussion** (10 minutes)

- What was challenging about developing your IDP?
- Did you use any resources (including people) while developing your IDP?
- What are some of the goals that you have for your research experience and how do these goals support your long-term goals?
- How will your mentor be involved in supporting you to attain these goals?
- Have you already discussed these goals with your mentor? If not, when and how do you plan to do so?

► **Wrap-up** (1 minute)

- Encourage trainees to use SMART goal setting as a tool to identify and refine professional development goals throughout their career.

Optional Follow-up: If you will be working with trainees over the course of a semester or longer, you may wish to occasionally revisit their professional development plans at key points in the course or workshop series (at mid-term, at the end of the semester or summer session, etc.) and invite trainees to revise their plans as necessary.

Contributed by A. R. Butz and C. Pfund with information from Pfund, Brace, Branchaw, Handelsman, Masters, and Nanney. (2012). *Mentor Training for Biomedical Researchers.* New York: W.H. Freeman & Co and from Pfund, C., Branchaw, J. L., and Handelsman, J. (2014), *Entering Mentoring: A Seminar to Train a New Generation of Scientists,* 2nd ed. New York: W.H. Freeman & Company.

PROFESSIONAL DEVELOPMENT PLANS

Learning Objectives

Trainees will:

- Develop a plan to guide career and professional development.

An Individual Development Plan (IDP) will help you to set goals and identify strategies to reach your goals. It is a self-tracking tool that can also be used to facilitate communication and alignment of expectations with your mentor. Annual (or more frequent) review of the plan will provide opportunities to celebrate your achievements, incorporate revisions, and ensure you're making progress toward your goals.

Setting goals can be challenging. Use the "Questions to Guide IDP Development" handout to guide development of your IDP and use the "IDP Planning Worksheet" to draft your IDP. Once you have some ideas about your goals, make sure they are "SMART:" Specific, Measurable, Action-oriented, Realistic, and Time-bound. Discussions with your mentors, instructors, peers, and family members may help you to refine your goals and your timeline.

A SMART goal is:

- **S**pecific—The goal is clear and focused.
- **M**easurable—You can easily determine whether or not you have achieved it.
- **A**ction-oriented—The action plan for achieving your goal is clear and logical.
- **R**ealistic—The goal is attainable given the difficulty of the task and the timeframe in which you have to complete it.
- **T**ime-bound—You have specified a deadline.

As you develop your goals and plans to achieve them, you will start to see where you can benefit from mentoring. The "Identifying Mentors" worksheet will help you to align the skills and competencies you hope to gain with potential mentors so that you can have a clear understanding of what you are seeking from each mentor.

Bring a copy of your draft IDP to the next session for discussion.

Contributed by A. R. Butz and C. Pfund with information from Pfund, Brace, Branchaw, Handelsman, Masters, and Nanney. (2012). *Mentor Training for Biomedical Researchers.* New York: W.H. Freeman & Co and from Pfund, C., Branchaw, J. L., and Handelsman, J. (2014), *Entering Mentoring: A Seminar to Train a New Generation of Scientists,* 2nd ed. New York: W.H. Freeman & Company.

QUESTIONS TO GUIDE IDP DEVELOPMENT

1. **What are your goals?**
   - **Ultimate goal** e.g., (*I will be a professor of neuroscience at a research university.*)

   - **Long-term (5–10 years)** e.g., (*I will be a postdoctoral fellow studying the genetic basis of neurological disorders.*)

   - **Intermediate-term (2–5 years)** e.g., (*I will earn my Ph.D. degree in Neuroscience.*) (*I will contribute to the discovery of the genetic basis of Alzheimer's disease.*)

   - **Short-term (1–2 years)** e.g., (*I will earn my B.S. degree in Genetics.*) (*I will publish my undergraduate research project in a peer-reviewed journal.*)

   - **Immediate (6 months–1 year)** e.g., (*I will earn an "A" in Biochemistry class.*) (*I will learn brain slice immunohistochemical staining techniques.*) (*I will participate in a summer research program to experience another university.*)

2. **What competencies and skills will you need to successfully reach your goals?** (See the list at the end of this document for specific ideas.)
   - Disciplinary knowledge
   - Research and technical skills
   - Professional and interpersonal skills
   - Management and leadership skills

3. **What activities and experiences will you engage in to gain the competencies and skills?**
   - Taking classes
   - Tutoring, study groups
   - Technique training
   - Research experiences
   - Scientific meeting attendance
   - Professional development workshops

4. **How will you assess your progress in mastering these competencies and skills?**
   - Mastery of coursework
   - Mentor/instructor feedback
   - Successful experimental outcomes
   - Peer review

5. **Who will help you reach your goals and how?**
   - Teachers
   - Mentors
   - Peers
   - Family members

Contributed by A. R. Butz and C. Pfund with information from Pfund, Brace, Branchaw, Handelsman, Masters, and Nanney. (2012). *Mentor Training for Biomedical Researchers.* New York: W.H. Freeman & Co and from Pfund, C., Branchaw, J. L., and Handelsman, J. (2014), *Entering Mentoring: A Seminar to Train a New Generation of Scientists,* 2nd ed. New York: W.H. Freeman & Company.

EXAMPLES OF SKILLS

Research and Technical

- Critical reading (research literature)
- Experimental design
- Experimental techniques
- Computer skills
- Documentation/Laboratory notebook
- Problem solving and trouble-shooting
- Data and statistical analysis
- Critical analysis
- Responsible conduct of research
- Identification of new research directions and next steps

Professional and Interpersonal

- Reliability and follow-through
- Communication (oral and written)
- Writing (manuscript, grant, fellowship)
- Teaching
- Mentoring
- Collaborating and working in teams
- Giving/receiving constructive feedback
- Collegiality
- Networking

Management and Leadership

- Time management (meeting deadlines)
- Prioritizing and organizing work
- Leading and motivating others
- Research project management
- Budget management
- Supervising/managing people
- Delegating responsibility

Contributed by A. R. Butz and C. Pfund with information from Pfund, Brace, Branchaw, Handelsman, Masters, and Nanney. (2012). *Mentor Training for Biomedical Researchers.* New York: W.H. Freeman & Co and from Pfund, C., Branchaw, J. L., and Handelsman, J. (2014), *Entering Mentoring: A Seminar to Train a New Generation of Scientists,* 2nd ed. New York: W.H. Freeman & Company.

INDIVIDUAL DEVELOPMENT PLAN

An Individual Development Plan is a professional tool that outlines objectives that you and your mentor/supervisor have identified as important for your professional development. A comprehensive review of your career goals and objectives identified at the beginning of your appointment and during regular check-ins provide constructive feedback from your mentor/supervisor that can help you become an independent investigator.

Use this table to record your goals and how you will achieve your goals through engagement in learning activities and experiences that support development of competencies and skills. Identify how you will assess progress toward your goals and who can support you as you work to achieve your goals.

Goals	Competencies and Skills	Activities and Experiences	Assessment of Progress	Mentors and Their Roles
Long-term **1.**				
Intermediate-term **1.** **2.** **3.**				
Short-term **1.** **2.** **3.**				
Immediate **1.** **2.** **3.**				

Contributed by A. R. Butz and C. Pfund with information from Pfund, Brace, Branchaw, Handelsman, Masters, and Nanney. (2012). *Mentor Training for Biomedical Researchers.* New York: W.H. Freeman & Co and from Pfund, C., Branchaw, J. L., and Handelsman, J. (2014), *Entering Mentoring: A Seminar to Train a New Generation of Scientists,* 2nd ed. New York: W.H. Freeman & Company.

Trainee Materials

If applicable: Please describe your transition plan from your current position to the next stage in your career (from undergraduate to graduate study, from graduate study to postdoctoral study).

Additional Comments:

Contributed by A. R. Butz and C. Pfund with information from Pfund, Brace, Branchaw, Handelsman, Masters, and Nanney. (2012). *Mentor Training for Biomedical Researchers.* New York: W.H. Freeman & Co and from Pfund, C., Branchaw, J. L., and Handelsman, J. (2014), *Entering Mentoring: A Seminar to Train a New Generation of Scientists,* 2nd ed. New York: W.H. Freeman & Company.

IDENTIFYING AND APPROACHING POTENTIAL MENTORS

Identifying Mentors

Identify people who can assist you in achieving your goals by developing the skills and competencies you listed in your plan. For each potential mentor, identify objectives, develop a list of what you can offer, and propose outcomes. Put your initial thoughts down on paper before you approach a mentor, and then revise it as your relationship changes.

Approaching Mentors

First approach potential mentors by sending an email that includes a request for a meeting, a brief summary of your goals, and why you think there would be a good fit between you and the mentor. Let them know how you are hoping to work with them, such as one-on-one, as one of many mentors, or as part of a mentoring team or committee. You might want to let them know how you think they would be able to contribute to your professional development.

Mentorship Plan[2]					
Mentor	**Long- or Short-Term Goal(s)**	**Skill or Competency**	**Activities and Experiences**	**What I can offer**	**Outcome**

Managing Relationships with Your Mentors

Relationships should be nurtured and respected. If you and your proposed mentor develop a working relationship, have some guidelines for how you will work together. Here are some tips:

- Schedule standing meetings ahead of time and keep them.
- Give your mentor(s) plenty of time to review drafts of grants and manuscripts.
- Don't be a black hole of need; limit the number of requests you make of any given mentor.
- Develop authorship protocols so that expectations are clear.
- Saying thank you is priceless.

[2] Information from Ann J. Brown, MD MHS, Vice Dean for Faculty, Duke University School of Medicine. Accessed 5/28/10 at http://facdev.medschool.duke.edu

Contributed by A. R. Butz and C. Pfund with information from Pfund, Brace, Branchaw, Handelsman, Masters, and Nanney. (2012). *Mentor Training for Biomedical Researchers.* New York: W.H. Freeman & Co and from Pfund, C., Branchaw, J. L., and Handelsman, J. (2014), *Entering Mentoring: A Seminar to Train a New Generation of Scientists,* 2nd ed. New York: W.H. Freeman & Company.

REFLECTING ON YOUR MENTORING RELATIONSHIP

Learning Objectives

Trainees will:

- Revisit the goals and expectations established with a research mentor.
- Identify and address any issues that have arisen in the mentor–trainee relationship.

Trainee Level:

undergraduate or graduate trainees
novice, intermediate, or advanced trainees

Activity Components and Estimated Time for Completion

- Trainee Pre-Assignment Time: 30 minutes
- In Session Time: 30 minutes

Total time: 1 hour

When to Use This Activity

This activity guides trainees to reflect on their relationship with their research mentor. Other activities that may be used with this activity include:

- Aligning Mentor and Trainee Expectations

Inclusion Considerations

Discuss how cultural, gender, race/ethnicity, and other differences between mentors and trainees have impacted their relationships. What challenges have these differences presented? How have they addressed those challenges, or do they need support in figuring out how to address them?

What benefits have arisen from the differences? How have their different backgrounds enriched the relationship and the contributions that each have made to it?

Areas of Trainee Development

- Professional and Career Development Skills
  - Develop confidence in pursuing a research career.
- Research Comprehension and Communication Skills
  - Develop effective interpersonal communication skills.
- Equity and Inclusion Awareness and Skills
  - Develop skills to deal with personal differences in the research environment.
- Researcher Confidence and Independence
  - Develop confidence as a researcher.

Contributed by B. Montgomery with information from Branchaw, J. L., Pfund, C., and Rediske, R. (2010). *Entering Research: A Facilitator's Manual.* New York: W.H. Freeman & Co.

Implementation Guide

Trainee Pre-Assignment (30 minutes)

► **Introduction**

- Maintaining a positive relationship between a trainee and their research mentor is very important and can be achieved through frequent, open, and candid communication.
- To facilitate this communication, trainees will schedule a meeting with their research mentor prior to the class session (1 week to several weeks ahead of time) to discuss several topics related to their mentor–trainee relationship using the guiding questions/topics below.
- Instruct trainees to answer the questions, then meet with their mentor to discuss the questions. Trainees should be encouraged to provide a copy of the questions to their mentor to reflect upon prior to the in-person meeting.

► Guiding questions/topics for trainee reflection:

- What seems to be working well for you in the mentor–trainee relationship?
- What is not working so well for you? What might you do differently?
- Review the goals and expectations you established with your mentor at the beginning of your relationship. Do you still agree that these goals and expectations are appropriate for your research experience, or do they need to be adjusted? Are you satisfied with the rate of progress you have made toward reaching the goals? If not, what might you do differently?
- What has the relationship you have with your mentor taught you about what you must do to be successful as a researcher?
- What aspects of mentoring do you need to get from someone other than your direct mentor? Who can provide this mentoring?

► Have trainees write a paragraph summarizing the conversation they had with their mentor and bring it to the next session.

Workshop Session (30 minutes)

► Discuss the questions as a group during the session. Offer to meet with trainees one-on-one to discuss any issues that have arisen in their relationships with their mentors that they are not comfortable sharing with the group.

► Discussion questions:

- Was it difficult to have this conversation with your mentor? Why or why not?
- As a result of this conversation, have you and your mentor decided to change the way you are working together or the goals for your work? If so, how?
- Ideally, how frequently would you like to have conversations like this with your mentor? Why?

Contributed by B. Montgomery with information from Branchaw, J. L., Pfund, C., and Rediske, R. (2010). *Entering Research: A Facilitator's Manual.* New York: W.H. Freeman & Co.

REFLECTING ON YOUR MENTORING RELATIONSHIP

Learning Objectives

Trainees will:

- Revisit the goals and expectations established with a research mentor.
- Identify and address any issues that have arisen in the mentor–trainee relationship.

Maintaining a positive relationship with your research mentor is very important and can be achieved through frequent, open, and candid communication. To facilitate this communication, answer the questions below, then meet with your mentor to discuss them. Prior to the in-person meeting, you should provide a copy of the questions to your mentor to reflect upon. Schedule a time to meet with your research mentor to discuss your mentoring relationship.

- **Questions to guide discussion with your mentor:**
  - What seems to be working well for you in the mentor-trainee relationship?
  - What is not working so well for you? What might you do differently?
  - Review the goals and expectations you established with your mentor at the beginning of your relationship. Do you still agree that these goals and expectations are appropriate for your research experience, or do they need to be adjusted? Are you satisfied with the rate of progress you have made toward reaching the goals? If not, what might you do differently?
  - What has the relationship you have with your mentor taught you about what you must do to be successful as a researcher?
  - What aspects of mentoring do you need to get from someone other than your direct mentor? Who can provide this mentoring?
- **After meeting with your research mentor, write a paragraph summarizing your conversation.** Come to the next session prepared to discuss the following questions:
  - Was it difficult to have this conversation with your mentor? Why or why not?
  - As a result of this conversation, have you and your mentor decided to change the way you are working together or the goals for your work? If so, how?
  - Ideally, how frequently would you like to have conversations like this with your mentor? Why?

Contributed by B. Montgomery with information from Branchaw, J. L., Pfund, C., and Rediske, R. (2010). *Entering Research: A Facilitator's Manual.* New York: W.H. Freeman & Co.

RESEARCH ARTICLES 1: INTRODUCTION

Learning Objectives

Trainees will:

- Learn the basic structure of a research article.

Trainee Level:

undergraduate or graduate trainees
novice trainees

Areas of Trainee Development

- Research Comprehension and Communication Skills
  - Develop research communication skills.
  - Develop logical/critical thinking skills.
- Researcher Confidence and Independence
  - Develop independence as a researcher.

Activity Components and Estimated Time for Completion

- In Session Time: 1 hour

Total time: 1 hour

When to Use This Activity

This activity may be used with novice research trainees who have very limited or no experience reading research articles. In particular, trainees who are beginning to read research articles in order to explore potential research topics and identify potential research mentors.

Inclusion Considerations

Novice trainees, especially those who have traveled nontraditional academic pathways, may have had limited exposure or opportunity to read research articles. Reassure them that they will learn strategies to break down the articles into manageable chunks and that their reading skills will improve with practice. If trainees have challenges with reading (e.g., disabilities), talk about these challenges and encourage them to seek help from professionals and to share this information with their research mentor, so that they are aware and can help.

Contributed by B. Montgomery with information from Branchaw, J. L., Pfund, C., and Rediske, R. (2010). *Entering Research: A Facilitator's Manual.* New York: W.H. Freeman & Co.

Implementation Guide

Workshop Session (1 hour)

- Bring a variety of research articles from different journals and different disciplines to the session or distribute them prior to the session to preview. Ask trainees to look over the articles and to use the trainee materials to identify the major sections and kinds of information presented in each, and to compare the differences in articles from different journals and disciplines.
- Either the facilitator or a trainee volunteer from the group should document the group's description of each section for future reference.
- If used with trainees who are already working in research groups, the facilitator may solicit suggestions of research articles from the trainees' primary mentors.

Contributed by B. Montgomery with information from Branchaw, J. L., Pfund, C., and Rediske, R. (2010). *Entering Research: A Facilitator's Manual.* New York: W.H. Freeman & Co.

RESEARCH ARTICLES 1: INTRODUCTION

Learning Objectives

Trainees will:

► Learn the basic structure of a research article.

---

1. Review the research articles provided by the facilitator.
2. List the major sections of the articles and describe the kind of information presented in each section.

Section	Description

3. When comparing articles from the various journals and disciplines, what differences do you notice in the format of the sections?

Contributed by B. Montgomery with information from Branchaw, J. L., Pfund, C., and Rediske, R. (2010). *Entering Research: A Facilitator's Manual.* New York: W.H. Freeman & Co.

RESEARCH ARTICLES 2: GUIDED READING

Learning Objectives

Trainees will:

- ► Learn strategies to effectively and efficiently read research articles in the discipline.

Trainee Level:

undergraduate or graduate trainees
novice trainees

Activity Components and Estimated Time for Completion

- ► Trainee Pre-Assignment Time: 2 hours
- ► In Session Time: 1 hour

Total time: 3 hours

When to Use This Activity

This activity is appropriate for use with novice researchers who have not yet been matched with a mentor or who are beginning a research experience. The activity is used for guiding novice trainees through the process of reading and understanding a research article in their discipline.

Inclusion Considerations

Novice trainees, especially those who have traveled nontraditional academic pathways, may have had limited exposure or opportunity to read research articles. Reassure them that they will learn strategies to break down the articles into manageable chunks and that their reading skills will improve with practice. If trainees have challenges with reading (e.g., disabilities), talk about these challenges and encourage them to seek help from professionals and to share this information with their research mentor, so that they are aware and can help.

Areas of Trainee Development

- ► Research Comprehension and Communication Skills
  - Develop logical/critical thinking skills.
  - Develop research communication skills.
- ► Researcher Confidence and Independence
  - Develop independence as a researcher.

Contributed by B. Montgomery and D. Wassarman with information from Branchaw, J. L., Pfund, C., and Rediske, R. (2010). *Entering Research: A Facilitator's Manual.* New York: W.H. Freeman & Co.

Implementation Guide

Trainee Pre-Assignment (2 hours)

- Prior to the session, identify a research article and distribute it with the "Research Article Worksheet."
- The article should be from the discipline in which the trainees are doing research (chemistry, physics, engineering, etc.), but need not be directly related to the trainees' specific areas of research. Trainees should read the selected research article and complete the worksheet before the meeting.

Workshop Session (1 hour, 5 minutes)

- **Introduction**
  - The goal of this activity is to introduce trainees to strategies to effectively and efficiently read research articles in their discipline(s).
- **Discussion** (1 hour)
  - Lead the group through a discussion of each part of the article using the worksheet as a guide. If the group is large, trainees can be organized into groups of two or three to go through the worksheet first. The whole group can then be brought together to share the small-group discussions.
  - In addition to the specific questions on the worksheet, the following general questions can be used to promote discussion:
    - In what ways did the worksheet help you understand the article?
    - Which section of the paper helped you the most in understanding the research presented?
    - Which section of the article was the most challenging to understand?
    - Based on your understanding of the research article, what are logical next steps or future directions for the research?
  - During the discussion, the facilitator may encourage development of a critique of the article by asking trainees to begin to evaluate what they are reading. Additionally, the question about next steps may be used to help students apply the knowledge gained from reading the article.
- **Wrap-up** (5 minutes)
  - Conclude the discussion by asking trainees to identify:
    - Questions they still have about the research presented in the article and information or resources that they used or could have used to help them develop a deeper understanding of the article.
    - How the research described may connect with their emerging or current research ideas or projects. (This step may be particularly helpful for more advanced trainees.)

Contributed by B. Montgomery and D. Wassarman with information from Branchaw, J. L., Pfund, C., and Rediske, R. (2010). *Entering Research: A Facilitator's Manual.* New York: W.H. Freeman & Co.

RESEARCH ARTICLES 2: GUIDED READING

Learning Objective

Trainees will:

- Learn strategies to effectively and efficiently read research articles in the discipline.

Preparation for the Research Article Discussion

Read the research article carefully, using the "Research Article Worksheet" to guide your reading. Attempt to answer all of the questions on the worksheet. Bring your completed worksheet to the discussion.

Contributed by B. Montgomery and D. Wassarman with information from Branchaw, J. L., Pfund, C., and Rediske, R. (2010). *Entering Research: A Facilitator's Manual.* New York: W.H. Freeman & Co.

RESEARCH ARTICLE WORKSHEET

Title:

Authors:

Journal:

Year:

The Basics:

1. What is the general topic that is being researched?
2. Why is it important to carry out research on this topic?
3. What hypotheses or research questions does the paper address?
4. What experiments were done to test the hypothesis or investigate the research question?
5. If a hypothesis was tested, did the authors support or reject the hypothesis?
6. What are the major conclusions?
7. What evidence supports each of the conclusions?
8. What is the next step in this research?

The Critique:

1. Is the paper well written? How do you know?
2. Do the conclusions seem logical given the data presented? Why or why not?
3. Why are the conclusions in this paper important?
4. What were the best aspects of the research presented, and how could it be improved?

Additional Resources:

1. What are the basic concepts that you need to know to understand the science presented in your paper?
2. Identify a chapter or section in a textbook that outlines these basic concepts. Is reading this helpful to your understanding?
3. What other information or resources might help you to better understand the paper?

Contributed by B. Montgomery and D. Wassarman with information from Branchaw, J. L., Pfund, C., and Rediske, R. (2010). *Entering Research: A Facilitator's Manual.* New York: W.H. Freeman & Co.

Further Questions:

Write *at least* five comments or questions about the article.

1.

2.

3.

4.

5.

Contributed by B. Montgomery and D. Wassarman with information from Branchaw, J. L., Pfund, C., and Rediske, R. (2010). *Entering Research: A Facilitator's Manual.* New York: W.H. Freeman & Co.

RESEARCH ARTICLES 3: PRACTICAL READING STRATEGIES

Learning Objectives

Trainees will:

- Develop critical reading skills.

Trainee Level:

undergraduate or graduate trainees
novice or intermediate trainees

Activity Components and Estimated Time for Completion

- Pre-Assignment Time: 2 hours
- In Session Time: 1 hour

Total time: 3 hours

When to Use This Activity

This activity should be used with novice or intermediate trainees who have begun working in a research group and have a specific research topic to guide their selection of a relevant research article from the literature. Other activities that may be used with this activity include:

- Research Articles 1: Introduction
- Research Articles 2: Guided Reading

Inclusion Considerations

Novice trainees, especially those who have traveled nontraditional academic pathways, may have had limited exposure or opportunity to read research articles. Reassure them that they will learn strategies to break down the articles into manageable chunks and that their reading skills will improve with practice. If trainees have challenges with reading (e.g., disabilities), talk about these challenges and encourage them to seek help from professionals and to share this information with their research mentor, so that they are aware and can help.

Areas of Trainee Development

- Research Comprehension and Communication Skills
  - Develop logical/critical thinking skills.
  - Develop research communication skills.
- Researcher Confidence and Independence
  - Develop independence as a researcher.

Contributed by B. Montgomery with information from Branchaw, J. L., Pfund, C., and Rediske, R. (2010). *Entering Research: A Facilitator's Manual.* New York: W.H. Freeman & Co.

Implementation Guide

Trainee Pre-Assignment (2 hours)

- Before the meeting, trainees should identify and read at least one research article covering important background information that they will need to plan and conduct their research project. This could be an article published by their own research team or one from another team in the field.
- Distribute the "Guide to Reading Research Articles" (included in the Trainee Materials) as a reference to support trainees as they read their article.
- In addition, facilitators may also refer students to other published resources such as: Ruben, A. (2016). How to read a scientific paper. *Science.* doi:10.1126/science.caredit.a1600012, http://www.sciencemag.org/careers/2016/01/how-read-scientific-paper

Workshop Session (1 hour)

- **Activity** (30 minutes)
  - Trainees work in small groups of two to four to share and generate a list of reading strategies that they used. Some strategies that students have suggested include:

> - Read the abstract, introduction, and conclusion first to get an overview of the main points.
> - Skip the methods section on the first read, since the methods may be confusing to readers with limited experience doing experiments.
> - Take notes as you read.
> - Write out questions as you read.
> - Search an introductory textbook or the web for terms or ideas that are difficult to understand. These sources often use simple language in their explanations.

- **Discussion** (30 minutes)
  - Bring the small groups together for the second half of the meeting to share strategies.
  - Generate a common list on a whiteboard or flipchart that can be distributed to students as a follow-up after the session.
  - Discuss the pros and cons of each strategy and which types of strategies work best for different types of articles.
- **Optional Activity: Mentor Interview** (15 minutes)
  - As an optional follow-up activity, facilitators can ask trainees to interview their mentors about reading and reviewing research literature in their discipline. Some potential interview questions include:
    - What are your three "go-to" journals and why?
    - How much time do you spend reading research literature on a weekly basis?
    - What strategies do you use when efficiently reading and deciphering research literature?
    - What advice would you give to a trainee who is just beginning to read and search research literature?
  - To debrief the interviews, trainees can discuss their interviews in small groups and share new reading strategies that they learned from their mentors.

Contributed by B. Montgomery with information from Branchaw, J. L., Pfund, C., and Rediske, R. (2010). *Entering Research: A Facilitator's Manual.* New York: W.H. Freeman & Co.

RESEARCH ARTICLES 3: PRACTICAL READING STRATEGIES

Learning Objectives

Trainees will:

- Develop critical reading skills.

---

1. Work with your mentor to identify a research article that covers important background information that you will need to plan and conduct your research project. This could be an article published by your mentor's research team or one from another team in the field.
2. As you read the article, use the "Guide to Reading Research Articles" and attempt to answer the questions posed in the guide. Take notes and write questions about each section of the article as you read. Be prepared to share the strategies that you use to understand the article at the next meeting.
3. Develop a set of questions about the article to discuss with your research mentor(s). These questions might be to clarify your understanding of the research presented in the article, or about how the research in the article relates to your research project.

Contributed by B. Montgomery with information from Branchaw, J. L., Pfund, C., and Rediske, R. (2010). *Entering Research: A Facilitator's Manual.* New York: W.H. Freeman & Co.

GUIDE TO READING RESEARCH ARTICLES

The BIG Picture

Before trying to understand the details presented in a research paper, it is wise to get an overview. Consider the following questions when first scanning papers:

- What is the central question/hypothesis the author is proposing?
- What assumptions are made both when proposing hypotheses and when evaluating them in light of the data collected?
- What data do they collect to assess their question/hypothesis?
- What is their conclusion given the data?

Basic Understanding

Research articles are typically organized in sections as outlined below. Knowing what types of information are present in each section allows one to more efficiently and effectively find information.

Title:

Paper titles are usually succinct, stand-alone overviews of a paper's contents. Authors usually include keywords that abstracting services could use in indexing the article. So, if you are new to a field and/or subject, it is useful to take note of the words used in the title as they may provide keywords to use in any future literature searches.

Abstract:

The purpose of the abstract is to provide the reader with a succinct summary of the article. Thus, the abstract should provide information about the specific research problem being investigated, the methods used, the results obtained, and what the results of the study mean in the larger context of the research study and, in some cases, the field of study. This means that the abstract is a good place to look first if you are trying to decide whether or not the paper is relevant to your work.

Introduction:

The introduction section generally provides an overview of the research problem being studied—why it is a worthy problem, what work has already been done by others to solve it, and what the authors may have already done in this area. Introductions are a good place to go if you are new to the subject.

- What is the main question they are interested in pursuing?
- What background research/pattern/theoretical prediction motivates this question?
- Why is this question interesting in light of the background they discuss?
- Do they offer one hypothesis or more than one?
- What assumptions are made when proposing the hypotheses?

Contributed by B. Montgomery with information from Branchaw, J. L., Pfund, C., and Rediske, R. (2010). *Entering Research: A Facilitator's Manual.* New York: W.H. Freeman & Co.

Methods:

The experimental section will provide detailed information on how the authors accomplished the experiments described in their paper. Such information typically includes sources for all reagents and/or materials used, names and models of all instrumentation used, methods for synthesizing any reagents, and provide quantitative information on the characterization of any new materials synthesized.

- Do their proposed methods critically test their hypotheses?
- Are any of their methods confounded?
- Did the authors use a creative method to assess their question/hypothesis?
- Are their methods simple and elegant or complicated and convoluted?
- Did they come up with a new technique to better evaluate a problem that others have struggled with?

Results:

Some articles will distinguish between "Results" and "Discussion," while others will combine this information into one section, "Results and Discussion." In papers that contain two distinct sections ("Results" and "Discussion"), the data obtained from the study are introduced in the "Results" section and their interpretation is delayed until the "Discussion" section. In papers that contain one section ("Results and Discussion"), results are introduced and interpreted experiment-by-experiment.

- What does the data say about the question/hypotheses?
- Is there only one interpretation of the data?
- Are there any big surprises/unexpected results?

Discussion:

Keep the following in mind:

- Does the author say that the results support or reject the hypothesis?
- Do you agree with the author's interpretation of the data?
- What novel insights are gained from the results?
- What do the results imply more generally for the field of interest? For other fields?
- What will the authors do next?

Sophisticated Understanding

With experience, reading the research literature in a given field will come more easily. This includes the ability to better evaluate what is being presented, and the ability to ask more sophisticated questions.

- When reading papers be critical, but also pay attention to exciting findings, novel insights, and creative ideas. It's easy to criticize, but hard to praise!
- What critical experiment would *you* do to evaluate the proposed hypothesis?
- Form an opinion after looking at the data, *before* reading the authors' interpretation and conclusions.
- Do you agree with the authors' interpretation or are there others?
- If more than one hypothesis is offered, is each exclusive, meaning that it proposes a distinctly alternative explanation that is incompatible with the others, or could some of the hypotheses operate simultaneously?
- Are there compelling alternatives given the data?
- What assumptions are made about the effectiveness of the experiments or the accuracy of the data?

Contributed by B. Montgomery with information from Branchaw, J. L., Pfund, C., and Rediske, R. (2010). *Entering Research: A Facilitator's Manual.* New York: W.H. Freeman & Co.

RESEARCH CAREERS: THE INFORMATIONAL INTERVIEW

Learning Objectives

Trainees will:

- Explore possible research careers and consider how the skills learned by doing research may be transferable to other types of careers.

Trainee Level:

undergraduate or graduate trainees
novice, intermediate, or advanced trainees

Areas of Trainee Development

- Professional and Career Development Skills
  - Explore and pursue a research career.
- Research Comprehension and Communication Skills
  - Develop effective interpersonal communication skills.
  - Develop research communication skills.

Activity Components and Estimated Time for Activity

- Trainee Pre-Assignment Time: 1 hour
- In Session Time: 45 minutes

Total Time: 1 hour, 45 minutes

When to Use This Activity

Though this activity is suitable for trainees at any career stage, it is helpful if they've had at least one semester of research. Other activities that may be used with this activity include:

- The Next Step in Your Career

Inclusion Considerations

Trainees from different cultural, socioeconomic, and other backgrounds may be influenced by family members who see doing research as unfamiliar and not in alignment with familial notions of success (i.e., working during the summer, getting a job right after college, etc.). This may have bearing upon a trainee's understanding and expectations of a research career. Invite them to share their perspective and discuss questions they might ask during their interview to better understand research careers.

Contributed by K. Spencer, J. Branchaw, and J. Gleason with information from Branchaw, J. L., Pfund, C., and Rediske, R. (2010) *Entering Research: A Facilitator's Manual.* New York: W.H. Freeman & Co.

Implementation Guide

Trainee Pre-Assignment (1 hour)

► **Research Career Interview** (approximately 4 weeks in advance of the due date)

- Tell trainees that many of the skills learned doing research are important not only for preparing for a research career but for many related careers as well.
- Distribute the list of careers and ask trainees to select a career on the list, identify an individual with this career, and conduct an email interview with that person. Trainees may also choose a career that is not on the list. However, the career should incorporate the use of research skills.
- Discuss with trainees the importance of establishing interpersonal relationships and appropriate email etiquette. Explain that an email template for the interview is provided and that they should contact the person at least 2 weeks before this assignment is due. Ask trainees to submit the interviewee's name in advance.
- Once the interview is complete, trainees should complete the reflection essay assignment that appears in the trainee materials.

Workshop Session (45 minutes)

► **Interview Presentations** (30 minutes)

- Have each trainee share what they learned from the interview using these questions as a guide:
  - Why did you explore your chosen career?
  - What, if any, preconceived ideas about the career have changed based on what you learned?
  - What specific research thinking and/or technical skills are required for the career you explored?

► **Large-Group Discussion Questions** (15 minutes)

- In what nondiscipline specific careers might research skills be useful or required?
  - Leadership, management, decision-making positions, civic engagement
- If a trainee were interested in the chosen career, what kinds of experiences, other than research training, should they pursue to prepare for it?
- Has reflecting on the skills learned in research and the potential for transferring skills to multiple different careers changed what you hope to get out of your research experience or graduate degree?
- *If the activity is used with undergraduate trainees:* Do you think that learning to do research should be a required part of the undergraduate curriculum in your discipline? Why or why not?
  - Yes—it is the best way to understand the discipline; everyone needs an appreciation for how new knowledge is generated.
  - No—not everyone needs these skills to get a job; it is not practical to require this of everyone; lab classes are enough.

Contributed by K. Spencer, J. Branchaw, and J. Gleason with information from Branchaw, J. L., Pfund, C., and Rediske, R. (2010) *Entering Research: A Facilitator's Manual.* New York: W.H. Freeman & Co.

RESEARCH CAREERS: THE INFORMATIONAL INTERVIEW

Learning Objectives

Trainees will:

- Explore possible research careers and consider how the skills learned by doing research may be transferable to other types of careers.

The skills learned doing research are important when preparing for a research career, but also for many other types of careers. Below is a list of careers for which research training is important.

Select a career on the list (or one that is not on the list), identify an individual with this career, and do an informational interview via email or in person with that person.

Possible Careers

- Museum professional
- Consultant
- "Big data" analyst
- Entrepreneur
- Editor
- Science writer
- Patent lawyer
- Government scientist
- Science policy advisor
- Private industry scientist
- Research university professor
- Research university professional staff (researcher, instructor)
- Teaching university/college professor
- Outreach coordinator (private or academic)
- Clinical researcher (e.g., clinical chemist, hospital clinic manager)
- Academic or private administrator/leader
- Others ___________

Use the email template on the next page to contact the interviewee. **Send the email at least 2 weeks before this assignment is due and email the name of the interviewee to your facilitator.** It may be useful to contact more than one individual to ensure that you receive a response.

Once a response is received, write a summary and **reflection essay** about what you learned using the question prompts on the next page.

Contributed by K. Spencer, J. Branchaw, and J. Gleason with information from Branchaw, J. L., Pfund, C., and Rediske, R. (2010) *Entering Research: A Facilitator's Manual.* New York: W.H. Freeman & Co.

INFORMATIONAL INTERVIEW EMAIL TEMPLATE

Dear Dr./Mr./Ms.____________,

I am a student at ________________ and am writing to request an email interview. We are studying different careers that research training can prepare us for, and I am interested in your career as a ____________. If you are willing to answer the few questions listed below, I would really appreciate it. If you do not have the time, perhaps you could forward this to a colleague who might?

Sincerely, ____________

Interview Questions

1. What do you do in your job?
2. What kind of education or training is needed for your career? Is research training needed? Why or why not?
3. What is a typical starting salary in your career?
4. How much time do you have for personal, noncareer interests?
5. What advice do you have for young people interested in pursuing your career?

Summary and Reflection Essay Prompt

After you receive a response to your interview questions, write a brief summary of the responses. Then reflect on and answer the following questions:

- Why did you explore your chosen career?
- What, if any, preconceived ideas about the career have changed based on what you learned?
- What specific research thinking and technical skills are required for the career you explored?
- Based on the informational interview, are you interested in pursuing this career? Why or why not?

Contributed by K. Spencer, J. Branchaw, and J. Gleason with information from Branchaw, J. L., Pfund, C., and Rediske, R. (2010) *Entering Research: A Facilitator's Manual.* New York: W.H. Freeman & Co.

RESEARCH DOCUMENTATION: CAN YOU DECIPHER THIS?

Learning Objectives

Trainees will:

- Develop an appreciation for the need to keep detailed notes.
- Recognize the ethical implications of poor note taking.

Trainee Level:

undergraduate or graduate trainees
novice, intermediate, or advanced trainees

Activity Components and Estimated Time for Completion

- In Session Time: 15 minutes

Total time: 15 minutes

When to Use This Activity

This activity may be used with undergraduate or graduate trainees at any level of research experience and at any stage in the mentoring relationship. Other activities that may be used with this activity include:

- Research Documentation Process

Inclusion Considerations

Consider learning styles, differences, and disabilities when discussing best practices in research documentation. Ask whether trainees have concerns about traditional best practices to empower them to talk about any of these barriers or concerns with their mentor or another advisor.

Areas of Trainee Development

- Practical Research Skills
  - Develop ability to conduct a research project.
- Research Ethics
  - Develop responsible and ethical research practices.
- Research Comprehension and Communication Skills
  - Develop disciplinary knowledge.
  - Develop research communication skills.

Contributed by J. Gleason with information from Branchaw, J. L., Pfund, C., and Rediske, R. (2010). *Entering Research: A Facilitator's Manual.* New York: W.H. Freeman & Co.

Implementation Guide

Trainee Pre-Assignment

► Trainees should bring copies of pages of their laboratory or field notebook to this session.

Workshop Session (15 minutes)

► **Small-Group Review** (5 minutes)
- Trainees exchange copies of pages from their laboratory or field notebook in groups of three.
- If trainees are not allowed to share copies from their notebook, then distribute copies of notebook pages from a research group that has approved use of these pages in the workshop. These may be from the facilitator's or a colleague's research group. Additional examples may be found on the Internet, particularly using an image search.
- To ensure that both good and poor examples of notebook documentation are available for this activity, the facilitator may want to bring some specific examples.
- Trainees should be instructed to try to explain one another's experiments based on what is written in the notebook.

► **Large-Group Discussion**—Explaining the experiment
- Ask trainees to share how easy or difficult it was to explain their peers' experiments based solely on what was documented in their notebook. (5 minutes)
  - Could you repeat this experiment accurately? If not, why is that a problem?
  - Do you understand the purpose of the experiment?
  - What information might improve the notes?
  - Is there potential for inaccurate data recording?
  - If the results or conclusions drawn from the experiment described were called into question, is there enough information in the notebook to support them?
- Generate a list of best practices for note keeping. Record ideas on a whiteboard or flipchart. (5 minutes) For example:
  - The date on every page
  - Numbered pages
  - A detailed description of the procedures followed
  - Legible writing in pen (permanent)
  - Correcting mistakes in a manner that the original entry can be seen
  - An explanation of the purpose or rationale of the procedures used
  - Inclusion of all data, including "failed" experiments
  - Interpretation of successes and failures

► **Wrap-up** (2–3 minutes)
- Trainees should reflect on their laboratory documentation practices:
  - Could their current notebook documentation withstand a challenge to the results? If not, how do they need to change their practice?
  - Based on today's exercise and discussion, trainees should identify two or three things they plan to modify in how they document their research.

Contributed by J. Gleason with information from Branchaw, J. L., Pfund, C., and Rediske, R. (2010). *Entering Research: A Facilitator's Manual.* New York: W.H. Freeman & Co.

RESEARCH DOCUMENTATION: CAN YOU DECIPHER THIS?

Learning Objectives

Trainees will:

- Develop an appreciation for the need to keep detailed notes.
- Recognize the ethical implications of poor note taking.

Pre-Assignment

Bring a copy of pages from your lab or field notebook describing an experiment to the next session.

In groups of three, exchange copies of notebook pages, review them, and try to explain the experiments based on what is written in the notebook to the third person.

- Are you able to follow the flow of experimental work described in the notebook? Why or why not?

- What parts of the notebook were useful in explaining the experiments?

- What pieces of information were missing that made it difficult to explain the experiments?

Contributed by J. Gleason with information from Branchaw, J. L., Pfund, C., and Rediske, R. (2010). *Entering Research: A Facilitator's Manual.* New York: W.H. Freeman & Co.

RESEARCH DOCUMENTATION PROCESS

Learning Objectives

Trainees will:

- Explain why it is important to accurately document research.
- Identify key elements in research documentation.
- Identify commonalities and differences in documentation associated with different research fields.
- Understand the ethical implications of documenting research.

Trainee Level:

undergraduate or graduate trainees novice trainees

Areas of Trainee Development

- Practical Research Skills
  - Develop ability to conduct a research project.
- Research Ethics
  - Develop responsible and ethical research practices.
- Research Comprehension and Communication Skills
  - Develop disciplinary knowledge.
  - Develop research communication skills.

Activity Components and Estimated Time for Completion

- Trainee Pre-Assignment Time: 1 hour
- In Session Time: 1 hour

Total time: 2 hours

When to Use This Activity

This activity should be used very early in the novice trainee's research career, preferably as soon as they start working with a research group to establish good research documentation practices. It can be implemented before or after a trainee selects a mentor.

Inclusion Considerations

Consider learning styles, differences, and disabilities when discussing best practices in research documentation. Ask whether trainees have concerns about traditional best practices to empower them to talk about any of these barriers or concerns with their mentor or another advisor.

Contributed by J. Gleason with information from Branchaw, J. L., Pfund, C., and Rediske, R. (2010). *Entering Research: A Facilitator's Manual.* New York: W.H. Freeman & Co.

Implementation Guide

Trainee Pre-Assignment (1 hour)

- Have trainees complete the "Research Documentation Process" assignment, which requires them to discuss with their mentor how research is documented in their group. Trainees should bring their outlines to the session.
- Before the discussion, you may also distribute *Guidelines for SCIENTIFIC RECORD KEEPING in the Intramural Research Program at the NIH*, which is available as a PDF and is a good resource on scientific notebook keeping. (https://oir.nih.gov/sites/default/files/uploads/sourcebook/documents/ethical_conduct/guidelines-scientific_recordkeeping.pdf)

Workshop Session (1 hour)

- Research activities are diverse; thus the documentation of those activities can vary, including both written and electronic forms. What is outlined below may include documentation of activities that none of the trainees in the cohort are doing. Other trainees may have documentation needs that are not addressed. However, **the underlying principle of documenting all work done so that it can be repeated should come through in the discussion.**
- If this activity is used with graduate students in a program that has rotations, include a discussion about watching for similarities and differences as they rotate among groups. Graduate students may also discuss comparisons to groups in which they worked as undergraduates.

- **Activity: What to document?**
  - Ask each trainee to name one thing that should be included in each research notebook entry. Generate a comprehensive list for the subsequent discussion. (5 minutes)

    This list might include:
    - Date
    - Hypothesis
    - Explanation of goals/rationale for the experiment
    - Detailed procedures identifying experimental and control treatments
    - Reagents
    - Key for labeling and identifying tubes, animals, etc.
    - Data, both successful and unsuccessful results
    - Analyses of data
    - Interpretation and thoughts about what to do next
    - Computer scripts for data mining and data analysis
    - Locations of transects and other field notes
    - References to locations of specimens and electronic data
    - Citations for methods, reagents, analyses, etc.
  - Using the list of items generated above, discuss the commonalities and differences across trainees' research groups. (20 minutes)

Contributed by J. Gleason with information from Branchaw, J. L., Pfund, C., and Rediske, R. (2010). *Entering Research: A Facilitator's Manual.* New York: W.H. Freeman & Co.

- **Discussion Questions**
  - Why is it important to keep a research notebook?

    - to be able to repeat the experiment
    - to be able to write up the results for publication
    - to document for patents
    - to defend against accusations of fraud

  - What format is required?

    - Does the lab use paper copies or electronic copies?
    - If paper copies, what happens to any photo, video, or other computer output?
    - If electronic copies, how is the electronic copy maintained? How are the data backed up?

  - Who owns the data? Where is the notebook to be kept?

    - All data belongs to the research group and university or company.

  - How will the notebook be used in the future? How does planning for the future influence how notes are kept?

    - The notebook needs to be detailed for you or other lab members to do follow-up experiments.
    - Discuss obligations for sharing data post publication (both the scientific ethics of sharing and publication requirements).

  - What elements of research need to be documented? This could lead to a discussion of:

    - procedures, including descriptions of experiments, observations, and computer scripts
    - raw data
    - data analysis, including the procedures used to analyze the data
    - processed data

  - Are there elements that are used in other research groups that you may find helpful to your own research?
  - How do your group's research documentation protocols reflect the culture in your research group? How do they reflect the communication style in your research group?

► **Wrap-up** (5 minutes)

- Summarize key points of the discussion with trainees. Encourage trainees to clarify with their mentor any parts of the documentation process that are unclear to them.

Contributed by J. Gleason with information from Branchaw, J. L., Pfund, C., and Rediske, R. (2010). *Entering Research: A Facilitator's Manual.* New York: W.H. Freeman & Co.

RESEARCH DOCUMENTATION PROCESS

Learning Objectives

Trainees will:

- Explain why it is important to accurately document research.
- Identify key elements in research documentation.
- Identify commonalities and differences in documentation associated with different research fields.
- Understand the ethical implications of documenting research.

Meet with your mentor to go over the protocol you must follow when documenting your research. Aspects of research that need to be documented may include a description of the methods followed, the raw data results, the analysis used, and the results of analyses. Discuss the level of detail expected in the documentation for your research group and ask your mentor to identify a research team member who keeps an exemplary notebook, or to show you another excellent example. The specifics of what and how to document research will vary depending on the nature of the research (e.g., field or lab based, computational, library).

Write an outline of the documentation protocol that you are to follow when doing research and identify the parts of the process that are common to your entire research group and the parts that are specific to your project. In addition, address the following questions:

- What kinds of notes are kept? Are they hard-copy documents or electronic files?
- Where are the notes kept?
- What parts of the research are documented? What level of detail is needed in documenting experiments?
- How are data to be recorded?

Contributed by J. Gleason with information from Branchaw, J. L., Pfund, C., and Rediske, R. (2010). *Entering Research: A Facilitator's Manual.* New York: W.H. Freeman & Co.

RESEARCH EXPERIENCE REFLECTIONS 1: ENTERING RESEARCH?

Learning Objectives

Trainees will:

- Explore realistic expectations for working in a research group.
- Self-evaluate readiness for research or *graduate* research.

Trainee Level:

undergraduate trainees
novice or advanced trainees

Areas of Trainee Development

- Researcher Identity
  - Develop an identity as a researcher.
- Research Comprehension and Communication Skills
  - Develop an understanding of the research environment.

Activity Components and Estimated Time for Completion

- In Session Time: 45 minutes

Total time: 45 minutes

When to Use This Activity

This activity is designed for trainees who are looking for a research experience, both as a novice undergraduate trainee new to research or an advanced undergraduate trainee who is considering a graduate research experience. This activity can help trainees to develop the self-awareness needed to select an appropriate mentor. Other activities that may be used with this activity include:

- Research Experience Reflections 2: Reflections Exercise

Inclusion Considerations

Encouraging trainees to share openly about their expectations in this activity may reveal unique perspectives that trainees from diverse backgrounds bring. For example, those from different cultural and/or socioeconomic backgrounds may be influenced by family members who see doing research as unfamiliar and not in alignment with familial notions of success (i.e., working during the summer, getting a job right after college, etc.). This may have a bearing upon a trainee's understanding and expectations of a research experience. Invite trainees to share their perspective and discuss strategies they might use to help others better understand research, why they are motivated to do it, and what success in research looks like.

Contributed by E. Frazier with information from Branchaw, J. L., Pfund, C., and Rediske, R. (2010). *Entering Research: A Facilitator's Manual.* New York: W.H. Freeman & Co.

Implementation Guide

Workshop Session (45 minutes)

- **Introduction** (15 minutes)
  - The goal of this activity is to familiarize trainees with realistic expectations for their research experience.
  - Distribute the trainee materials and give trainees 10 minutes to answer the questions individually. Once they have drafted their answers, give trainees 5 minutes to pair and share with a peer.
- **Discussion** (20–30 minutes)
  - Ask trainees to share their answers with the whole group. If the group is large (more than 10), trainees may be divided into two smaller groups. Encourage participation from everyone. Begin by asking for volunteers to share, but be prepared to call on someone if necessary.
  - Guiding questions to facilitate overall discussion:
    - What goals do you have for your research experience?
    - What are your expectations for working with a research group?
    - What skills are needed to be an effective researcher? Do you have those skills? If not, what can you do to develop those skills?
  - It is important to make note of the areas of concern that trainees share during this discussion so that you can make sure that they are addressed. Often, there will be variability in the level of sophistication of the trainees' knowledge about research. If so, encourage the group to discuss the differences and to learn from one another. As the facilitator, provide information or clarification only when necessary.
  - Discussion of Expectations:
    - Address any unrealistic expectations that trainees bring up in the large-group discussion. If possible, invite an experienced trainee or a panel of trainees to facilitate this discussion.
    - Trainees may have varying levels of knowledge about the research process, depending on their career stage and previous research experience. In particular, undergraduate trainees at institutions that do not have a strong tradition of undergraduate research may lack an understanding of what will be expected of them and what they can expect from the research team. For example, they may expect:
      - to be given an independent project on their first day;
      - that faculty members and graduate student mentors work for them and should offer their full attention when the trainee is around;
      - that they will be able to do research in short discontinuous periods of time like other jobs (e.g., during the 30 minutes they have between classes);
      - that they will find the cure for cancer in two months;
      - that they will get a publication after working for just one semester;
      - that the research lab is a place to socialize with friends, not necessarily a place of work.
  - Discussion of Abilities/Skills:
    - Frequently, novice trainees identify intelligence, the ability to talk or write about research, and excellent time management skills as traits they must possess in order to participate in a research experience. They may be under the impression that if they do not already possess these traits, then they are not good enough to do research. Therefore, it is important to emphasize that a research experience provides the opportunity to develop these abilities and skills through hands-on training; it does not require that they are already developed.

Alternative Implementation

This activity can also be implemented in a group with new researchers and those who have one or more semesters of experience with research. This implementation will allow novice researchers to learn from experienced researchers and will encourage experienced researchers to build upon what they have learned in previous semesters. This implementation strategy may be useful in programs with peer mentoring.

Contributed by E. Frazier with information from Branchaw, J. L., Pfund, C., and Rediske, R. (2010). *Entering Research: A Facilitator's Manual.* New York: W.H. Freeman & Co.

RESEARCH EXPERIENCE REFLECTIONS 1: ENTERING RESEARCH?

Novice Undergraduate

Learning Objectives

Trainees will:

- Explore realistic expectations for working in a research group.
- Self-evaluate readiness for research.

Outline your expectations for your research experience by writing brief answers to the following questions. Once you've answered all of the questions, pair with someone in the group to share.

- Why do you want to do research?
- What specific goals do you hope to achieve in your research experience?
- What are your expectations of working with a research group? Please list them below.
- What do you think will be expected of you as an undergraduate student conducting research on a "real" research team? Please list them below.
- What contributions will you bring to your research team?
- What is your greatest concern, and what are you excited about the most?
- Explain your understanding of the research process as you see it today. How does one approach a research question?
- What do you think are important abilities/skills for an individual to have to be able to conduct research?
- Which of those abilities/skills do you have?
- Which of those abilities/skills do you lack? What can you do to develop the abilities that you think you may lack?

Contributed by E. Frazier with information from Branchaw, J. L., Pfund, C., and Rediske, R. (2010). *Entering Research: A Facilitator's Manual.* New York: W.H. Freeman & Co.

RESEARCH EXPERIENCE REFLECTIONS 1: ENTERING RESEARCH?

Advanced Undergraduate

Learning Objectives

Trainees will:

- Explore realistic expectations for working in a research group.
- Self-evaluate readiness for *graduate* research.

Outline your expectations for your graduate training experience by writing brief answers to the following questions. Once you've answered all of the questions, pair with someone in the group to share.

- Why do you want to do research and earn a graduate degree?
- What specific goals do you hope to achieve with your graduate research?
- What are your expectations of working with a research group?
- What do you think will be expected of you as a graduate student conducting research as part of a thesis advisor's research team?
- What contributions will you bring to your research team?
- What is your greatest concern, and what are you excited about the most?
- Explain your understanding of the research process as you see it today. How does one approach a research question?
- What do you think are important abilities/skills for an individual to have to be able to conduct research at the graduate level?
- Which of those abilities/skills do you have?
- Which of those abilities/skills do you lack? What can you do to develop the abilities that you think you may lack?

Contributed by E. Frazier with information from Branchaw, J. L., Pfund, C., and Rediske, R. (2010). *Entering Research: A Facilitator's Manual.* New York: W.H. Freeman & Co.

RESEARCH EXPERIENCE REFLECTIONS 2: REFLECTION EXERCISE

Learning Objectives

Trainees will:

- Critically analyze and articulate their learning and growth through guided written reflections.

Trainee Level:

undergraduate or graduate trainees
novice, intermediate, or advanced trainees

Areas of Trainee Development

- Determined by reflection prompts used.

Activity Components and Estimated Time for Completion

- Trainee Pre-Assignment Time: 1 hour
- In Session Time: 1 hour

Total time: 2 hours

When to Use This Activity

This reflection essay assignment may be administered at any time throughout the course of a research experience. Other activities that may be used with this activity include:

- Research Experience Reflections 1: Entering Research?
- Research Experience Reflections 3: Research Experience Exit Interview

Inclusion Considerations

Encouraging trainees to share openly about their research experiences in this reflection essay may reveal unique issues that trainees from diverse backgrounds are experiencing. Those from different cultural, socioeconomic, and other backgrounds might have family members who see research activities and experiences as unfamiliar and not in alignment with familial notions of success (i.e., working during the summer, getting a job right after college, etc.). This may have bearing upon the trainee's research experience reflection. Be open to diverse interpretations of what is happening in the research environment and invite trainees to explain their perspective. Emphasize that no single interpretation of a research experience is "right" or "wrong." Invite trainees to discuss how their perspectives vary and how this can lead to challenges as well as benefits in the research environment.

Reflection exercise contributed by A. R. Butz with information from J. TeSlaa. *Exploring Service in Science.* University of Wisconsin–Madison.

Implementation Guide

General prompts are provided for pre-, mid-, and post-research experience reflections, "any time reflection," and specific prompts relating to each area of trainee development. Prompts can be mixed and matched to create customized reflection assignments based on the areas of trainee development addressed in the implementation. All of the prompts and the accompanying rubric align to the *Entering Research* conceptual framework.

Trainee Pre-Assignment (varies)

- Ask trainees to write one or two paragraphs to address the general prompt and an additional paragraph for each area of trainee development prompts assigned. Provide trainees with the assessment rubric to guide their writing.

Pre-research prompt: Describe your expectations for your research experience. Use the following questions to guide your reflection:

- Why do you want to do research?
- When and where will your research experience take place?
- What goals do you have for your research experience?
- Who are the individuals who will be involved in your research experience?
- What are your expectations for working with a research team? What will you do? What will others do?
- What skills do you need to be an effective researcher?
- What will you hear, see, smell, touch, as part of your research experience?
- How do you feel about you upcoming research experience?

Mid-research prompt: Revisit your pre-research reflection essay. How has your research experience met the expectations that you had at the beginning? Revise your essay as needed to accurately reflect your research experience. Use the following questions to help guide your reflection:

- Why do you want to do research?
- When and where is your research experience taking place?
- What goals do you have for your research experience?
- Who are the individuals involved in your research experience?
- What are your expectations for working with a research group? What do you do? What do others do?
- What skills do you need to be an effective researcher?
- What have you heard, seen, smelled, touched, as part of your research experience?
- How do you feel about your research experience so far?

End of research prompt: Revisit your mid-research reflection essay. Would you still describe your research experience in the same way? Revise your essay to reflect what you have learned in this [class/seminar] and in your research experience. Use the following questions to help guide your reflection:

- Why do you want to do research?
- When and where did your research experience take place?
- What goals do you have for your research experience?
- Who are the individuals who have been involved in your research experience?
- What are your expectations for working with a research group? What do you do? What do others do?
- What skills did you need to be an effective researcher?
- What did you hear, see, smell, touch, as part of your research experience?

Reflection exercise contributed by A. R. Butz with information from J. TeSlaa. *Exploring Service in Science.* University of Wisconsin–Madison.

- How do you feel about your research experience? What goals do you have for your next or continued research experience moving forward?
- What advice would you give to new students interested in doing a research experience?

Any time reflection prompt: Describe something that happened in your research experience that was particularly memorable for you. This can be a positive or negative experience. Use the following questions to help guide your reflection:

- When and where did this experience take place?
- Who was there? Who wasn't there?
- What did you do? What did others do? What actions did you or others take?
- Who did not speak or act?
- What did you hear, see, smell, touch?
- Did you or others laugh, cry, make a face, etc.?
- How did you feel about this experience?

Reflection exercise contributed by A. R. Butz with information from J. TeSlaa. *Exploring Service in Science.* University of Wisconsin–Madison.

REFLECTION PROMPTS SPECIFIC TO AREAS OF TRAINEE DEVELOPMENT

Research Comprehension and Communication

- ► Interpersonal Communication Skills
  - What have you learned about communicating with members of your research group? your primary research mentor?
  - In what ways has your relationship with your primary research mentor and members of your research team developed?
- ► Knowledge About Your Research Discipline
  - What have you learned about your area of research?
- ► Research Communication Skills
  - In what ways have you had the opportunity to communicate about your research to members of your research team, other researchers, family members, or the general public? What have you learned from these experiences?
- ► Logical/Critical Thinking Skills
  - What opportunities have you had to critically reflect upon your own research, the research process, or the research of others? What was that experience like?
- ► Understanding of the Research Environment
  - In what ways has your experience in this [course/seminar] and/or working with your primary research mentor helped you understand the research process and the "culture" of research? What are some of the unspoken rules or assumptions about research that you have encountered? Do you agree or disagree with them?

Practical Research Skills

- ► Ability to Design a Research Project
- ► Ability to Conduct a Research Project
  - Describe your experience designing and conducting a research project. What was the process like? Did anything unexpected happen? What did you learn?

Research Ethics

- ► Responsible and Ethical Research Practices
  - What have you learned about conducting research in a responsible and ethical way? Have you encountered any ethical situations in your research that made you feel uncomfortable? If so, how did you handle them?

Researcher Identity

- ► Identity as a Researcher
  - Describe a researcher. What do they look like? What do they study? What does their schedule and list of daily tasks look like? Describe the extent to which your experience aligns with that of the researcher you described.

Reflection exercise contributed by A. R. Butz with information from J. TeSlaa. *Exploring Service in Science.* University of Wisconsin–Madison.

Researcher Confidence and Independence

- Independence as a Researcher
  - What opportunities have you had to work independently on your research project? How did the opportunity to work on research independently make you feel? Were you successful in those opportunities?
- Confidence as a Researcher
  - What is something that has happened that made you feel more or less confident as a researcher? Describe the event and the people, places, and things involved. How did this experience impact your confidence to do research in the future?

Equity and Inclusion Awareness and Skills

- Skills to Deal with Personal Differences in the Research Environment
  - What opportunities have you had to work with individuals in the context of your research experience whose background (e.g., race/ethnicity, religion, socioeconomic status, prior training, parents' education level) is different than your own? Describe the ways in which you differ from these individuals, and any benefits/challenges you have encountered as a result of working with them. What assumptions did you have about these individuals?
- Advancing Equity and Inclusion in the Research Environment
  - Have you encountered or witnessed exclusion or unfair treatment (e.g., bias, stereotype threat, or imposter syndrome) in your research experience? How did you and others react to the incident? How did this experience impact your identity as a researcher, your feelings about your research experience, and your feelings about being a researcher?
  - What can you do as a researcher to make your research group and the research environment in general more inclusive?

Professional and Career Development Skills

- Explore and Pursue a Research Career
- Confidence in Pursuing a Research Career
  - What opportunities have you had to explore possible careers in research? Have you learned about any new careers? What type of research career do you think would be the best fit for you? Why? Have these experiences made you more or less confident in your ability to pursue a research career?

Note: There are no trainee materials provided for this activity. Facilitators should generate custom materials based on the reflection prompts they choose to use.

Reflection exercise contributed by A. R. Butz with information from J. TeSlaa. *Exploring Service in Science.* University of Wisconsin–Madison.

RESEARCH EXPERIENCE REFLECTION ESSAY

Assessment Rubric

	Lacking (1)	In Development (2)	Good (3)	Excellent (4)
Integration	Provides no clear connection between the experience and the learning.	Provides minimal and/or unclear connection between the experience and the learning.	Provides adequate and reasonably clear connection between the experience and the learning.	Provides thorough and very clear connections between the experience and the learning.
Clarity	Consistently makes inaccurate statements and/or fails to provide supporting evidence for claims.	Makes several inaccurate statements and/or supports few statements with evidence.	Usually but not always makes statements that are accurate and well supported with evidence.	Consistently makes statements that are accurate and well supported with evidence.
Precision	Consistently fails to provide specific information, descriptions, or data.	Only occasionally provides specific information, descriptions, or data.	Usually but not always provides specific information, descriptions, or data.	Consistently provides specific information, descriptions, or data.
Writing	Consistently makes typographical, spelling, and/or grammatical errors.	Makes several typographical, spelling, and/or grammatical errors.	Makes few typographical, spelling, and/or grammatical errors.	Makes very few or no typographical, spelling, and/or grammatical errors.
Depth	Fails to address salient questions that arise from statements being made; consistently oversimplifies when making connections; fails to consider any of the complexities of the issue.	Addresses few of the salient questions that arise from statements being made; often oversimplifies when making connections; considers little of the complexity of the issue.	Addresses some but not all of the salient questions that arise from statements being made; rarely oversimplifies when making connections; considers some but not all of the full complexity of the issue.	Thoroughly addresses salient questions that arise from statements being made; avoids oversimplifying when making connections; considers the full complexity of the issue.
Breadth	Ignores or superficially considers alternative points of view and/or interpretations.	Gives minimal consideration to alternative points of view and/or interpretations and makes very limited use of them in shaping the learning being articulated.	Gives some consideration to alternative points of view and/or interpretations and makes some use of them in shaping the learning being articulated.	Gives meaningful consideration to alternative points of view and/or interpretations and makes very good use of them in shaping the learning being articulated.

Reflection exercise contributed by A. R. Butz with information from J. TeSlaa. *Exploring Service in Science.* University of Wisconsin–Madison.

	Lacking (1)	In Development (2)	Good (3)	Excellent (4)
Fairness	Consistently represents others' perspectives in a biased or distorted way.	Occasionally represents others' perspectives in a biased or distorted way.	Often but not always represents others' perspectives with integrity.	Consistently represents others' perspectives with integrity (without bias or distortion).
Growth	Provides no examples of growth in specified area(s) of trainee development.	Provides minimal examples of growth in specified area(s) of trainee development.	Provides examples and demonstrates moderate growth in specified area(s) of trainee development.	Provides examples and demonstrates substantial growth in specified area(s) of trainee development.

Rubric adapted from Ash, S. L., and Clayton, P. H. (2009). *Generating, deepening and documenting learning: The power of critical reflection in applied learning. J Appl Learning Higher Ed. 1*, 25–48.

Notes:

Reflection exercise contributed by A. R. Butz with information from J. TeSlaa. *Exploring Service in Science.* University of Wisconsin–Madison.

RESEARCH EXPERIENCE REFLECTIONS 3: RESEARCH EXPERIENCE EXIT INTERVIEW

Learning Objectives

Trainees will:

- Reflect on what was learned and the goals achieved during a research experience.

Trainee Level:

undergraduate or graduate trainees
novice, intermediate, or advanced trainees

Areas of Trainee Development

- Researcher Identity
  - Develop identity as a researcher.
- Researcher Confidence and Independence
  - Develop confidence as a researcher.
- Professional and Career Development Skills
  - Develop confidence in pursuing a research career.

Activity Components and Estimated Time for Completion

- Trainee Pre-Assignment Time: 30 minutes
- In Session Time: 30 minutes (1 hour if the pre-assignment is done in class)

Total time: 1 hour

When to Use This Activity

This activity can be implemented at the end of a trainee's research experience with a particular mentor, or at the mid- or end-point of an *Entering Research* course. Other activities that may be used with this activity include:

- Professional Development Plans

Inclusion Considerations

Acknowledge that trainees from different cultural, socioeconomic, and other backgrounds might have family members who see research activities and experiences as unfamiliar and not in alignment with familial notions of success (i.e., working during the summer, getting a job right after college, etc.). This may have bearing upon reflections of a research experience. Create space in the conversation for students to share their experience talking with friends and family about their research experience.

Contributed by A. R. Butz with information from Branchaw, J. L., Pfund, C., and Rediske, R. (2010). *Entering Research: A Facilitator's Manual.* New York: W.H. Freeman & Co.

Implementation Guide

Trainee Pre-Assignment (30 minutes)

► Ask trainees to respond to the reflection questions prior to coming to the session. If time permits, the reflections can also be completed during the session prior to discussion.

Workshop Session (30 minutes; 1 hour if the worksheet is completed as part of session)

► Worksheet discussion questions: (Note: Any of the reflection questions on the worksheet can be used for the class discussion.)
- What went well with your research experience?
- What did not go so well? How will you improve it moving forward in this experience or in your next research experience?
- What are your goals and expectations do you have for your research experience moving forward?
- What differences have you observed between the research done as part of a research group and research done as part of a course?

► Additional discussion questions:
- What advice would you give to new students interested in a research experience?

> - Be persistent when looking for a mentor, but not pushy.
> - Don't be afraid to ask your mentor questions.
> - Be creative with your schedule to allow blocks of time for research.
> - Go to research group meetings; attend conferences.
> - Find a group that is doing research that interests you.

- How has your research experience prepared you for your future career? Identify the skills you have learned that will be useful in your future career and explain how.

> - Problem-solving or troubleshooting skills.
> - Critical thinking skills.
> - Writing and presentation skills.
> - The importance of doing something right, paying attention to details.
> - Knowing general career options.

- Outline at least three differences between your first few months in the research group and where you are now.

> - Didn't know anything; now feel more comfortable.
> - Felt in the way; but can contribute to the research group now.
> - Now we can design our own experiments.

- What is the most important thing you learned in your research?

Contributed by A. R. Butz with information from Branchaw, J. L., Pfund, C., and Rediske, R. (2010). *Entering Research: A Facilitator's Manual.* New York: W.H. Freeman & Co.

RESEARCH EXPERIENCE REFLECTIONS 3: RESEARCH EXPERIENCE EXIT INTERVIEW

Learning Objectives

Trainees will:

- ► Reflect on what was learned and the goals achieved during a research experience.

---

1. Has your research experience been what you expected it to be? Why or why not?

2. What academic and personal goals have you already achieved in your research experience? How do those goals compare to what you aspired to achieve at the beginning of your research experience?

3. What values, experiences, and/or perspectives have you contributed to your research group? Have you been able to contribute in ways that you did not predict? If so, how?

4. What has been the most challenging aspect of your research experience and how did you deal with it?

5. What has been the most rewarding aspect of your research experience? Why was it so rewarding?

6. Did you have any negative experiences with your research experience? If so, what did you do to address those issues?

7. What goals do you have for your research experience moving forward?

8. What advice would you give to new students interested in doing a research experience?

9. How has your research experience prepared you for your future career? Identify the skills you have learned that will be useful in your future career and explain how.

10. Outline at least two differences between research done with a research group and research done as part of a laboratory course.

11. Can you identify at least two connections between your research project and what you are learning in your classes?

Contributed by A. R. Butz with information from Branchaw, J. L., Pfund, C., and Rediske, R. (2010). *Entering Research: A Facilitator's Manual.* New York: W.H. Freeman & Co.

RESEARCH GROUP DIAGRAM

Learning Objectives

Trainees will:

- Meet all of the people in a research group and learn about their areas of responsibility and research projects.
- Articulate how research group personnel differ in education, responsibilities, and contribution to the research team.

Trainee Level:

undergraduate or graduate trainees
novice trainees

Areas of Trainee Development

- Research Comprehension and Communication Skills
  - Develop effective interpersonal communication skills.
  - Develop an understanding of the research environment.
- Professional and Career Development Skills
  - Explore and pursue a research career.

Activity Components and Estimated Time for Completion

- Trainee Pre-Assignment Time: 1–2 hours
- In Session Time: 50 minutes

Total time: 1–2 hours, 50 minutes

When to Use This Activity

This activity is useful for undergraduate or graduate novice trainees who have recently joined a research group. By completing this activity, trainees will meet the individuals in their research group and learn about their different roles and responsibilities. Other activities that may be used with this activity include:

- Your Research Group's Focus
- Finding Potential Research Rotation Groups and Mentors

Inclusion Considerations

Consider that trainees who are members of underrepresented groups in research may feel out of place or marginalized in their research groups. This may make it difficult for them to initiate the "get to know you" discussions in this activity or impact their perceptions of those discussions. Encourage trainees to reach out to their mentors for help in meeting the rest of the members of their research group.

Contributed by A. R. Butz and D. Wassarman with information from Branchaw, J. L., Pfund, C., and Rediske, R. (2010). *Entering Research: A Facilitator's Manual.* New York: W.H. Freeman & Co.

Implementation Guide

Trainee Pre-Assignment (1–2 hours)

- Prior to the discussion, distribute the trainee materials. Trainees should first meet with their mentor to learn who is in the research group and about the various roles, educational backgrounds, and responsibilities of each member. They can track this information using the "Research Group Personnel Worksheet."
- Once trainees have learned about the members of their research group from their mentor, they should personally introduce themselves to each member of the group to learn more about who they are and the research they do.
- Using the information obtained from this worksheet and their one-on-one conversations, trainees should draw a diagram that conveys how the people and projects in their research group are connected. The research group's overall area of study should be represented, and ideally encompass all parts of the diagram. Importantly, trainees should include themselves in the diagram, showing how they and their project fit in and with whom in the group they see themselves collaborating.

Workshop Session (50 minutes)

- Discussion of Research Group Structure (30 minutes)
  - The goal of this discussion is to explore the variability in research group structures and research approaches. In particular, the discussion should highlight that research experiences depend a great deal on the research group, the people in it, and how it is structured, not just the research project.
  - **What do you like best about your research project and research group?**

> - I am very excited about the research my group does and my project.
> - I feel very important and useful with all of the training that I am to undergo before work even begins.
> - I like the freedom I have to do the project in the ways that I see necessary.
> - I like that my project is a long-term one—I wanted to do research this summer, so this research provides me with a summer job.

  - **What do you find most challenging about your research project and research group?**

> - The content is a little scary to me. Learning the jargon and protocol for this project is going to be very challenging, but I see it being worth it once I'm able to be very independent.
> - I think the most challenging part about being in my research group is going to be living up to the accomplishments of the other undergraduates who have been in the research group for a few years.
> - The most challenging part is that no one checks my work, so I spend a lot of time going over things so that my entries don't invalidate the research.
> - My mentor is very quiet, and English is not her native language. I'm worried that it will be difficult to communicate with her and that it will make my project harder.

Contributed by A. R. Butz and D. Wassarman with information from Branchaw, J. L., Pfund, C., and Rediske, R. (2010). *Entering Research: A Facilitator's Manual.* New York: W.H. Freeman & Co.

- **To whom would you go if you had a problem with your immediate mentor?**

> - I would go to other graduate students in the research group or the Principal Investigator.
> - I would ask other members of my research group who have been there for a while to give me advice. I would ask them what they thought of the problem and for their advice on how to solve it.
> - I don't really know whom I would go to about my mentor. I guess really the only thing would be to directly talk to him about any problems I might be having.

► Additional questions to stimulate discussion:

- Does your research group have regularly scheduled group meetings? If so, how often and what is discussed? Are you expected to attend these meetings?
- How do members of the research group interact and communicate with one another?

► **Research Group Diagram Presentations—Research Group Personnel** (20 minutes)

- Students should compare their research group diagrams. Dividing them into smaller groups of three or four works well. Distribute one copy of the "Research Group Personnel" handout included in the trainee materials to each group.
- Have each student describe his/her diagram and explain the relationships among the members represented in it.
- When presenting their diagrams, small groups should pool what they've learned about the personnel in their groups to begin to generate a comprehensive list of the types of positions on a research team using the "Research Group Personnel" blank table provided in the trainee materials.
- Gather the small groups together to create a comprehensive "Research Group Personnel" table. The facilitator can add positions to the table that may not be represented in the trainees' groups and modify the responsibilities to be in line with their specific institution. An example table is provided below.

► **Discussion Questions**

- Does the diagram show a hierarchical structure? Where is the Principal Investigator or professor on the diagram (at the top, in the middle)? Where are the undergraduate students?
- Does the diagram reflect your access to the Principal Investigator or professor?
- How are the diagrams in each group similar? How are they different?
- Does the diagram show how each person and project in the group is related to the trainee's project?
- Who is working in the biggest group? Who is working in the smallest group?
- What does your diagram tell you about the nature of your research group?
  - A group with many postdoctoral researchers, for example, may reflect a well-funded research group, a group led by a tenured faculty member, a PI with outside commitments (e.g., an M.D./Ph.D.), etc.

► *Additional questions to promote discussion (optional).* Research group diagrams also provide the opportunity to discuss different career paths and titles. The facilitator can use this opportunity to engage trainees in a discussion about different career pathways. Potential discussion questions include:

- What does it take to complete a Ph.D.?
- What are the benefits to completing a postdoc?
- What jobs are available for individuals who complete a bachelor's or master's level degree?
- What do different titles in industry mean (e.g., Scientist I vs. Scientist II)?

► Facilitators should highlight specific ways in which their institution is different from the example table with respect to the titles, roles and responsibilities. Encourage trainees to consider how roles might differ across different institutions.

Contributed by A. R. Butz and D. Wassarman with information from Branchaw, J. L., Pfund, C., and Rediske, R. (2010). *Entering Research: A Facilitator's Manual.* New York: W.H. Freeman & Co.

RESEARCH GROUP PERSONNEL EXAMPLES

Titles	Degree	Roles and Responsibilities
Professor (Assistant, Associate, Full)	M.S. or Ph.D.	Faculty member at an academic institution, who is the leader of the research group. Typically is the PI on grants funding the group's research. Senior mentor to all personnel in the research group. In addition to research, professors also have teaching and service responsibilities.
Principal Investigator (PI)	M.S. or Ph.D.	The lead person on a grant-funded project. Responsible for oversight of all aspects of the project. Often is a professor, but may be a senior researcher or a scientist who does not have teaching or service responsibilities.
Scientist	Ph.D.	A staff person whose primary responsibility is to do research. Although scientists are usually part of a professor's research group, core research facility, or research institute, they are highly independent and often provide leadership within the research group.
Researcher	M.S. or Ph.D.	A staff person who collaborates with the PI and scientists on the research team to carry out research projects.
Instructor	M.S. or Ph.D.	A staff person dedicated to teaching lecture and lab/field courses. Instructors may also serve as academic advisors.
Post-doc	Ph.D.	Postdoctoral fellows have earned their Ph.D. and are gaining additional training to prepare for the next step in their career. Post-docs are mentored by the PI, but work independently on the research team and may provide mentorship for more novice team members. Post-docs are usually focused on research, but some positions include teaching.
Technician	B.S. or M.S.	Technicians can perform a wide variety of tasks depending on their skill set and the needs of the research team. Technicians often are experts on particular techniques and, in addition to helping with research projects, they may supervise students, order supplies, and perform administrative tasks.
Graduate Student	B.A., B.S., M.S.	Graduate students have completed a bachelor's degree and are working toward a M.S. and/or Ph.D. degree. This includes taking courses and doing research. Graduate students are mentored by the PI as they learn research techniques, experimental design, data analysis, and publication skills with the goal of becoming an independent researcher. Graduate students will develop their own research question/hypothesis to investigate and must defend their research findings to earn their advanced degree.
Undergraduate Researcher	None	Undergraduate student working with a mentor on a research project either for academic credit or a stipend. Depending on the undergraduate researcher's level of experience, they may help their mentor with a project or work independently on a project of their own.
Lab Assistant	None	Undergraduate student working on the research team, usually paid by the hour, to support the team through maintenance tasks such as washing glassware/dishes, autoclaving, sorting samples, making common lab solutions, etc.

Contributed by A. R. Butz and D. Wassarman with information from Branchaw, J. L., Pfund, C., and Rediske, R. (2010). *Entering Research: A Facilitator's Manual.* New York: W.H. Freeman & Co.

RESEARCH GROUP DIAGRAM

Learning Objectives

Trainees will:

- Meet all of the people in a research group and learn about their areas of responsibility and research projects.
- Articulate how research group personnel differ in education, responsibilities, and contribution to the research team.

---

1. Meet with your mentor and ask him/her to give you an overview of the people in your research group and their roles and responsibilities. Use the worksheet on the next page to document what you learn.
2. Introduce yourself to the members of your research group one by one and ask them to tell you a little bit about themselves and their research. Use the "Personnel in Your Research Group" table to document what you learn about each group member. If the size of your research group makes it impossible for you to talk with each group member, consult with your mentor and your research group's website to obtain the information that you need.
3. Draw a diagram to identify the people and projects in your research group. The diagram should represent how the projects are connected to one another, how the people are connected to one another, and how the projects and people are connected. The research group's overall area of study should be represented, and ideally encompass all parts of the diagram. Specifically include how you and your project fit in, and with whom in the group you see yourself collaborating.

Contributed by A. R. Butz and D. Wassarman with information from Branchaw, J. L., Pfund, C., and Rediske, R. (2010). *Entering Research: A Facilitator's Manual.* New York: W.H. Freeman & Co.

PERSONNEL IN YOUR RESEARCH GROUP

Title (e.g., professor, grad student)	**Degree**	**Responsibilities and Specific Projects**

Contributed by A. R. Butz and D. Wassarman with information from Branchaw, J. L., Pfund, C., and Rediske, R. (2010). *Entering Research: A Facilitator's Manual.* New York: W.H. Freeman & Co.

RESEARCH GROUP PERSONNEL

Create a comprehensive table with your peers.

Titles	Degree	Roles and Responsibilities

Contributed by A. R. Butz and D. Wassarman with information from Branchaw, J. L., Pfund, C., and Rediske, R. (2010). *Entering Research: A Facilitator's Manual.* New York: W.H. Freeman & Co.

RESEARCH GROUP FUNDING

Learning Objectives

Trainees will:

- Explore how research groups are funded.
- Identify where research funds come from.
- Identify who is responsible for securing funding.
- Identify what funding is available to undergraduate or graduate research trainees.
- Gain an appreciation for the amount of funding necessary to sustain research.

Trainee Level

undergraduate or graduate trainees
novice, intermediate, or advanced trainees

Areas of Trainee Development

- Professional and Career Development Skills
  - Explore and pursue a research career.
- Research Comprehension and Communication Skills
  - Develop an understanding of the research environment.
- Researcher Confidence and Independence
  - Develop independence as a researcher.

Activity Components and Estimated Time for Completion

- Trainee Pre-Assignment Time: 30 minutes
- In Session Time: 35 minutes

Total time: 1 hour, 5 minutes

When to Use This Activity

This activity is appropriate for undergraduate or graduate trainees at any level who are currently in a research group. It can be implemented in several different ways to accommodate students at all levels of experience. Other activities that may be used with this activity include:

- Science Literacy Test

Inclusion Considerations

Include a discussion of funding opportunities designed to support individuals from underrepresented groups (e.g., Ford Fellowships, HHMI Gilliam Fellowships) in pursuing research careers as examples.

Contributed by C. Barta with information from Branchaw, J. L., Pfund, C., and Rediske, R. (2010). *Entering Research: A Facilitator's Manual.* New York: W.H. Freeman & Co.

Implementation Guide

Trainee Pre-Assignment (30 minutes)

- Distribute the worksheet "Research Group Funding" at least one week before the topic will be discussed. Ask the students to complete the worksheet by either talking with their PI or a knowledgeable group member in their research lab (graduate student, postdoctoral researcher, lab manager, program director, etc.).

Workshop Session

- **Introduction** (5 minutes)
  - This discussion about research funding will help trainees understand the responsibilities of researchers to apply for and secure research funding. It will also help trainees understand how their research is funded, which will emphasize the importance of being financially conscious when conducting research.
- **Discussion** (30 minutes)
  - Have students discuss what they learned about their research group's funding in small groups of three or four. (10 minutes)
  - Bring the group together for a larger discussion. (20 minutes)
    - What similarities and/or differences regarding funding did you discover between research groups?
    - Were you surprised by any of the aspects involved in funding research?
    - What is the structure of a research grant?
    - On average, how much funding is needed to sustain a research lab? How much does this amount vary depending on the type of research? The size of the research group?
    - How much funding is needed to support your research project for 1 week? 1 month? 1 year?
    - What can you do as a researcher to make sure the funding is used appropriately (i.e., not "wasted")?
    - What funding opportunities are currently available to you as a research trainee? (campus scholarships/research awards/fellowships, NSF GRFP, McNair Scholars, etc.)
    - What skills do you need to write a successful application for scholarships, fellowships, or research grants?
    - What are some resources you can utilize to strengthen your application writing skills? (campus writing centers, taking courses in scientific writing, asking researchers to share previous successful research grants/scholarship applications, attending seminars/workshops on campus, asking to help out with writing and/or submitting grants for their research group, asking program directors for examples of successful fellowship/scholarship applications, etc.).

Optional Activity #1: Discussion: Priorities for the Office of Science and Technology. (20 minutes) This activity can be used to discuss the links between government, public perception of science and technology and research funding. Look up the budget for the Office of Science and Technology and discuss the budgeting priorities as a group.

- What areas of science and technology do you believe should receive the most funding and why?
- Do these areas align with the priorities of the Office of Science and Technology? Why or why not?
- How much of the U.S. tax dollars are currently spent on Science and Technology? How much of U.S. tax dollars should be spent on Science and Technology research? (graphicacy.com/viz/budget/2015/index.html#/-/total-2015-discretionary-budget)
- Do you feel that this is a fair amount? What implications does the yearly fiscal budget have on research that is happening around the world? In the United States? On our campus?

Contributed by C. Barta with information from Branchaw, J. L., Pfund, C., and Rediske, R. (2010). *Entering Research: A Facilitator's Manual.* New York: W.H. Freeman & Co.

- How does the media affect research funding? How does public perception of science and technology affect research funding?
- What can individuals do to affect future funding rates?

Optional Activity #2: Research Funding at Your Institution (5- to 20-minute discussion). Discuss how grants are submitted, managed, and reviewed at your institution. Also, consider discussing commonly used terms such as overhead or indirect costs, matching funds, 9-month contract, etc.

Optional Activity #3: Invited Speaker: Funding Opportunities for Students (variable time limit). Invite the speaker to your course to talk about funding opportunities for trainees. Allow several minutes for questions.

Optional Activity #4: Panel of Funded Trainees (variable time limit).The panel could consist of three to five undergraduate, graduate, or post-doc students who have successfully applied for various funding opportunities. One variation to this panel is to invite three to five faculty members to talk about their success and failures in securing funding for their research and for their undergraduate or graduate students and/or postdoctoral trainees.

Contributed by C. Barta with information from Branchaw, J. L., Pfund, C., and Rediske, R. (2010). *Entering Research: A Facilitator's Manual.* New York: W.H. Freeman & Co.

RESEARCH GROUP FUNDING

Learning Objectives

Trainees will:

- ▶ Explore how research groups are funded.
- ▶ Identify where research funds come from.
- ▶ Identify who is responsible for securing funding.
- ▶ Identify what funding is available to undergraduate or graduate research trainees.
- ▶ Gain an appreciation for the amount of funding necessary to sustain research.

Arrange a meeting with either the Principal Investigator (PI) of your research group or another knowledgeable research group member (graduate student, postdoctoral researcher, lab manager, program director, etc.) and discuss the following questions regarding research funding.

1. What types of expenses does your research group have (supplies, chemicals, instrumentation, animals, salaries, etc.)?

2. On an annual basis, approximately how much does your research group spend on these expenses?

3. How are the activities in questions 1 and 2 funded? (federal grants, private grants, the state, the university or college, etc.)

4. Who is responsible for securing research funds?

5. How much time does this/these individual(s) spend on writing grants, searching for funding, etc.?

6. Briefly describe the structure of a research grant (i.e., what sections/reports are included in a research grant, what is the average length of a grant application, etc.).

Contributed by C. Barta with information from Branchaw, J. L., Pfund, C., and Rediske, R. (2010). *Entering Research: A Facilitator's Manual.* New York: W.H. Freeman & Co.

7. What skills do you need to write a successful research grant?

8. How can one learn/gain these skills?

9. What funding opportunities are available for trainees like me? Are there any helpful resources that I should be aware of to find potential funding opportunities (i.e., scholarships, fellowships, research grants for student projects, etc.)?

10. What kinds of requirements are there for these trainee funding opportunities? Deadlines?

Contributed by C. Barta with information from Branchaw, J. L., Pfund, C., and Rediske, R. (2010). *Entering Research: A Facilitator's Manual.* New York: W.H. Freeman & Co.

RESEARCH ROTATION EVALUATION

Learning Objectives

Trainees will:

- Learn about the factors that are important to consider when selecting a thesis research mentor and group.
- Reflect on and compare research group rotation experiences.

Trainee Level

graduate trainees
novice trainees

Areas of Trainee Development

- Professional and Career Development Skills
  - Explore and pursue a research career.
- Research Comprehension and Communication Skills
  - Develop effective interpersonal communication skills.

Activity Components and Estimated Time for Completion:

- In Session 1 Time: 10 minutes
- Trainee Pre-Assignment 2 Time: 1 hour
- In Session 2 Time: 30 minutes

Total time: 1 hour, 40 minutes

When to Use This Activity

This activity should be used with new graduate students who are participating in research group rotations at the beginning of their training program. The rubric for evaluating the student research rotation experiences should be distributed before rotations begin and completed throughout the time that the rotations are occurring. Other activities that may be used with this activity include:

- Case Study: "Whatever you do, don't join our lab."
- Finding Potential Research Rotation Groups and Mentors
- Prioritizing Research Mentor Roles
- Research Experience Reflections 1: Entering Research?

Inclusion Considerations

Trainees from diverse backgrounds may want to look for markers of attention to diversity and inclusion in their rotation research teams. Encourage all trainees, especially those from underrepresented groups, to ask questions of prospective mentors and research teammates about management style and their thoughts around equity, inclusion, and diversity. Encourage them to add these as "factors to consider" in their individual rubrics.

Contributed by J. Branchaw. (2018). *Research Rotation Evaluation.*

Implementation Guide

Workshop Session 1 (10 minutes)

► **Using the Research Rotation Rubric**

- Distribute a blank electronic copy of the research rotation rubric provided in the trainee materials and tell trainees that:
  - this is a tool to help them reflect on and compare their research rotation experiences.
  - each section should be completed at the end of each rotation, when the experience is fresh in their mind.
  - when taken as a whole at the end of all of their research rotation experiences, their scores and reflections will help them decide which mentor and research group is the best fit for their thesis research.

Trainee Pre-Assignment 2 (1 hour)

► Trainees should complete the research rotation rubric as they complete each rotation. The completed rubric will be used as part of Workshop Session 2.

Workshop Session 2 (30 minutes)

► **Introduction**

- Trainees will share their completed rubrics and discuss the factors that they consider most important and the process that they're using to decide which research group to join.
- Trainees should be encouraged to discuss their experiences openly and honestly. Consequently, it is important to stress that the conversations should be kept confidential.

► **Pair and Share** (20 minutes)

- Trainees pair with a peer, preferably one who did not rotate with the same research groups, to share their rubric scores and comments.
- Each trainee should focus on asking their partner questions that will help them determine how to weigh each factor in their decision-making process and that will help them develop follow-up questions to ask potential mentors or research groups before making a decision.

► **Wrap-up** (10 minutes)

- Ask each pair to summarize the main points of their discussion, focusing on the factors they discussed and the next steps each will take in their decision-making process.

RESEARCH ROTATION EVALUATION

Learning Objectives

Trainees will:

- Learn about the factors that are important to consider when selecting a thesis research mentor and group.
- Reflect on and compare research group rotation experiences.

The rubric outlines factors that are important to consider when selecting a thesis mentor and research group. At the end of each rotation, reflect on your experience and complete a column in the rubric by assigning a score and documenting your rationale for that score in the Notes column.

To what extent does each rotation meet your needs and/or expectations?
Score: 0 = does not meet needs/expectations 1 = meets some but not all needs/expectations
2 = meets needs/expectations 3 = exceeds needs/expectations 4 = greatly exceeds needs/expectations

	Rotation Score (0–4)			
Factors to Consider	**#1**	**#2**	**#3**	**Notes**
Mentorship provided by research team leader (PI)				
Collegiality and collaborative nature of the research team members				
Degree of alignment between the potential thesis research projects and my research interests				
Degree of alignment between average time to degree for trainees in research and my own expectations for time to degree				
The type of funding available (e.g., RA, TA, PA)				
The length of funding available (1 yr., 2 yrs., etc.)				
The degree of flexibility in how many and which hours of the day I am expected to be in lab each week				
Ability to be myself in this research group without fear of alienation or negative repercussions				

Contributed by J. Branchaw. (2018). *Research Rotation Evaluation.*

	Rotation Score (0–4)			
Factors to Consider	**#1**	**#2**	**#3**	**Notes**
The size of the research group and the infrastructure for communication in the group				
Access to resources beyond the group to support my success (e.g., equipment, professional development opportunities, community)				

Contributed by J. Branchaw. (2018). *Research Rotation Evaluation.*

RESEARCH WRITING 1: BACKGROUND INFORMATION AND HYPOTHESIS OR RESEARCH QUESTION

Learning Objectives

Trainees will:

- Identify and summarize the key background information needed to explain and justify a research project.
- Articulate the knowledge gap that the research will address.
- Formulate specific hypotheses or research questions to address the gap.

Trainee Level

undergraduate or graduate trainees
novice or intermediate trainees

Areas of Trainee Development

- Research Comprehension and Communication Skills
  - Develop disciplinary knowledge.
  - Develop research communication skills.
  - Develop logical/critical thinking skills.

Activity Components and Estimated Time for Completion

- Trainee Pre-Assignment Time: 2–3 hours
- In Session Time: 1 hour, 25 minutes

Total time: 3–4 hours, 25 minutes
Estimated time for activity may vary depending upon size of class.

When to Use This Activity

This activity is most valuable for trainees who have limited experience constructing hypotheses or research questions. Other activities that may be used with this activity include:

- Searching Online Databases
- Research Articles 1: Introduction
- Research Articles 2: Guided Reading
- Research Articles 3: Practical Reading Strategies
- Research Writing 3: Project Design
- Research Writing 6: Research Proposal

Sufficient time should be reserved for discussion so that trainees can talk through their challenges. Facilitators may wish to dedicate more time or multiple sessions to this topic to allow trainees to work through multiple versions of their framing funnel.

For more advanced trainees, this activity can be introduced as an approach to writing an introduction to a conference paper or article (see "Alternative Implementation" below).

Contributed by A. R. Butz, C. Barta, and A. Sokac with information from Branchaw, J. L., Pfund, C., and Rediske, R. (2010). *Entering Research: A Facilitator's Manual.* New York: W.H. Freeman & Co.

Inclusion Considerations

Learning and reading styles will vary among trainees. Invite them to share with you any learning accommodations they need or preferences they have and be flexible when setting reading and writing assignment deadlines. Encourage them to share with the group alternative ideas about how to approach reading research papers, organizing information, and constructing reviews.

Novice trainees, especially those who have traveled nontraditional academic pathways, may have had limited opportunity to practice research writing. Reassure them that their research writing skills will improve with practice and even the most experienced writers use an iterative process of review and revision when writing. If trainees have challenges with writing (e.g., disabilities), talk about these challenges and encourage them to seek help from professionals and to share this information with their research mentor, so that he or she is aware and can help.

Contributed by A. R. Butz, C. Barta, and A. Sokac with information from Branchaw, J. L., Pfund, C., and Rediske, R. (2010). *Entering Research: A Facilitator's Manual.* New York: W.H. Freeman & Co.

Implementation Guide

Trainee Pre-Assignment (2–3 hours)

- Prior to the session, trainees should use the "framing funnel" exercise provided in the trainee materials to identify and summarize the key background information needed to understand their research project.
- The "framing funnel" presents information in the following format:

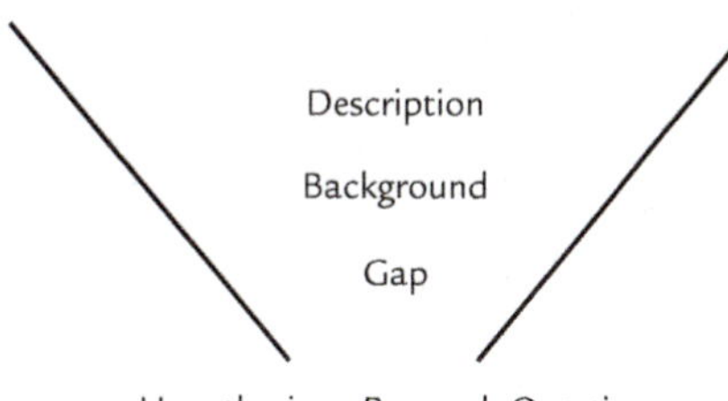

  - **Description**: In one sentence or less, describe what your research is about. What is the synopsis?
  - **Background**: What is already known? What does a reader need to know in order to understand your research?
  - **Gap**: What knowledge is still missing? What are the open questions that your research will address? Why is your research important?
- The framing funnel exercise can be simplified. Trainees can be asked to describe their research in "one word" instead of "a few words" and to use that word to start the first sentence of their presentation.

Workshop Session (1 hour 25 minutes)

- **Funnel Presentations and Discussion** (1 hour)
  - During the session, trainees present their framing funnels. Trainee presentations can occur either in small groups or in a large group. If in a large group, provide a way for trainees to project their funnel, or make copies of the funnel for everyone to follow along during the presentations. Using the board to give a "chalk talk" as their presentation is also very effective.
  - To encourage peer interaction and review, trainees should write at least one comment, question, or suggestion about each hypothesis/research question presented by each peer on an index card during the presentation. Require each trainee to verbally present their comment, question, or suggestion for at least two peers during the discussion. At the end of each presentation, collect the index cards and give them to the presenter. Alternatively, trainees can complete the peer-review rubric included in the trainee materials.
  - Discussion questions:
    - Does the hypothesis/research question "make sense" and follow from the background information?
    - Is it clear how testing this hypothesis/research question will contribute to the broader research mission of the trainee's research group?
    - What types of experiments could the trainee do to test the hypothesis or to derive the research question?
  - Trainees may find it challenging to briefly describe their work, identify what introductory/background information is critical, and articulate their hypothesis or research question. During this exercise, trainees frequently realize that their hypothesis or research question is not clearly defined. An effective follow-up exercise is to require them to revise it with their mentor and bring it back the next week.

Contributed by A. R. Butz, C. Barta, and A. Sokac with information from Branchaw, J. L., Pfund, C., and Rediske, R. (2010). *Entering Research: A Facilitator's Manual.* New York: W.H. Freeman & Co.

► **Discussion: Elements of a Good Hypothesis or Research Question** (20 minutes)

- The goal of this discussion is to capture the important features of the trainees' research project presentations to generate a list of the characteristics of a good hypothesis/research question.
- Ask trainees to reflect on the hypothesis/research question presentations just given and brainstorm a list of features of a good hypothesis/research question, either in the large or small groups. Solicit their ideas verbally and create a comprehensive list of features on the board, which they can use to determine whether and how they should revise their own hypotheses/research questions.
- Some features that trainees have identified include:

> - The hypothesis/research question follows logically from the background information.
> - The hypothesis/research question is narrowly defined and testable.
> - The hypothesis/research question addresses one aspect of a broader research investigation.

  - What is a hypothesis/research question?
  - How do you evaluate a hypothesis/research question? What makes it good or bad?
  - What is the difference between a hypothesis/research question and a research project?

► If you're planning to use this activity as part of a series leading up to the Research Project or Paper activities, you may want to revisit this discussion as students prepare their outlines. The goal would be to remind trainees of the elements of a good hypothesis/research question and to offer them the opportunity to reflect on, and possibly reconsider, their own hypothesis/research question. If trainees are presenting their outlines, this discussion can happen after the presentations for the day or be interwoven between presentations. Alternatively, facilitators can remind trainees to consider the questions as they continue to refine their outlines and hypotheses/research questions.

► **Wrap-up** (5 minutes)

- Summarize key points raised in the discussion.
- Encourage trainees to continue to refine their hypotheses/research questions in consultation with their mentors.

Alternative Implementations for Intermediate to Advanced Trainees:

Workshop Session

► **Introduction**

- Trainees with some experience developing hypotheses or research questions can benefit from this activity by using it as a way to create an outline and introduction to a conference paper or article or to develop their scientific paper critical reading skills.
- Explain that most introductions for research papers consist of the following:
  - Definitions and explanations of key concepts and ideas relevant to the research;
  - Background information/previous studies that inform the present study;
  - Identification of a knowledge gap which the paper will address;
  - Purpose of the study; and
  - Hypotheses or research questions.

Contributed by A. R. Butz, C. Barta, and A. Sokac with information from Branchaw, J. L., Pfund, C., and Rediske, R. (2010). *Entering Research: A Facilitator's Manual.* New York: W.H. Freeman & Co.

► **Alternative Assignment #1: Creating an Outline and Introduction to a Paper**

- Prior to the session, trainees use the framing funnel to create an outline of the introduction to their paper with the major headers as follows:
  - Describe the research
  - Important background information required to understand the research (i.e., "what is known?")
  - Knowledge gap that the research will address (i.e., "what is *not* known?")—Why is this research important?
  - Hypothesis/research question
  - Preview of experimental design
- Encourage trainees to use this as a guide when writing their introduction. Outlines can be presented during a class session as described in the facilitator notes above. If this session is used as part of a course or series of workshops that will result in a final proposal or paper, this activity can be used as a first step towards writing a paper or proposal. The peer-review rubric included in trainee materials can be used to collect and provide formative feedback to trainees.

► **Alternative Assignment #2: Reading Research Papers**

- Prior to the session, trainees read a well-written introduction of a published paper from their discipline and map it onto a framing funnel.
- **Discussion Questions**
  - Was the logic in the introduction easy to follow?
  - Did the introduction follow the framing funnel organization?
  - Are there stylistic items in this introduction that you might use in your own writing to construct a clear, logical introduction in your research papers?

Contributed by A. R. Butz, C. Barta, and A. Sokac with information from Branchaw, J. L., Pfund, C., and Rediske, R. (2010). *Entering Research: A Facilitator's Manual.* New York: W.H. Freeman & Co.

RESEARCH WRITING 1: BACKGROUND INFORMATION AND HYPOTHESIS OR RESEARCH QUESTION

Learning Objectives

Trainees will:

- Identify and summarize the key background information needed to explain and justify a research project.
- Articulate the knowledge gap that the research will address.
- Formulate specific hypotheses or research questions to address the gap.

A "framing funnel" can be used to describe and frame your research study. It presents information about your study in the following order:

1. **Description**: In one sentence or less, describe what your research is about. What is the synopsis?
2. **Background**: What is already known? What does a reader need to know in order to understand your research?
3. **Gap**: What knowledge is still missing? What are the open questions that your research will address? Why is your research important?
4. **Hypothesis or Research Question**

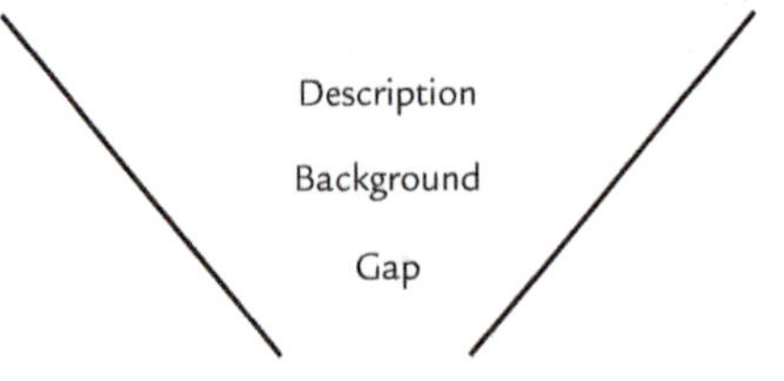

Use this worksheet to create your own framing funnel.

1. **Describe the research.** In one sentence or less, describe what your research is about. What is the synopsis?

2. **Identify and summarize the key background information needed to understand your research project.** In other words, what is already known about your topic? Outline the background information that leads to your hypothesis/ research question. Use the spaces below to write one or two sentences about each piece of information. Be sure to include the reference(s) you are citing!

   ►

   ►

Contributed by A. R. Butz, C. Barta, and A. Sokac with information from Branchaw, J. L., Pfund, C., and Rediske, R. (2010). *Entering Research: A Facilitator's Manual.* New York: W.H. Freeman & Co.

-
-
-
-

3. **Knowledge Gap.** What is not known about your topic? What are the open questions that your research will address? Why is your research important?

4. **Hypotheses/Research Question(s)**

Contributed by A. R. Butz, C. Barta, and A. Sokac with information from Branchaw, J. L., Pfund, C., and Rediske, R. (2010). *Entering Research: A Facilitator's Manual.* New York: W.H. Freeman & Co.

RESEARCH WRITING 1: BACKGROUND INFORMATION AND HYPOTHESIS OR RESEARCH QUESTION

Peer-Review Assessment Rubric

Name of the Presenter: ______________________________

Title/Topic: ______________________________

Use the scale to rate the extent to which the introduction/framing funnel met each objective. Provide comments in the space below each section.

1 = strongly disagree, 2 = disagree, 3 = somewhat agree, 4 = agree, 5 = strongly agree

1. **Describe the research:** A concise and clear summary of the research is presented. 1 2 3 4 5

2. **Key background information:** Key pieces of information about the topic reported in the literature are presented in a logical manner and support this research project. 1 2 3 4 5

3. **Knowledge Gap:** An issue about the topic that has not yet been addressed in the literature is identified for investigation. 1 2 3 4 5

4. **Hypotheses/Research Question(s):** A hypothesis or research question about the topic is posed to address the gap in the literature. 1 2 3 4 5

Overall, what are the strengths of this introduction/framing funnel?

Offer one specific suggestion for improvement of this introduction/framing funnel.

Contributed by A. R. Butz, C. Barta, and A. Sokac with information from Branchaw, J. L., Pfund, C., and Rediske, R. (2010). *Entering Research: A Facilitator's Manual.* New York: W.H. Freeman & Co.

RESEARCH WRITING 2: RESEARCH PROJECT OUTLINE AND ABSTRACT

Learning Objectives

Trainees will:

- Write an outline summarizing their research project.
- Write an abstract about their research project.

Trainee Level

undergraduate or graduate trainees
novice trainees

Areas of Trainee Development

- Practical Research Skills
  - Develop ability to design a research project.
- Research Comprehension and Communication Skills
  - Develop research communication skills.
  - Develop logical/critical thinking skills.

Activity Components and Estimated Time for Completion

- Trainee Pre-Assignment 1 Time: 1 hour
- In Session 1 Time: minimum of 1 hour, but can vary based on number of trainees
- Trainee Pre-Assignment 2 Time: 1 hour
- In Session 2 Time: 1 hour

Total Time: 4 hours, 10 minutes

When to Use This Activity

This activity is suitable for undergraduate and graduate novice researchers with limited to no experience developing outlines for research projects. Once the undergraduate or graduate students have a defined project, they may do this activity. Other activities that may be used with this activity include:

- Research Writing 1: Background Information and Hypothesis or Research Question

Inclusion Considerations

Novice trainees, especially those who have traveled nontraditional academic pathways, may have had limited opportunity to practice research writing. Reassure trainees that their research writing skills will improve with practice and even the most experienced writers use an iterative process of review and revision when writing. If trainees have challenges with writing (e.g., disabilities), talk about these challenges and encourage them to seek help from professionals and to share this information with their research mentor, so that he or she is aware and can help.

Contributed by E. Frazier with information from Branchaw, J. L., Pfund, C., and Rediske, R. (2010). *Entering Research: A Facilitator's Manual.* New York: W.H. Freeman & Co.

Implementation Guide

Trainee Pre-Assignment for Session 1 (1 hour)

- Trainees should prepare an outline of their research project and bring two hard copies to the next session. Provide a copy of the "Student Guidelines for a Research Project Outline" handout provided in the trainee materials and example outlines.

Workshop Session 1—Research Outline Presentations (minimum 1 hour)

- **Introduction**
  - This activity was designed to be implemented in two separate sessions and requires trainees to complete pre-work assignments. In the first session, trainees work on an outline of their research project. In the second session, they complete a scientific abstract. This activity requires trainees to explain terminology (jargon) that is specific to their area of research and builds confidence when they realize that they are able to accurately use this specific terminology.
- **Research Outline Presentations** (approximately 10 minutes per student)
  - Ask trainees to use their outlines to explain their research project to their peers in a 3- to 5-minute chalk talk. During the presentations, other trainees should complete a peer-review form (provided in trainee materials) and write at least one question or comment to share with the presenter. Trainees should give their forms to the presenter at the end of the session.
  - At the end of each presentation, the audience will be given 2–3 minutes to complete the peer-review form followed by time for one or two questions.
  - If there are a large number of participants, this activity can be done in small break-out groups of four or five trainees.
- **Wrap-up** (5 minutes)
  - Give trainees time to synthesize the comments from the other participants, identify the main changes they need to make to their outline, and consider the questions they were not able to answer.
  - Assign trainees to revise their outlines based on the feedback they received for the next session—OR—assign them to prepare an abstract for the next session.

Trainee Pre-Assignment for Session 2 (1 hour)

- Trainees should use their outline and peer feedback to write an abstract about their research and bring two hard copies to the next session. They should follow the instructions on the "Student Guidelines for Writing a Research Abstract" handout.

Workshop Session 2—Scientific Abstracts (1 hour)

- **Peer Review in Pairs** (30 minutes)
  - Put trainees in pairs and have them use the peer-review form provided in the trainee materials to review their peer's abstract. Spend 10 minutes to review and 5 minutes to discuss feedback. It is helpful if the facilitator indicates the halfway point in the discussion time to ensure both trainees have time to receive feedback.
  - Trainees repeat the process with a new partner.

Contributed by E. Frazier with information from Branchaw, J. L., Pfund, C., and Rediske, R. (2010). *Entering Research: A Facilitator's Manual.* New York: W.H. Freeman & Co.

► **Peer-Review Large-Group Discussion** (20 minutes)

- How did you feel when reviewing someone else's work? Was it difficult? Why?
- How did you feel when communicating your suggestions to the author? Was it difficult? Why?
- What are some lessons that you have learned from this activity that will help you with your own writing?

► **Wrap-up** (5–10 minutes)

- Ask if there are any questions, in general, about writing abstracts (e.g., depth of detail, addition of data or references).
- Ask trainees to revise their abstract using the feedback from their peers before the next class.

► **Optional Activity: Revisit Elements of a Good Hypothesis** (15 minutes)

- If you also implemented the "Research Writing 1: Background Information and Hypothesis/Research Question" activity, you may have a discussion to remind students of the elements of a good hypothesis and also to offer the opportunity to reflect on, and possibly reconsider, their own hypothesis. This discussion can happen after the presentations or the abstract review. Encourage trainees to revisit their hypotheses/research questions in light of the feedback received during their presentations to determine if they need to make additional changes to clarify their hypothesis or research question.
- **Discussion Questions**
  - What is a hypothesis/research question?
  - How do you evaluate one? What makes it good or bad?
  - What is the difference between a hypothesis, a research question, and a research project?
  - What is the significance or what are the implications of your hypothesis? Does it address a gap in the current knowledge of a field?

Contributed by E. Frazier with information from Branchaw, J. L., Pfund, C., and Rediske, R. (2010). *Entering Research: A Facilitator's Manual.* New York: W.H. Freeman & Co.

RESEARCH WRITING 2: RESEARCH PROJECT OUTLINE AND ABSTRACTS

Learning Objectives

Trainees will:

- Write an outline summarizing their research project.
- Write an abstract about their research project.

Researchers must be able to communicate their findings to colleagues in the research community. Presenting complex research questions in a clear and logical manner takes organization and practice. Over the next two sessions you will practice communicating about your research by writing a research project outline and an abstract.

Session 1: Research Project Outline

Materials:
Student Guidelines for a Research Project Outline
Example Research Project Outlines
Research Project Outline Peer Feedback Form

Pre-Work for Session 1: Create an outline of your research project using the guideline sheet and examples provided. Limit each section of your outline to one or two sentences. You will have 3–5 minutes to present your outline orally in a chalk-talk format in the next session. During the chalk talk you will have access to a whiteboard/chalkboard if you wish to write key terms or draw simple figures, so consider how you will use simple visuals in your presentation.

Session 2: Research Abstract

Materials:
Student Guidelines for Writing a Research Abstract
Examples of Research Abstracts
Research Abstract Peer-Review Form

Pre-Work for Session 2: Use the guidelines provided to write an abstract of your research project with no more than one or two sentences per topic. The abstract is a narrative of your work and should summarize your research in 200–300 words. Incorporate the feedback you received on your outline and presentation in your abstract. Print two hard copies and bring them to the session for peer review.

Contributed by E. Frazier with information from Branchaw, J. L., Pfund, C., and Rediske, R. (2010). *Entering Research: A Facilitator's Manual.* New York: W.H. Freeman & Co.

STUDENT GUIDELINES FOR WRITING A RESEARCH OUTLINE

1. Title of your research project and authorship.

2. Background of your research question: What is the gap in the knowledge that your project addresses?

3. Hypothesis or research question guiding the study.

4. Experimental design or methodology that will be used to collect data to address the hypothesis or research question.

6. Identify possible pitfalls or limitations of this study.

7. List the results (do not explain the results in this section). If you do not have results, list the expected results.

8. Conclusion/Significance: Do the results address the hypothesis? Was the hypothesis supported? Do the results address the gap in the knowledge? What is the significance?

Contributed by E. Frazier with information from Branchaw, J. L., Pfund, C., and Rediske, R. (2010). *Entering Research: A Facilitator's Manual.* New York: W.H. Freeman & Co.

RESEARCH PROJECT OUTLINE EXAMPLE

1. *Distribution and Habitat Use of the Gopher Tortoise in a Declining Southeast Florida Conservation Area.*
2. Gopher tortoise populations are declining throughout their range; therefore, it is important to study this species before they reach very low numbers that compromise their survival.
3. It is important to study the biology of species to provide information for their conservation.
4. Determine status, distribution, and habitat use of gopher tortoises in fragmented habitats.
5. Methods: burrow belt surveys, vegetation and soil analysis, tortoise counts and measurements (size, weight, etc.)
6. The information provided by this study could be useful to improve current conservation management practices for this species. (If you have results, skip this item and go to item 7.)

Results:

7. Tortoises avoid burrowing in densely vegetated areas and prefer open areas. Soil type was positively correlated with burrow density and high ratio of active to abandoned burrows and population study revealed lack of juveniles.
8. The high ratio of active to abandoned burrows indicates a healthy population, but the count of individual tortoises shows that the population includes mainly adult individuals with low recruitment of young. The utilization of burrow counts may not be an accurate way to evaluate the sustainability of gopher tortoise populations.

Outline based on article by J. T. Hindle and E. Frazier. 2012. Population Structure and Burrow of Gopherus polyphemus, in a small declining Southeast Florida Conservation Area. FAU Undergraduate Research Journal *1*(1): 23–36.

Contributed by E. Frazier with information from Branchaw, J. L., Pfund, C., and Rediske, R. (2010). *Entering Research: A Facilitator's Manual.* New York: W.H. Freeman & Co.

Research Project Outline Peer-Review Form

Presenter: ______________________________ Reviewer: ______________________________

1. Based on the project outline and oral presentation, how well do you understand this research project?
   ____ no understanding at all
   ____ weak understanding
   ____ good understanding
   ____ very good understanding
2. In your own words, write one or two sentences to describe the main focus of this research project.

3. Write at least one question or comment for the presenter.

Contributed by E. Frazier with information from Branchaw, J. L., Pfund, C., and Rediske, R. (2010). *Entering Research: A Facilitator's Manual.* New York: W.H. Freeman & Co.

STUDENT GUIDELINES FOR WRITING A RESEARCH ABSTRACT

A research abstract is a summary of your research in 200–300 words. The abstract is typically in paragraph format written in the sequence outlined below:

- Title your research project and list the authors.
- The first and second sentences introduce the larger question of your research project or gap in the knowledge.
- The third and fourth sentences provide a very short background to introduce the importance of your research question and the knowledge gap it addresses.
- The fifth sentence states the hypothesis or research question.
- The sixth and seventh sentences are a brief description of the experimental design or methodology that was used to address your hypothesis or research question.
- The eighth sentence addresses any pitfalls or predicted shortcomings of the methodology and what was done to address them.
- The ninth through eleventh sentences are a brief description of the results.
- The last sentences address whether the results support the hypothesis or address the research question, the significance of your research, and how it addresses a gap in the current knowledge in the field.

Contributed by E. Frazier with information from Branchaw, J. L., Pfund, C., and Rediske, R. (2010). *Entering Research: A Facilitator's Manual.* New York: W.H. Freeman & Co.

EXAMPLES OF RESEARCH ABSTRACTS

Example of Abstract without Results

Distribution and Habitat Use of the Gopher Tortoise in a Declining Southeast Florida Conservation Area. Joshua Scholl[1,2], Evelyn Frazier[2], and Tobin Hindle. 2012. *FAU Undergraduate Research Journal* (1): pages 23–36

Gopher tortoises have been declining throughout their range over the last few decades due mostly to urbanization, which often leads to the creation of island habitats. This confines populations and eliminates natural management by wildfires resulting in degraded island habitats. To maximize conservation efforts in rapidly developing regions it is critically important to investigate not only the natural ecology of native species, but specifically how they are affected in confined and degraded habitats. We studied a gopher tortoise population to determine its status, distribution, and habitat use in a confined, degraded ecosystem on the Florida Atlantic University campus in Boca Raton, Florida. We conducted complete burrow surveys using belt transects, directly captured tortoises, and performed vegetation and soil analyses through aerial photos and U.S. Geological Survey data, respectively. The status of the population was assessed directly based on carapace length measurements and indirectly through ratios of active to abandoned burrow categories. The information obtained in this study can be used to improved current conservation management practices for this species.

Example of Abstract with Results

Distribution and Habitat Use of the Gopher Tortoise in a Declining Southeast Florida Conservation Area. Joshua Scholl[1,2], Evelyn Frazier[2], and Tobin Hindle. 2012. *FAU Undergraduate Research Journal* (1): pages 23–36

Gopher tortoises have been declining throughout their range over the last few decades due mostly to urbanization, which often leads to the creation of island habitats. This confines populations and eliminates natural management by wildfires resulting in degraded island habitats. To maximize conservation efforts in rapidly developing regions it is critically important to investigate not only the natural ecology of native species, but specifically how they are affected in confined and degraded habitats. We studied a gopher tortoise population to determine its status, distribution, and habitat use in a confined, degraded ecosystem on the Florida Atlantic University campus in Boca Raton, Florida. We conducted complete burrow surveys using belt transects, directly captured tortoises, and performed vegetation and soil analyses through aerial photos and U.S. Geological Survey data, respectively. The status of the population was assessed directly based on carapace length measurements and indirectly through ratios of active to abandoned burrow categories. Tortoises burrowed densely in areas of low vegetation and completely avoided areas with closed canopies, which comprised about 15% of the habitat. Soil types had a significant correlation to the spatial distribution of burrows. We found a high ratio of active to abandoned burrows, which could indicate an active and healthy population; however, demographic data compiled from captured tortoises revealed a lack of juveniles, suggesting an unsustainable population.

Contributed by E. Frazier with information from Branchaw, J. L., Pfund, C., and Rediske, R. (2010). *Entering Research: A Facilitator's Manual.* New York: W.H. Freeman & Co.

Abstract Peer-Review Form

Author: ______________________ Reviewer: ______________________

CONTENT

Does the abstract include the:

1. Title and Authors

   YES SORT OF NO

   Explain:

2. Purpose of the project, context, and/or relevance

   YES SORT OF NO

   Explain:

3. Hypothesis/Research Question

   YES SORT OF NO

   Explain:

4. Research methods/approach taken to answer the research question

   YES SORT OF NO

   Explain:

5. Results (or preliminary results) and Conclusions

   YES SORT OF NO

   Explain:

6. Significance/importance of results/new understandings that come as a result of the project that the public will appreciate

   YES SORT OF NO

   Explain:

Contributed by E. Frazier with information from Branchaw, J. L., Pfund, C., and Rediske, R. (2010). *Entering Research: A Facilitator's Manual.* New York: W.H. Freeman & Co.

STYLE (The 4 C's)

Is the abstract:

1. **Complete** (covers major parts of project)

   YES SORT OF NO

   Explain:

2. **Concise** (no excess wordiness or unnecessary information without being too abbreviated)

   YES SORT OF NO

   Explain:

3. **Clear** (grammar and spelling are correct, readable, well organized, lacks jargon and overly technical terms, accessible to the target audience)

   YES SORT OF NO

   Explain:

4. **Cohesive** (sentences and information flow smoothly)

   YES SORT OF NO

   Explain:

LENGTH

Does the abstract fit within the character/word limit?

YES NO, TOO LONG NO, TOO SHORT

ANY ADDITIONAL COMMENTS:

Contributed by E. Frazier with information from Branchaw, J. L., Pfund, C., and Rediske, R. (2010). *Entering Research: A Facilitator's Manual.* New York: W.H. Freeman & Co.

RESEARCH WRITING 3: PROJECT DESIGN

Learning Objectives

Trainees will:

- Design activities or experiments to test a hypothesis or investigate a research question.

Trainee Level

undergraduate or graduate trainees
novice trainees

Areas of Trainee Development

- Practical Research Skills
  - Develop ability to design a research project.
- Research Comprehension and Communication Skills
  - Develop logical/critical communication skills.
  - Develop research communication skills.

Activity Components and Estimated Time for Completion

- Trainee Pre-Assignment Time: 1 hour
- In Session Time: 1 hour, 25 minutes

Total time: 2 hours, 25 minutes

When to Use This Activity

This activity is used to guide novice trainees in developing a sequence of experiments or activities to conduct a research project. The development of a project design should occur after a trainee has chosen a mentor and a research topic, but prior to initiation of a research project. Other activities that may be used with this activity include:

- Research Writing 1: Background Information and Hypothesis or Research Question
- Research Writing 6: Research Proposal

Inclusion Considerations

Novice trainees, especially those who have traveled nontraditional academic pathways, may have had limited opportunity to practice research writing. Reassure them that their research writing skills will improve with practice and even the most experienced writers use an iterative process of review and revision when writing. If trainees have challenges with writing (e.g., disabilities), talk about these challenges and encourage them to seek help from professionals and to share this information with their research mentor, so that he or she is aware and can help.

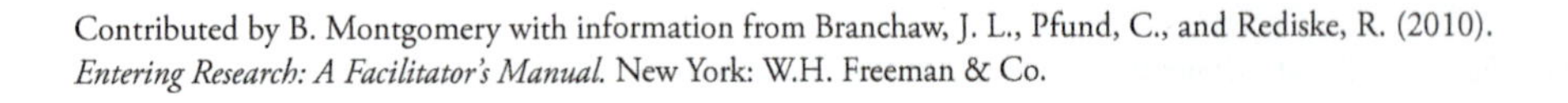
Contributed by B. Montgomery with information from Branchaw, J. L., Pfund, C., and Rediske, R. (2010). *Entering Research: A Facilitator's Manual.* New York: W.H. Freeman & Co.

Implementation Guide

Trainee Pre-Assignment (1 hour)

► Trainees complete the project design assignment and arrive prepared to present their project designs in small groups. In this assignment, trainees are guided to design the experiments or activities to test their hypothesis or investigate their research question.

Workshop Session (1 hour, 25 minutes)

► **Presentation of Project Designs** (1 hour)

- Trainee presentations can occur either in small or large groups. If in a large group, either provide a way for trainees to project their outlines or make copies of the outlines for everyone to follow during the presentations.
  - Each trainee should state his/her (revised) hypothesis or research question, and then present the project design. If the "framing funnel" activity was used prior to this activity as part of either the "Research Writing 1: Background Information and Hypothesis or Research Question" or "Article Organization, Comprehension, and Recall" activities, this may have motivated a revision of the original hypothesis or research question.
  - To encourage discussion, distribute index cards to the trainees and ask them to write at least one comment, question, or suggestion about the design during the presentation. Require each trainee to verbally present his or her comment, question, or suggestion to at least two peers during the discussion. At the end of each presentation, collect the index cards and give them to the presenter.
- **Discussion Questions**
  - Do the experiments or activities described address the hypothesis or research question that was stated?
  - Can the trainee predict potential outcomes of the experiments and explain whether each of those outcomes would support the stated hypothesis?
  - Will the data allow an assessment of the validity of the hypothesis or research question, independent of the outcomes?
  - Are the proposed experiments or activities and analysis achievable in the project timeframe?
  - Is the project design based on peer-reviewed literature? Is there precedent in the literature for the proposed design?

► **Discussion of Elements of a Good Experimental/Project Design** (20 minutes)

- The goal of this discussion is to generate a list of the characteristics of a good research design.
- Have trainees reflect on the research design presentations just given and brainstorm a list of characteristics of a good research design, either in the large or small groups. Summarize the discussion by generating a comprehensive list of characteristics, which can be used by trainees to determine whether they should revise their own research designs.
- Some characteristics that students have outlined in previous offerings include:

> - The experiments directly address the stated hypothesis.
> - Regardless of the outcome, the data will shed light on the validity of the hypothesis.
> - The experimental techniques can be completed in the allotted time frame.
> - The experimental design is based on established procedures and approaches in the literature.

Contributed by B. Montgomery with information from Branchaw, J. L., Pfund, C., and Rediske, R. (2010). *Entering Research: A Facilitator's Manual.* New York: W.H. Freeman & Co.

► **Wrap-up** (5 minutes)
- Summarize key points of the discussion with trainees.
- Encourage trainees to revisit their project design with their mentors based on the feedback that they received during their presentation.

Contributed by B. Montgomery with information from Branchaw, J. L., Pfund, C., and Rediske, R. (2010). *Entering Research: A Facilitator's Manual.* New York: W.H. Freeman & Co.

RESEARCH WRITING 3: PROJECT DESIGN

Learning Objectives

Trainees will:

- Design activities or experiments to test a hypothesis or investigate a research question.

Develop a research project design by outlining the experiments or activities (e.g., field observations) that you will conduct to test your hypothesis or investigate your research question. For each experiment or activity, address the questions below. Develop a timeline or flowchart with target dates for all experiments or activities.

For each experiment or activity explain the following:

1. What techniques will be used and what is the justification for selecting each technique? If there are control experiments, include them in your description.
2. What type of data will be collected and why is this data relevant for testing your hypothesis or investigating your research question?
3. What are the potential results of the planned experiments or activities? How would each result support (or not support) your hypothesis? Alternatively, what information would each result provide to address your research question?
4. Where possible, generate an illustration, graph, or table of the possible results.

Contributed by B. Montgomery with information from Branchaw, J. L., Pfund, C., and Rediske, R. (2010). *Entering Research: A Facilitator's Manual.* New York: W.H. Freeman & Co.

RESEARCH WRITING 4: RESEARCH LITERATURE REVIEW AND PUBLISHING PROCESS

Learning Objectives

Trainees will:

- Learn how peer-reviewed research papers are published.

Trainee Level

undergraduate or graduate trainees
intermediate or advanced trainees

Areas of Trainee Development

- Research Comprehension and Communication Skills
  - Develop disciplinary knowledge.
- Professional and Career Development Skills
  - Explore and pursue a research career.

Activity Components and Estimated Time for Completion:

- Trainee Pre-Assignment Time: 5 minutes
- In Session Time: 30 minutes

Total time: 35 minutes

When to Use This Activity

This activity is recommended for beginning graduate students, especially those in research rotations. Other activities that may be used with this activity include:

- Article Organization, Comprehension, and Recall
- Searching Online Databases
- Importance of Reading in Graduate School

Inclusion Considerations

Learning and reading styles will vary among trainees. Invite trainees to share with you any learning accommodations they need or preferences they have and be flexible when setting reading and writing assignment deadlines. Encourage them to share with the group alternative ideas about how to approach reading research papers, organizing information, and constructing reviews.

Contributed by A. Sokac. (2018). *Research Writing 4: Research Literature Review and Publishing Process.*

Implementation Guide

Trainee Pre-Assignment (5 minutes)

► Read "Journal Jargon" handout. Note any questions or points of confusion.

Workshop Session (30 minutes)

► **Journal Jargon Activity** (5 minutes)

- Based on your reading of the "Journal Jargon" handout, what questions do you have about the terms associated with the journal publishing process?
- Facilitators may wish to highlight specific journals in their discipline(s) or the disciplines of the trainees to show what impact factors look like in different research fields. Encourage trainees to talk with their mentors or with advanced trainees in their research group or lab to identify top-tier journals.

► **Peer-Reviewed Literature Activity** (25 minutes)

- Gather printouts of the author instructions from several journals, specifically the few paragraphs describing the editorial process. For an example, see the Information for Authors page on the Developmental Cell website (*http://www.cell.com/developmental-cell/authors*)
- Distribute the author instructions from several different journals (one per trainee). Allow trainees time to read the author instructions individually. (3–5 minutes)
- In the large group, ask for volunteers to describe the editorial process of their particular journal and discuss the similarities and differences of the editorial processes. (20 minutes)
    - Which processes are similar among these journals?
    - Which processes are different?
    - Has anyone published a paper? Was your experience different or similar to the editorial processes described?
    - How might differences in peer review impact the publications?

RESEARCH WRITING 4: RESEARCH LITERATURE REVIEW AND PUBLISHING PROCESS

Learning Objectives

Trainees will:

- Learn how peer-reviewed research papers are published.

Pre-Assignment Reading: Journal Jargon

Primary literature: These are journal articles that represent new research findings with data and methods included.

Peer reviewed: This describes an article or journal with content that is evaluated and refined by experienced colleagues in the field.

Impact factor: This is a "score" that is calculated for academic journals. It represents the average number of citations received per article in a given journal. The company, Thomson Reuters, calculates impact factors for a given journal as follows[1]:

A = the number of times articles published in 2013–2014 were cited in indexed journals during 2015
B = the number of articles, reviews, proceedings or notes published in 2013–2014 Impact factor 2015 = A/B

Top-tier journals: These journals are considered to be the most prestigious in the field, and typically have the highest impact factors.

Broad interest journals: Articles in these journals are meant for a broad audience. In *Nature*, *Cell*, and *Science*, these articles are often touted as "seminal" or "paradigm shifting."

Specialist journals: Articles in these journals will have a narrower audience. While it varies by journal, these articles often report more incremental advances.

Letters/Communications: These are short articles that are used to report experimental findings very quickly. These articles include limited experimental results and methods.

Reviews: These are compilations of the primary literature. Different reviews will have different agendas. Some reviews are written for the purpose of simply summarizing or giving a broad overview of the vast literature within a field of study. Other reviews are written to address some dispute or a gap within a field.

Methods articles: These are journal articles that describe a method in detail, often including a full protocol as well as troubleshooting advice. These articles are often found in "method" or "protocol" journals.

Publication costs: These are fees paid by the authors to the journal to cover publication expenses. Depending on the journal, these costs may be a flat fee or may be calculated based on the number of color figures.

Open access: These are journals or journal articles that are provided online, for free, for everyone to view, and download. The costs of open access publishing are offset by a sponsor or by fees that the author covers.

[1]http://www.sciencegateway.org/impact/

Contributed by A. Sokac. (2018). *Research Writing 4: Research Literature Review and Publishing Process.*

ACTIVITY: PEER-REVIEWED LITERATURE

Read the "For Authors" information from one journal. Outline the editorial process of this journal here and report back to the group.

Discussion Questions

- What is common among the peer-review processes of different journals?

- Are there any differences among the peer-review processes of different journals?

- How might differences in peer-review impact the publications?

- What are some examples of non-peer-reviewed reference materials?

Contributed by A. Sokac. (2018). *Research Writing 4: Research Literature Review and Publishing Process.*

RESEARCH WRITING 5: THE PEER-REVIEW PROCESS

Learning Objectives

Trainees will:

- ▶ Learn about the peer-review process in STEM.
- ▶ Categorize reviewer comments and use this framework to review peers' research.

Trainee Level

undergraduate or graduate trainees
novice or intermediate trainees

Areas of Trainee Development

- ▶ Research Ethics
  - Develop responsible and ethical research practices.
- ▶ Research Comprehension and Communication Skills
  - Develop research communication skills.
  - Develop logical/critical thinking skills.
- ▶ Professional and Career Development Skills
  - Explore and pursue a research career.
  - Develop confidence in pursuing a research career.

Activity Components and Estimated Time for Completion

- ▶ In Session Time: 1 hour

Total time: 1 hour

Alternative Implementation: This activity can be broken up into two sessions as follows:

- ▶ Session One:
  - Think-Pair-Share Activity (15 minutes)
  - Reviewer Critique Activity (20 minutes)
- ▶ Session Two:
  - Peer Review and Publishing Discussion (20 minutes)
  - Wrap-up (5 minutes)

When to Use This Activity

This activity introduces trainees to the peer-review process. To provide just-in-time support for trainees, implement this activity after trainees attempt to do a peer review. Ideally, this activity is implemented the first time that peer review is used in a workshop series and it is inserted between two rounds of review, allowing trainees to reflect on and improve their reviews. Other activities that may be used with this activity include:

- ▶ Research Writing 6: Research Proposal
- ▶ Mini-Grant Proposal

This activity defines the intellectual expectations of peer reviewers and builds confidence in trainees who may not believe that they are qualified to evaluate their peers' research. In addition, it conveys that peer review is an important part of pursuing a research career and that trainees need to develop their peer-review skills to become successful researchers. This activity can be used as part of a series of activities that address peer review, or it can be used as a stand-alone activity in a shorter workshop that trainees attend to receive

Contributed by A. Bramson with information from Branchaw, J. L., Pfund, C., and Rediske, R. (2010). *Entering Research: A Facilitator's Manual.* New York: W.H. Freeman & Co.

feedback on a proposal or paper draft. Incorporating the first part of this activity (the think-pair-share exercise) into discussions about publishing, the scientific literature, funding, and/or the writing and evaluating of research proposals is recommended.

Inclusion Considerations

Discuss the potential implications or perceptions of cross-cultural/cross-gender feedback. How might the similarities or differences between the backgrounds of the reviewer and the person whose work is being reviewed impact how criticism is intended and how it is perceived? Encourage trainees to ask questions of their mentors and other reviewers, who might be reluctant to provide honest feedback for fear of being discouraging—especially to those from historically underrepresented backgrounds. Encourage trainees to always ask, "What else would be helpful for me to know about being successful in this lab/on this project?"

Contributed by A. Bramson with information from Branchaw, J. L., Pfund, C., and Rediske, R. (2010). *Entering Research: A Facilitator's Manual.* New York: W.H. Freeman & Co.

Implementation Guide

Workshop Session (1 hour)

► **Think-Pair-Share (15 minutes).** Organize trainees into pairs or groups of three to discuss the questions presented below for about 5 minutes. Bring trainees together in a large group to share the main points of their discussions. Potential responses are provided here:

- **What is meant by "peer" in the context of peer review?**

> - Peers are typically experts in the same field as the author [i.e., of a paper or a grant proposal] and are therefore qualified to evaluate the disciplinary merits (competence, significance, originality, etc.) of the work.
> - Peers are involved in research and publishing in the same discipline and are knowledgeable on the prior work in the field, the methods used, the limits of the data, etc.
> - Sometimes researchers are called upon to review work in a field in which they are not expert. It is assumed that all researchers in a disciplinary field possess the basic knowledge and skill set to critically evaluate any research in that field.

- **How is the peer-review process used in STEM?**

> - Peer review is the main process by which the quality of research is maintained. It allows the community to establish and enforce minimal standards.
> - The major areas that peer reviewers consider are:
>   - The validity of the project (errors or problems in the data, experimental design, methodology, interpretation of results, etc.).
>   - The importance of the findings and the originality of the work (how much it advances the field).
>   - The clarity of the presentation (a reader from the same field should be able to replicate the work; the conclusions, experimental design, and hypothesis should all follow logically)
> - Researchers write grant proposals to get money to fund research and publish their research results in journals. Grant proposals and journal articles are reviewed by peers to determine whether the grant should be funded, or an article should be published.
> - Program officers at funding agencies and editors of journals are responsible for identifying qualified researchers to review grant proposals and journal articles. Authors can also suggest possible reviewers.
> - Peer review can contribute to the quality of grant proposals or journal articles by providing new interpretations of data or broader perspectives that the authors may not have considered.
> - Peer review can serve as a way to identify and address possible misconduct.

- **Why is it important for the general public to understand the role of peer review?**

> - Peer review provides a way for the general public to evaluate the validity of reported research results and conclusions. If the research results have been peer reviewed, then the public (as well as the research community) can consider them valid and trustworthy.
> - Frequently, public funds are used to support research. Therefore, it is important that the public understand how research funding decisions are made, which relies heavily on peer review, and that the process for making those decisions is transparent.

Contributed by A. Bramson with information from Branchaw, J. L., Pfund, C., and Rediske, R. (2010). *Entering Research: A Facilitator's Manual.* New York: W.H. Freeman & Co.

- Mention to trainees that reviewing can be time consuming and researchers don't get paid to do it. However, responsible researchers are expected to participate in the peer-review process and generally benefit from this participation.

► **Reviewer Critique Activity (20 minutes)** Distribute the trainee materials and introduce the categories of reviewer comments (Adapted from *Peer Review and the Acceptance of New Scientific Ideas*, https://archive.senseaboutscience.org/data/files/resources/17/peerReview.pdf).

- Categories of reviewer comments:
  - **Significance:** Are the findings original? Is the paper suitable for the subject focus of this journal? Is it sufficiently significant?
  - **Presentation:** Is the paper clear, logical, and understandable?
  - **Scholarship:** Does it take into account relevant current and past research on the topic?
  - **Evidence:** Are the methodology, data, and analyses sound? Is the statistical design and analysis appropriate? Are there sufficient data to support the conclusions?
  - **Reasoning:** Are the logic, arguments, inferences, and interpretations sound? Are there counterarguments or contrary evidence to be taken into account?
  - **Theory:** Is the theory sufficiently sound, and supported by the evidence? Is it testable? Is it preferable to competing theories?
  - **Length:** Does the article justify its length?
  - **Ethics:** In papers describing work on animals or humans, is the work covered by appropriate licensing or ethical approval? (Many biological and medical journals have their own published guidelines for such research.)
- Give trainees 5–10 minutes to read the example reviewer comments and individually identify which category(ies) each represents. Review the examples as a large group by asking trainees to read the examples aloud, share the categories that they identified, and explain why they selected those categories.

► **Presentation: Journal Article Publishing Process (10 minutes)**

- Reviewers' comments go to the journal editor, who makes a final decision on whether the article is suitable for publication.
- The editor communicates the details of the reviewers' comments to the author, along with one of the following decisions on the manuscript:
  - Paper is accepted as is. (very rare)
  - Paper is accepted with minor revisions. (editor checks that the revisions have been made after re-submission)
  - Major revisions needed. Acceptance is dependent on whether the author can satisfactorily deal with flaws identified by reviewers. (may need to go through peer review again)
  - Paper is rejected. (but author is advised to try to publish elsewhere)
  - Paper is rejected; further re-submission to same or other journals is not advised because the work is seriously flawed.

Contributed by A. Bramson with information from Branchaw, J. L., Pfund, C., and Rediske, R. (2010). *Entering Research: A Facilitator's Manual.* New York: W.H. Freeman & Co.

- Show the example flowchart below on the process an article goes through to get published (provided on the first page of trainee materials). It can be printed out and handed to trainees or projected onto a screen.

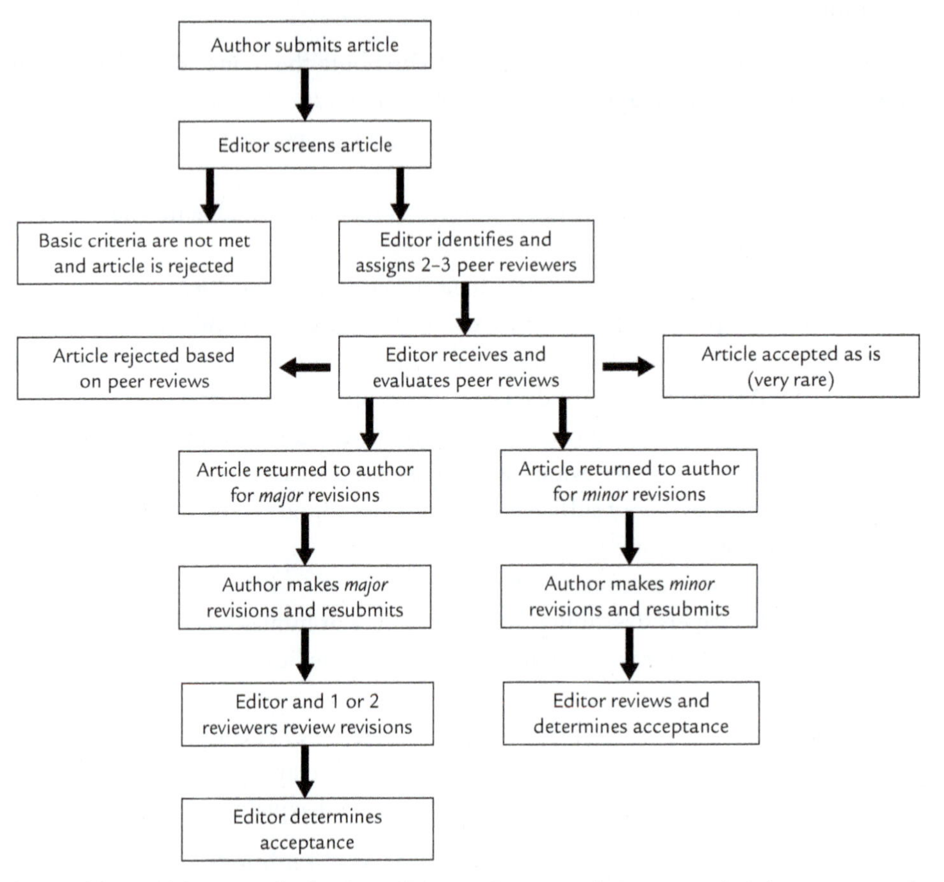

► **Discussion of Possible Problems with the Peer-Review Process.** (10 minutes) Ask trainees to brainstorm possible problems or flaws that they see with the peer-review system. Discuss how the problems or flaws they identify might be addressed. For example:

- ***Competition*** A reviewer could try to delay a competitor's paper from being accepted.
- ***Personal Prejudice*** Reviewers could reject articles or grant proposals based on personal grounds rather than flawed science, OR reviewers recommend that lower quality work be published because it supports their own work.
- ***Abuse of Privileged Information*** Reviewers are not allowed to use any part of a paper they are reviewing or to keep copies or show it to anyone else, but it happens.

How to Avoid These Problems: Reviewers are obligated to tell editors or program officers whether they have a potential conflict of interest with a paper or grant that they are asked to review. In general, the scientific community aspires to high standards of quality and ethical conduct, which discourages abuse. Editors scrutinize referees' reviews and multiple reviewers are used in order to prevent many of these problems.

► **Wrap-up** (5 minutes)

- Remind trainees that the goal of peer review is not to simply edit grammar and spelling, but rather to evaluate the validity of the research, the relevance of the findings, and the clarity of the presentation.
- Acknowledge any concerns that trainees may have about their ability to effectively contribute as a peer reviewer and reassure them that it is a skill that they will develop over time with practice. It is not necessary to understand all the details of proposals when reviewing, but trainees should at least be able to evaluate the big-picture goals and relevance of the proposal or article, if it is well written.

Contributed by A. Bramson with information from Branchaw, J. L., Pfund, C., and Rediske, R. (2010). *Entering Research: A Facilitator's Manual.* New York: W.H. Freeman & Co.

RESEARCH WRITING 5: THE PEER-REVIEW PROCESS

Learning Objectives

Trainees will:

- Learn about the peer-review process in STEM.
- Categorize reviewer comments and use this framework to review peers' research.

The Path of an Article from Submission to Publication

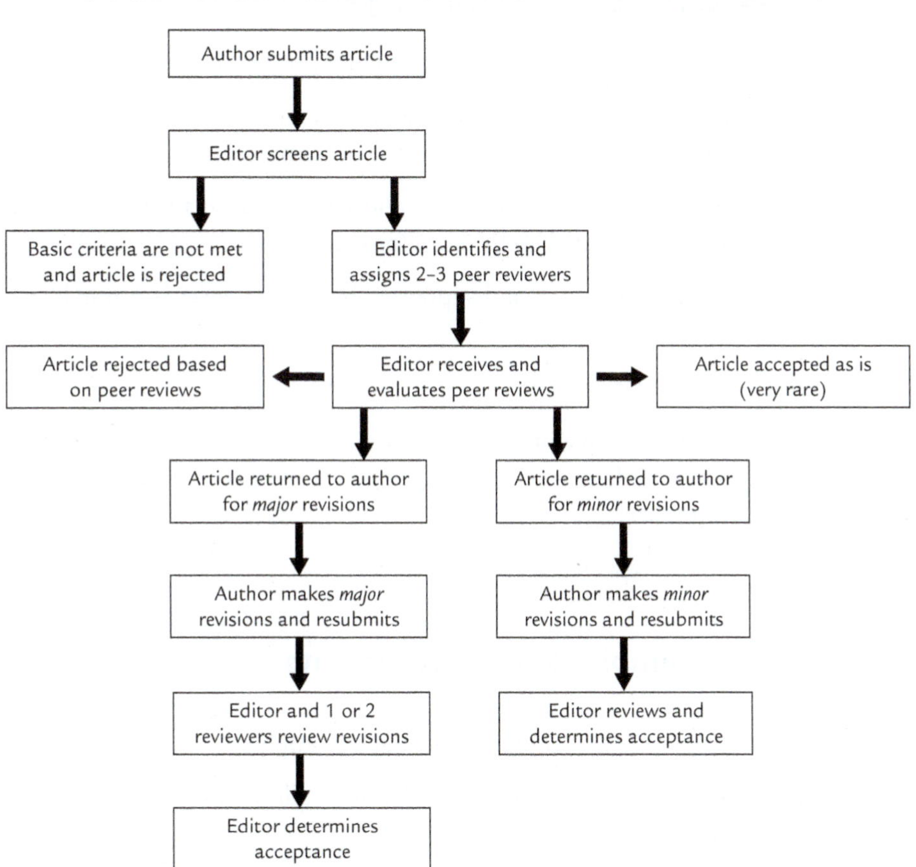

Contributed by A. Bramson with information from Branchaw, J. L., Pfund, C., and Rediske, R. (2010). *Entering Research: A Facilitator's Manual.* New York: W.H. Freeman & Co.

Trainee Materials

ACTIVITY: REVIEWER CRITIQUE

Categories of Referee Comments

(Adapted from *Peer Review and the Acceptance of New Scientific Ideas,* https://archive.senseaboutscience.org/data/files/resources/17/peerReview.pdf)

1. **Significance:** Are the findings original? Is the paper suitable for the subject focus of this journal? Is it sufficiently significant?
2. **Presentation:** Is the paper clear, logical, and understandable?
3. **Scholarship:** Does it take into account relevant current and past research on the topic?
4. **Evidence:** Are the methodology, data, and analyses sound? Is the statistical design and analysis appropriate? Are there sufficient data to support the conclusions?
5. **Reasoning:** Are the logic, arguments, inferences, and interpretations sound? Are there counter-arguments or contrary evidence to be taken into account?
6. **Theory:** Is the theory sufficiently sound, and supported by the evidence? Is it testable? Is it preferable to competing theories?
7. **Length:** Does the article justify its length?
8. **Ethics:** In papers describing work on animals or humans, is the work covered by appropriate licensing or ethical approval? (Many biological and medical journals have their own published guidelines for such research.)

Example Referee Comments

Identify the category/categories of referee comments for each example.

_______ "The experiment has not been repeated sufficiently to allow this statistical test to be used."

_______ "In general, the experiments to address these hypotheses are carefully done, but I believe the data are greatly over-interpreted and the authors neglect alternative explanations for their results as well as precedents in the literature that would provide a different model."

_______ "The explanation of the null hypothesis is not clear, and this is a very important point: what is the baseline?"

_______ "How do the authors exclude the possibility that most of the two proteins are sorted to protein storage vacuoles in endosperm by another mechanism?"

_______ "The statistical analysis, which underpins the major conclusions, is flawed. The authors state that the difference in plasma insulin levels between the experimental and control groups reached statistical significance (i.e., $p < 0.05$). However, the small print in Table 2 shows

Contributed by A. Bramson with information from Branchaw, J. L., Pfund, C., and Rediske, R. (2010). *Entering Research: A Facilitator's Manual.* New York: W.H. Freeman & Co.

that the *t*-test that they applied was one-tailed, which would be appropriate only if there were clear reasons to expect a deviation from equality in one particular direction. Since there was no such hypothesis, a two-tail test should have been used. By my calculation, this raises the probability of the observation to $p = 0.07$ and the result can rightly be described as no more than a trend. Since this finding is pivotal to the paper, I strongly advise that the authors should extend the study, presumably with a new, much larger sample, in order to test properly whether this result is secure, before publishing it."

_______ "This reagent will give poor resolution and therefore the claims are not justified."

_______ "This experimental design will not detect false positives."

_______ "A delicate subject is the data presented in Figure 4. It seems to me that I have seen these exact data before, albeit in the form of tables. If this is true (I do hope it isn't), such a work philosophy is highly unethical."

_______ "It is unacceptable that the authors do not refer to the extensive work in this field from researcher(s). Since those papers describe very similar results, they are not only relevant, but they also render the present study much less original than the authors claim. With no special reason to justify the publication of a replication, I think that the paper should not be published, certainly not in a journal for which there is much competing material."

_______ "The authors cite the earlier paper of A (2001), as the basis of their experimental design and interpretation, but do not refer to the widely accepted failures of B (2002) and C et al. (2002) to replicate those findings. This contradictory evidence should be cited, and, unless the authors can adduce a convincing argument for rejecting these contradictory results, I cannot see how their own paper can be accepted, since it rests so fundamentally on the results of A."

_______ "... the paper is extremely dense and data-rich, and is much longer than is usual for this journal. In my opinion the latter part of the results should be removed, together with Figs. 11–13. These data are rather preliminary, perhaps not as conclusive as the authors imply, and in several respects unsatisfactory."

_______ "This study is topical, highly original and technically impressive. Although the results are unexpected, and not entirely easy to interpret, I think that the paper should be published with high priority. It will have considerable impact in the field."

Contributed by A. Bramson with information from Branchaw, J. L., Pfund, C., and Rediske, R. (2010). *Entering Research: A Facilitator's Manual.* New York: W.H. Freeman & Co.

Implementation Guide

RESEARCH WRITING 6: RESEARCH PROPOSAL

Learning Objectives

Trainees will:

- Develop a logical progression of ideas.
- Develop research writing skills.

Trainee Level

undergraduate or graduate trainees
novice or intermediate trainees

Areas of Trainee Development

- Research Comprehension and Communication Skills
  - Develop research communication skills.
  - Develop logical/critical thinking skills.
  - Develop disciplinary knowledge.

Activity Components and Estimated Time for Completion:

- In Session 1: 35 minutes
- In Session 2: 50 minutes

Total time: 1 hour, 25 minutes

When to Use This Activity

To successfully implement this activity, trainees should have a research mentor, have a well-defined research question and methods, and have conducted a literature search to acquire knowledge to write the introduction and methods. Facilitators working with trainees who have no prior experience writing research proposals should ensure that trainees have had the opportunity to draft the separate components of a research proposal before using this activity. Other activities that may be used with this activity include:

- Research Writing 1: Background Information and Hypothesis or Research Question
- Research Writing 2: Research Project Outline and Abstract
- Research Writing 3: Project Design
- Research Writing 4: Research Literature Review and Publishing process
- Research Writing 5: The Peer-Review Process
- Technical Writing Tips
- Communicating Research Findings 1: Poster Presentations
- Communicating Research Findings 2: Oral Presentations

This activity is designed to generate a written research proposal. A template for the research proposal is provided in the trainee materials.

Contributed by E. Frazier with information from Branchaw, J. L., Pfund. C., and Rediske, R. (2010). *Entering Research: A Facilitator's Manual.* New York: W.H. Freeman & Co.

Inclusion Considerations

Novice trainees, especially those who have traveled nontraditional academic pathways, may have had limited opportunity to practice research writing and may be overwhelmed at the prospect of writing a research proposal. Reassure trainees that their research writing skills will improve with practice and even the most experienced writers use an iterative process of review and revision when writing. The goal of this assignment is to help them develop their research writing and critical thinking skills by integrating several earlier pieces of their writing into a comprehensive proposal. Remind trainees that they are not starting from scratch, but are refining and connecting pieces that have already been reviewed into a compelling, complete proposal.

Contributed by E. Frazier with information from Branchaw, J. L., Pfund. C., and Rediske, R. (2010). *Entering Research: A Facilitator's Manual.* New York: W.H. Freeman & Co.

Implementation Guide

Workshop Session 1: Introduction to Research Proposal Assignment (35 minutes)

- Research proposals can be used to apply for research funding and are also a requirement in most graduate training programs. In this activity, trainees compile and integrate the components of their research proposal, which they have previously drafted, into one document.

- **Components of a Research Proposal** (15 minutes)
  - Review with trainees the components of a research proposal. Facilitators may wish to provide copies of complete proposals in the appropriate format for trainees to review. Note that the template provided in the trainee materials covers the major components of proposals, but the order/formatting of these components may vary depending upon the discipline and audience for which the proposal is written.
  - Proposal instructions from federal funding agencies such as the National Science Foundation or the National Institutes of Health may be used to guide trainees. If trainees are drafting proposals for specific grants, facilitators may wish to alter the template or encourage trainees to develop a proposal that is aligned with the requirements of their particular funding agencies.

- **Assignment: Proposal Draft and Peer Review** (10 minutes)
  - Introduce the research proposal and peer-review assignments.
    - Trainees should combine drafts of their abstract, research question, introduction, literature review, project design, and reference list into a full proposal using the template provided, or one that is specific to their needs. If these components have not been drafted as part of prior activities (see "When to Use This Activity"), then facilitators should allow time either in or outside of the class session for trainees to prepare these components of the research proposal. Encourage trainees to schedule time with their mentors to review their proposal prior to sharing it with their peers.
    - Assign trainees to review draft proposals of two of their peers.
  - Trainees should share first drafts of their proposals with their assigned peer reviewers at least 48 hours before Session 2. We recommend the proposal be two or three pages in length. If the proposal will instead be in the form of a scientific poster, or an oral presentation, please see "Communicating Research Findings 1: Poster Presentations" and "Communicating Research Findings 2: Oral Presentations," respectively.
  - In addition to preparing their research proposals, trainees should complete reviews of their assigned research proposals prior to Session 2 using the rubric provided in the trainee materials.

- **Wrap-up** (5 minutes)
  - Reiterate the proposal draft deadline and make sure trainees know their reviewing assignments.

Workshop Session 2: Peer Review of Research Proposal (50 minutes)

- **Introduction** (5 minutes)
  - Debrief the proposal writing experience with trainees. What part of the proposal was most challenging for them to write and why?

- **Activity: Peer Review** (30 minutes)
  - Invite trainees to meet with their peer reviewers to go over the review comments in general, and specifically the three aspects on which they requested feedback.

Contributed by E. Frazier with information from Branchaw, J. L., Pfund. C., and Rediske, R. (2010). *Entering Research: A Facilitator's Manual.* New York: W.H. Freeman & Co.

- Reviews can take place in one session or, if time permits, you may wish to dedicate additional meeting times for trainees to receive feedback on subsequent versions of their proposals as they work towards a final draft.
- Encourage trainees to ask questions and discuss specific strategies for how to improve the draft. Trainees should leave with a list of items from the review to discuss with their mentor.

► **Discussion Questions** (5–10 minutes)

- What are the major issues you need to address in your proposal?
- How will you address these issues in your final draft? To whom, if anyone, will you go for help in revising your research proposal?

► **Wrap-up** (5 minutes)

- Outline the next steps in the revision process with your trainees. Indicate when trainees should be prepared to turn in a final draft of their research proposal. If you choose to dedicate additional meetings to peer review, develop and hand out a schedule of revision due dates and review sessions.

Contributed by E. Frazier with information from Branchaw, J. L., Pfund. C., and Rediske, R. (2010). *Entering Research: A Facilitator's Manual.* New York: W.H. Freeman & Co.

RESEARCH WRITING 6: RESEARCH PROPOSAL

Learning Objectives

Trainees will:

- ▶ Develop a logical progression of ideas.
- ▶ Develop research writing skills.

Research Proposal Format

The research proposal is a form of writing in which researchers present a potential research project to their peers for review. The future tense is used since the work being proposed will be done in the future.

For this assignment, each research proposal will be evaluated by research mentors and peers to ensure that the research question is relevant, the project is feasible and that the methodology is appropriate. The proposals may be submitted to funding agencies such as NSF or NIH to request funding for the research project and are also an integral part of graduate training programs where trainees submit a thesis research proposal to their faculty committee for approval. If the research proposal is not approved by the committee, the trainee cannot start the research project.

The exact format and sections of research proposals vary depending on the audience for whom they are written (e.g., funding agency, institution, graduate program committee). Generally, however, they include the following components:

1. **Cover page**
   a. **Title of research project**
   b. **Trainee/Researcher name**
   c. **Research advisor/mentor name**
   d. **Date**
   e. **Institution name**
2. **Project Summary/Abstract** (approximately 150–300 words). The abstract summarizes the project in a way that is easily read and understood by a disciplinarily literate reader. It should address the relevance and implications of the study.
3. **Project Narrative/Research Plan.** The project narrative includes the following components:
   a. **Introduction/Background:** The goal of the introduction section is to provide the reader with enough background information to explain why the research question is important, what is the gap in the knowledge that will be addressed, and what will be the significance or impact of the research. List at least three or four key previous findings from the literature that lay the foundation for the project.
   b. **Hypotheses or Research Question/Specific Project Aims:** This should flow logically from the background information provided. Consider the following:
      i. Can you break down this hypothesis/question/objective/specific project aim or goal into smaller questions or aims/goals? If so, what would these smaller questions or aims/goals be? Are these smaller questions independent of each other (meaning that one can exist without the other)?
      ii. Do your questions follow a logical progression of ideas?
      iii. Can you identify potential problems with your hypotheses?

Trainee Materials

Contributed by E. Frazier with information from Branchaw, J. L., Pfund. C., and Rediske, R. (2010). *Entering Research: A Facilitator's Manual.* New York: W.H. Freeman & Co.

c. **Relevance/Significance and Implications:** Why is this research important? What may be the potential implications of your results? How will it benefit basic research, human health, or development of a commercial product?
d. **Project Design/Approach:** What experiments/activities will be completed in order to test your hypotheses or explore your research questions/specific aims or goals?
e. **Expected Results:** What kind of data would support the hypothesis? What kind of data would not support it? Have you or others in your research group generated preliminary results that could be included?
f. **Acknowledgments:** Who is supporting you in this research (e.g., mentor, other research group members, funding sources)?

4. **References.** All references cited have to be of peer-reviewed publications and have to be mentioned in the introduction and/or methods section. If a reference was not cited in the introduction or methods, delete it from this section. The most common format used is American Psychological Association format (http://www.apastyle.org), but it varies by field and funding agency. Follow the instructions of the funding agency.

Assignment: Prepare a complete draft of your research proposal using the outline above. Identify at least three aspects of your proposal about which you would like peer-reviewed feedback and list them below:

1.

2.

3.

Contributed by E. Frazier with information from Branchaw, J. L., Pfund. C., and Rediske, R. (2010). *Entering Research: A Facilitator's Manual.* New York: W.H. Freeman & Co.

PEER REVIEW OF RESEARCH PROPOSAL

Assessment Rubric

Use the rubric below to do an in-depth review of two peers' proposals. Be sure to provide ***specific feedback*** on the three aspects for which the author has requested feedback.

Reviewer: ______________________ Author: ______________________

	0	1	2	3
Title and Authors	Absent	Title is lengthy and unclear.	Title is lengthy but clear.	Title is concise and clear.
Background	Absent	The background information presented lacks the content needed to understand the scientific basis of the hypothesis or research question.	The relevant background information is presented, but poorly organized. Therefore, the hypothesis or research question does not follow logically from it.	The relevant background information is presented and organized such that the hypothesis or research question follows logically from it.
Hypothesis or Research Question/ Specific Aims	Absent	A statement is made, but it is neither a hypothesis nor a research question.	A hypothesis or research question statement is made, but it is neither concise nor follows logically from the background information.	A clear and concise hypothesis or research question statement is made that follows logically from the background information.
Relevance/ Significance and Implications	Absent	The stated relevance and implications of the research are general and do not specifically address issues relevant to the disciplinary field or society at large.	The stated relevance and implications of the research address only issues relevant to the disciplinary field or society at large.	The stated relevance and implications of the research address both issues relevant to the disciplinary field and society at large.
Experimental/ Project Design	Absent	Experiments are listed but lack detail and are not connected to the stated hypothesis or research question.	Experiments are listed, and either well explained or connected to the stated hypothesis or research question, but not both.	Experiments are listed, well explained, and connected to the stated hypothesis or research question.
Expected Results	Absent	Potential results are described, but lack a figure to represent them or a statement of whether they would support the stated hypothesis or research question.	Potential results are described, but lack either a figure to represent them or a statement of whether they would support the stated hypothesis or research question.	Potential results are described in a figure and a statement about whether they would support the stated hypothesis or research question is made.
Writing	Absent	Choppy sentence fragments with grammatical and spelling errors.	Mostly clear sentences. Some work needed on clarity and flow. Some grammatical or spelling errors.	Clear. Each sentence deals with one topic and flows logically. No grammatical or spelling errors.

Contributed by E. Frazier with information from Branchaw, J. L., Pfund. C., and Rediske, R. (2010). *Entering Research: A Facilitator's Manual.* New York: W.H. Freeman & Co.

PEER-REVIEW INSTRUCTIONS

Please evaluate each component according to the rubric guidelines. Offer ***specific*** suggestions for how to improve the components.

Title and Authors	0	1	2	3
Comments/Suggestions:				
Background	0	1	2	3
Comments/Suggestions:				
Hypothesis or Research Question	0	1	2	3
Comments/Suggestions:				
Relevance and Implications	0	1	2	3
Comments/Suggestions:				
Experimental/Project Design	0	1	2	3
Comments/Suggestions:				
Expected Results	0	1	2	3
Comments/Suggestions:				
Acknowledgments	0	1	2	3
Comments/Suggestions:				
Writing (sentence structure, grammar, spelling)	0	1	2	3
Comments/Suggestions:				

Comment on the three aspects in the proposal about which the author has requested feedback.

1.

2.

3.

Contributed by E. Frazier with information from Branchaw, J. L., Pfund. C., and Rediske, R. (2010). *Entering Research: A Facilitator's Manual.* New York: W.H. Freeman & Co.

RESEARCH WRITING 7: RESEARCH PAPER

Learning Objectives

Trainees will:

- Identify the characteristics of effective research papers.
- Practice technical writing skills to communicate research.
- Recognize that research writing provides opportunities for reporting research findings.

Trainee Level

undergraduate or graduate trainees
novice or intermediate trainees

Areas of Trainee Development

- Research Comprehension and Communication Skills
  - Develop research communication skills.
  - Develop disciplinary knowledge.
  - Develop logical/critical thinking skills.

Activity Components and Estimated Time for Completion

- In Session 1 Time: 10 minutes
- In Session 2 Time: 50 minutes

Total time: 1 hour

Alternative Implementation: Weekly Review

- 30 minutes during the first session plus 10 minutes per subsequent session

When to Use This Activity

This activity should be implemented when trainees have been working in their research groups for several months and are preparing to present the results of their research. Other activities that may be used with this activity include:

- Research Articles 1: Introduction
- Research Writing 1: Background Information and Hypothesis or Research Question
- Research Writing 2: Research Project Outline and Abstract
- Research Writing 3: Project Design
- Research Writing 4: Research Literature Review and Publishing Process
- Research Writing 5: The Peer-Review Process
- Research Writing 6: Research Proposal
- Tips for Technical Writers

Inclusion Considerations

Novice trainees, especially those who have traveled nontraditional academic pathways, may have had limited opportunity to practice research writing. Reassure trainees that their research writing skills will improve with practice and even the most experienced writers use an iterative process of review and revision when writing. If trainees have challenges with writing (e.g., disabilities), talk about these challenges and encourage them to seek help from professionals and to share this information with their research mentor, so that he or she is aware and can help.

Contributed by D. Wassarman. (2018). *Research Writing 7: Research Paper.*

Implementation Guide

The writing and feedback process for the research paper is highly malleable and can be adjusted to accommodate the structure of the workshop or course. In general, trainees should have the opportunity to develop a draft, receive feedback from peers and the facilitator, and incorporate the feedback into the final paper.

Workshop Session 1: Full Paper Review (10 minutes)

- **Introduction** (10 minutes)
  - Distribute and briefly introduce the Research Paper Template in the trainee materials. Facilitators may wish to provide word limits on each section. Assign trainees to write a complete research paper using this template.
  - Pair trainees to do a peer review of one another's research papers using the Assessment Rubric. Direct them to exchange papers at least 1 week before the due date to allow enough time to complete the review.

Workshop Session 2: Full Paper Peer Review (50 minutes)

- Assign trainees to prepare a complete draft of their research paper. Trainees will exchange and review drafts using the Research Paper Assessment Rubric before the session. During the session, trainees will provide feedback to their partner. The facilitator can use the rubric to provide feedback at the individual level and to discuss common challenges with the whole group.

- **Peer Review in Pairs** (30 minutes)
  - Put trainees in pairs and allow 20 minutes for them to share their Research Paper Assessment Rubric reviews. It is helpful if the facilitator indicates the halfway point in the discussion time to ensure both trainees have time to receive feedback.

- **Peer-Review Large-Group Discussion** (20 minutes)
  - How did you feel when reviewing someone else's work? Was it difficult? Why?
  - How did you feel when communicating your suggestions to the author? Was it difficult? Why?
  - What are some lessons that you have learned from this activity that will help you with your own writing?

- Trainees may make revisions to their papers based on the reviews and the entire review process can be repeated the following week with new peer-review partners.

Alternative Implementation: Weekly Review

- **Weekly Feedback by Research Paper Section**
  - Assign one section (e.g., introduction, methods) of the paper to be completed each week. Trainees will bring a hard copy of their writing to class and receive feedback from their peers and/or facilitator.
  - **Weekly Peer Review in Pairs** (10 minutes)
    - Put trainees in pairs and have them use the Research Paper Assessment Rubric to review their peer's paper. Spend 5 minutes to review and 5 minutes to discuss feedback. It is helpful if the facilitator indicates the halfway point in the discussion time to ensure both trainees have time to receive feedback.
    - Consider changing peer-review partners each week so trainees benefit from feedback from multiple perspectives.
  - **Peer-Review Large-Group Discussion** After the first peer-review session only (20 minutes)
    - How did you feel when reviewing someone else's work? Was it difficult? Why or why not?
    - How did you feel when communicating your suggestions to the author? Was it difficult? Why or why not?
    - What are some lessons that you have learned from this activity that will help you with your own writing?
    - In each subsequent week, the students should incorporate the feedback and draft prior to the next section.

Contributed by D. Wassarman. (2018). *Research Writing 7: Research Paper.*

RESEARCH WRITING 7: RESEARCH PAPER

Learning Objectives

Trainees will:

- Identify the characteristics of effective research papers.
- Practice technical writing skills to communicate research.
- Recognize that research writing provides opportunities for reporting research findings.

Research Paper Template

1. **Manuscript Title, Authors, Affiliations.** Refer to papers from your research group for formatting of these parts.
2. **Abstract:** The paragraph should provide a succinct summary of the whole article. Begin with one or two sentences that provide information about the research problem to be investigated (ideas that are expanded upon in the INTRODUCTION), followed by one or two sentences that describe the methods used (ideas that are expanded upon in the MATERIALS AND METHODS) and the results obtained (ideas that are expanded upon in the RESULTS). End the paragraph with one or two sentences that discuss the potential implications of the results (ideas that are expanded upon in the DISCUSSION).
3. **Introduction:** The introduction should narrow down from a broad topic to the specifics of your research. Begin by introducing your topic for researchers in vastly different fields from your own (for example, for a physicist if you are a biologist), followed by sentences that narrow in on your area of interest. Then, clearly state the question or goal that you addressed. End the paragraph by describing the overall conclusions of your manuscript, emphasizing the implications and significance of your findings for a general scientific audience.
4. **Materials and Methods:** Provide a separate paragraph for each method. Refer to papers from your research group to determine the level of detail needed.
5. **Results:** Begin with a paragraph that introduces the specific research question/goal your manuscript will address, explaining any terms that may be unfamiliar to researchers in other fields. Next, expand on the materials and methods section to provide a very general overview of your experimental design, taking particular care to maintain a conversational tone, avoid extensive technical detail, and cite previous publications as necessary. Finally, describe your overall findings, providing sample sizes as appropriate, and outline any study limitations. Refrain from describing experimental techniques in detail; focus instead on what the experiment is designed to determine.
6. **Discussion:** Briefly and objectively discuss the potential implications of your findings with regard to your field at large, using language that a nonspecialist would appreciate. For example, you might focus on the specific advance that your study contributes to the field by pairing it with a single sentence that states what was previously known. Alternatively, if your study has clinical, engineering, political, economic, or social implications, state them in a sentence.
7. **References:** Look at a journal in your research area for formatting suggestions or ask your research mentor for the preferred citation style. All references should be in the same format.
8. **Figure/Table Legends:** Provide a concise and clear caption for each figure, diagram, or table. Data should be summarized in a logical format in the appropriate table or graph types. Graph axes are appropriately labeled, and scales and captions are informative and complete.

Contributed by D. Wassarman. (2018). *Research Writing 7: Research Paper.*

PEER REVIEW OF RESEARCH PAPER

Assessment Rubric

	0	1	2	3
Title, Authors, Affiliations	Absent	Title is lengthy and unclear; authors and/or affiliations are not included.	Title is lengthy but clear; authors and affiliations may be included.	Title is concise and clear; authors and affiliations are included and clear.
Abstract	Absent	Does not summarize the research and results. Provides only basic background information.	Summarizes research and results, but provides too much information, is not concise, or uses jargon without definitions.	Concise descriptive summary of research and results; provides only relevant information with limited jargon.
Introduction	Absent	The background information presented lacks the content and support needed to understand the scientific basis of the hypothesis or research question and the gap of knowledge the research seeks to address is not apparent.	The relevant background information is presented, but poorly supported by peer-reviewed literature and/or organized such that the hypothesis or research question does not follow logically from the background. The gap of knowledge the research seeks to address may be vague and not explicitly stated.	The relevant background information is presented, supported by primary and secondary research papers, and is organized such that the hypothesis or research question follows logically from it. The gap of knowledge the research seeks to address is readily apparent to the reader.
Materials and Methods	Absent	The materials and methods used for experiments are described in little detail. The procedures are not outlined and/or are difficult to follow.	Most of the materials and the setup used in the experiment are accurately described. Some ambiguity in procedures still remains OR too much unnecessary detail is included.	All materials and methods used for experiments presented are described in the paper. All steps are written in a precise and concise manner such that the experiments can be repeated by another laboratory.
Results	Absent	Results are stated, though too disorganized or poorly recorded to make sense of.	Data are summarized and presented in a mostly organized and logical format. Figure tags are absent or are not used appropriately.	Data are summarized in a logical format and organized so it is easy for the reader to see trends. Figures are referenced in the text using figure tags (e.g., Figure 1) when appropriate.

Contributed by D. Wassarman. (2018). *Research Writing 7: Research Paper.*

	0	1	2	3
Discussion	Absent	Implications are not discussed. Paper ends abruptly and lacks an overall conclusion.	Experiments are listed, but implications are discussed only in vague terms and are not connected to the stated hypothesis or research question. Paper ends abruptly and lacks an overall conclusion.	A logical chain of reasoning from hypothesis to data to conclusions is clearly and persuasively explained. Conflicting data, if present, are adequately addressed. Alternative explanations are considered. Limitations of the data and/or experimental design and corresponding implications for data interpretation are discussed. Paper gives a clear indication of the significance and direction of the research in the future.
References	Absent	Does not meet minimum number of peer-reviewed references OR references are not cited correctly in the body of the text and at the end of the document using appropriate formatting guidelines.	Peer-reviewed references are listed, but not cited in text, or there are references cited in the text that are not listed.	Minimum number of peer-reviewed references are included and cited correctly in the body of the text and at the end of the document using appropriate formatting guidelines.
Figures	Absent	Potential results are described, but lack either a figure or table to represent them, OR captions are not complete.	Data are summarized in a logical format, but either does not display all of the data and/or the table or graph types are inappropriate. Graph axes may or may not be appropriately labeled and scaled. Captions may be incomplete.	Clear, accurate diagrams are included and make the experiment easier to understand. Diagrams are labeled neatly and accurately. Data are summarized in a logical format. Table or graph types are appropriate. Graph axes are appropriately labeled and scales and captions are informative and complete.
Writing	Absent	Writing is disjointed and does not flow from one idea to the next. Many grammatical and spelling errors.	Mostly clear sentences. Some work needed on clarity and flow. Few grammatical or spelling errors.	Clear. Each sentence deals with one topic and flows logically. No grammatical or spelling errors. Writing quality, grammar, word usage, and organization facilitate the reader's understanding of the paper.

Contributed by D. Wassarman. (2018). *Research Writing 7: Research Paper.*

CASE STUDY: RESPONDING TO FEEDBACK

Learning Objectives

Trainees will:

- Understand the role of constructive feedback.
- Communicate effectively across diverse backgrounds and cultures.

Trainee Level

undergraduate or graduate trainees
novice or intermediate trainees

Activity Components and Estimated Time for Completion

- In Session Time: 20 minutes

Total time: 20 minutes

When to Use This Activity

This activity is best suited for trainees who have given or will soon give an oral presentation of their research. Other activities that maybe used with this activity include:

- The Power of Social Persuasion
- Messages Sent and Received

Inclusion Considerations

Discuss the cross-cultural/cross-gender implications of feedback by asking trainees to consider how an individual from a similar background of the person providing criticism might process the feedback compared to someone from a different background. Encourage trainees to ask questions of their mentors and others who might be reluctant to provide honest feedback for fear of discouraging them—especially those from historically underrepresented backgrounds, "What else would be helpful for me to know about being successful in this research group/on this project?"

Areas of Trainee Development

- Research Comprehension and Communication Skills
  - Develop effective interpersonal communication skills.
- Equity and Inclusion Awareness and Skills
  - Develop skills to deal with personal differences in the research environment.
- Professional and Career Development Skills
  - Develop confidence in pursuing a research career.

Contributed by C. Pfund and A. R. Butz with information from Pfund, C., Brace, C., Branchaw, J., Handelsman, J., Masters, K., and Nanney, L. (2012). *Mentor Training for Biomedical Researchers.* New York: W.H. Freeman & Co.

Implementation Guide

Workshop Session (20 minutes)

- Distribute the case study and have someone read it aloud or have trainees read the case individually. (2–3 minutes)
- Alternative: Facilitators may have the trainees role-play this case study.

► **Discussion** (15 minutes)

- Invite the trainees to discuss the questions posed at the end of the case study in small groups or as a large group. Facilitators may want to record the ideas generated in this discussion on a whiteboard or flipchart.
- **Additional Discussion Questions**
  - What are the characteristics of positive, negative, and constructive feedback? Can you give examples? How can constructive feedback improve your performance?
  - Should your mentor try to achieve a balance between positive and negative feedback? What reasons might result in you having difficulty receiving negative feedback? How can you uncover these reasons and address them?
  - What are the characteristics of good communication? What does it look like? Does it change depending on audience? Don't forget nonverbal communication.
  - How can you tell if you heard a comment the way it was intended to be heard?
  - Discuss the role of trust in giving and receiving feedback.
  - What additional things would the trainee have to consider or do if they were using a sign or language interpreter during the presentation?
  - What if English were the mentor's second language and speaking fluently was a challenge? What if English were the trainee's second language and speaking fluently was a challenge? Would you handle the situation differently?
  - Does a difference in gender affect communication in this case?
  - How does power play into this interaction? Would the feedback be received differently if it were coming from a peer instead of your mentor?

Contributed by C. Pfund and A. R. Butz with information from Pfund, C., Brace, C., Branchaw, J., Handelsman, J., Masters, K., and Nanney, L. (2012). *Mentor Training for Biomedical Researchers.* New York: W.H. Freeman & Co.

CASE STUDY: RESPONDING TO FEEDBACK

Learning Objectives

Trainees will:

- ▶ Understand the role of constructive feedback.
- ▶ Communicate effectively across diverse backgrounds and cultures.

As she leaves the crowded conference room, Jessica's mentor, Dr. Tariq, says that he'll see her in a few minutes in his office. Though she has been with the research team for over a year, she tries to avoid speaking one-on-one with Dr. Tariq. She has a hard time understanding his accent and is anxious about meeting in his office to discuss her talk without someone else from the research group present. What will she do if she misunderstands what he is saying?

When she arrives, he meets her gaze, smiles, and says with a heavy accent "Thanks for coming by. I wanted to make sure we could review your talk since the conference is in a week and I'm out of town next week." Beyond her anxiety about Dr. Tariq's accent, Jessica knows she needs to practice her talk many more times. However, they scheduled the lab practice talk early, so Dr. Tariq could attend. She is very nervous about her performance and stares at him without comment.

Dr. Tariq says, "As you know, I think this research is really important. You'll be representing the research team when you give this presentation, so we need to be sure that you're ready to do a good job." Jessica nods slightly, and shifts in her seat, the weight of her responsibility as a representative of the lab sinking in.

"I think there are a few things that could tighten your presentation." *Here it comes…* She braces herself and stares at Dr. Tariq. "For example, you had some long sentences, and even whole paragraphs on your slides. You need to shorten those slides." Jessica looks at the floor, wondering how she can shorten the slides without losing information? How will she know what to say if she shortens them? This is her first research presentation and she needs the notes. She doesn't want Dr. Tariq to think that she doesn't understand her research or won't be able to represent the lab, but she is very nervous and fears she may forget things that are not on the slides.

Dr. Tariq continues, "To cut back on the time, I think you could cut the four slides on the background and just briefly summarize those." She worked several days on those slides and now she is supposed to just cut them!? Jessica begins to panic.

"What do you think?" says Dr. Tariq. She wants to say everything that is running through her mind and tell him how hard she worked on the background slides and that she is really scared to give this presentation, but instead says "I can look at it" and keeps her face expressionless.

There is a long moment of silence. "Do you have any questions for me?" says Dr. Tariq. "No, not right now." says Jessica. "Ok then, well, good luck!" he says.

She smiles feebly and says, "Thanks." At least she understood what he was saying this time.

Questions to Consider

1. What are the main issues raised in this case study?

2. What could Jessica have done to engage with Dr. Tariq differently?

3. How could this situation have been handled differently by Dr. Tariq?

4. What should Jessica do now?

5. As a trainee, how can you tell if you heard your mentor's comments the way they were intended to be heard?

Contributed by C. Pfund and A. R. Butz with information from Pfund, C., Brace, C., Branchaw, J., Handelsman, J., Masters, K., and Nanney, L. (2012). *Mentor Training for Biomedical Researchers.* New York: W.H. Freeman & Co.

SAFETY TRAINING CHECKLIST

Learning Objectives

Trainees will:

- Become familiar with the measures necessary to safely engage in research in a discipline and participate in formal safety training as required.

Trainee Level:

undergraduate or graduate trainees
novice, intermediate, or advanced trainees

Areas of Trainee Development

- Research Ethics
  - Develop responsible and ethical research practices.
- Practical Research Skills
  - Develop ability to conduct a research project.

Activity Components and Estimated Time for Completion

- In Session Time: 5 minutes
- Trainee Pre-Assignment Time: 15 minutes for worksheet; time to complete training will vary based upon the requirements of each research lab and institution

Total time: varies

When to Use This Activity

This activity should be implemented after trainees have selected a research group, but prior to beginning research. It is most useful for novice undergraduate and graduate trainees; however, it can be implemented with all trainees regardless of prior research experience.

Inclusion Considerations

Consider learning styles, differences, and disabilities when discussing safety training. Ask whether trainees have concerns about the accessibility of the training materials and/or courses they need to complete. Some may need accommodations to complete their required training, but not know how to make the request.

Contributed by J. Svaren. (2018). *Safety Training Checklist.*

Implementation Guide

Workshop Session (5 minutes to introduce assignment)

► **Trainee Assignment**

- Trainees should discuss with their mentor the safety or research training sessions they need to complete prior to beginning research. Trainees can use the checklist included in the trainee materials to guide their conversation and keep track of training sessions they have completed.

Contributed by J. Svaren. (2018). *Safety Training Checklist.*

SAFETY TRAINING CHECKLIST

Learning Objectives

Trainees will:

- Become familiar with the measures necessary to safely engage in research in a discipline and participate in formal safety training as required.

Work in many research environments requires some type of standardized safety training that the research organization (e.g., university or company) provides. The type of training required will depend upon the type of research that you do. It is the responsibility of the research group to provide specific training in various aspects of safety (yours and the groups') and in ethical conduct. Ultimately, however, it is your responsibility to take care of your own personal safety. Thus, it is imperative that you complete all required safety training sessions and ask about any risks and/or precautions associated with your research that you may be unsure of.

Your research organization may provide learning modules (either online or in-person) that are required to engage in certain types of research. Ask your research mentor if you will be doing research in any of the following areas, and if you should participate in any safety training prior to engaging in research. Indicate the training sessions that you need and record the date each training session is successfully completed in the table provided on the next page. If you are required to complete a training that is not listed, use the blank spaces provided to list the training and when it was completed.

Ask your mentor the following questions:

- Is any documentation needed to confirm completion of training (e.g., certificates of completion)? *If so, make sure to keep a copy for your records.*
- Do you need to become recertified in any of the training sessions during your research experience? If so, how frequently will you need to redo the training?

Contributed by J. Svaren. (2018). *Safety Training Checklist.*

SAFETY TRAINING CHECKLIST

Safety Training Topic	Required? (Yes/No)	Date Scheduled	Date Completed
Molecular biology (includes working with viruses, bacteria)			
Human or primate samples (working with tissues, such as blood and culture of human cell lines)			
Working with chemicals			
Working with radioactive materials			
Working with hazardous materials			
Working with research animals (mice, rats, etc.)			
Working with sharps (e.g., syringes, razor blades, scalpels, tools, saws, knives)			
Human Subjects Research (e.g., CITI/IRB Training)			
HIPAA/Protected Health Information Training (e.g., medical records)			
Fire safety			
Safety training specific to your research lab (storage of chemicals/reagents, cleanliness policies, location of safety showers, etc.). List here or in space provided below:			
International Trade in Arms Regulations (ITAR)			
Working with cryogens			
Shop safety			
Safe lifting practices			
Electrical safety			
Electrostatic discharge (ESD prevention)			
Driving safety and registration			
Working in a potentially asphyxiating environment			
Vacuum training			
Laser training			
Transporting materials			
Other Required Training:			
Other Required Training:			
Other Required Training:			
Other Required Training:			

Contributed by J. Svaren. (2018). *Safety Training Checklist.*

Implementation Guide

SCIENCE AND SOCIETY

Learning Objectives

Trainees will:

- Learn how science can be perceived differently by the scientific community and the general public.
- Become aware of how science and society interact.
- Consider the social implications of research.
- Recognize the responsibility to communicate research to the general public.
- Develop strategies to translate research to the general public.

Trainee Level

undergraduate or graduate trainees
novice or intermediate trainees

Areas of Trainee Development

- Research Comprehension and Communication Skills
  - Develop research communication skills.

Activity Components and Estimated Time for Completion

- Trainee Pre-Assignment Time: 20 minutes
- In Session Time: 35 minutes

Total time: 1 hour, 5 minutes

When to Use This Activity

This activity is suitable for trainees at any career stage, but may be most effective when used with trainees who have limited experience communicating their science to a broad audience (i.e., novice to intermediate trainees). Other activities that may be used with this activity include:

- Science Literacy Test
- Discussion of the Nature of Science
- Science or Pseudoscience
- What Happens to Research Results?

Inclusion Considerations

Trainees from nonscience backgrounds may be particularly adept at connecting science and society, especially if they are the first in their family to pursue a career in research. Encourage them to share how they became interested in research, including where and how they learned about it.

Contributed by L. Adams and A. R. Butz with information from Branchaw, J. L., Pfund, C., and Rediske, R. (2010). *Entering Research: A Facilitator's Manual.* New York: W.H. Freeman & Company.

Implementation Guide

Trainee Pre-Assignment (20 minutes)

- Have trainees read a short article on communication in science prior to the session. Suggested readings:
  - Hendrix, M. J. C., and Campbell, P. W. (2001). Communicating science: From the laboratory bench to the breakfast table. *Anat. Rec., 265*: 165–167. doi:10.1002/ar.1150
  - Bearzi, M. (2013, October 11). *5 Simple tips for communicating science.* Retrieved from http://voices.nationalgeographic.com/2013/10/11/5-simple-tips-for-communicating-science/
- Distribute the handout included in the trainee materials and have them consider their responses to each of the questions prior to the session. Explain that the discussion in this session will focus on their answers to each of these questions.

Workshop Session (35 minutes)

- **Activity:** Why is it important for the general public to understand your research? (10 minutes)
  - Show video: "Shrimp on treadmills? Or your benefits?" (AARP): http://blog.aarp.org/2011/06/16/shrimp-on-treadmills-or-your-benefits/
  - Explain that the benefits of some studies may not be immediately apparent or seen as relevant to the general public. If studies are taken out of context, they may be seen as not important or a waste of money.
  - The critique offered in this video shows differences in how the scientific community and the public may perceive research. Time permitting, facilitators may wish to pull up research from the Pew Research Center on differences in views between the public and scientists from the American Association for the Advancement of Science (AAAS) on questions about scientific research that have arisen over the past few years:
    - Major Gaps Between the Public, Scientists on Key Issues (Pew Research Center): http://www.pewinternet.org/interactives/public-scientists-opinion-gap/
  - Trainees may say that the public does not need to know about their research because it is so basic. If this comes up in your session, encourage them to take a different perspective and try to make a connection between the research that they are doing and the general public.
  - *Note:* Facilitators may wish to use this session to talk about federally funded research and how studies are funded. Federally funded research is administered through agencies like the National Science Foundation (NSF) and the National Institutes of Health (NIH) and paid for with taxpayer dollars. Congress sets the budget for these agencies, who then award grants based on a peer-review process (i.e., other scientists determine whose research gets funded). The studies referenced in the video are taken from the report *NSF: Under the Microscope*, by Senator Tom Coburn in 2011 (see http://lcweb2.loc.gov/service/gdc/coburn/2014500020.pdf for the full report). For more activities relating to research funding, see the activity "Research Funding."
- **Discussion Questions** (20 minutes)
  - What should people know about science?

> - Basic science concepts
> - The process of research

Contributed by L. Adams and A. R. Butz with information from Branchaw, J. L., Pfund, C., and Rediske, R. (2010). *Entering Research: A Facilitator's Manual.* New York: W.H. Freeman & Company.

- Why is it important for the general public to understand your research?

> - The research may be funded with taxpayer dollars.
> - The specific relevance will vary by research topic, but each trainee should be able to answer this question.

- How will you make your research presentation accessible to the general public?

> - I will avoid using scientific jargon.
> - I will use images and drawings whenever I can to explain my research.
> - I will focus on the "BIG" research question and not the specific details of my experiments.

Contributed by L. Adams and A. R. Butz with information from Branchaw, J. L., Pfund, C., and Rediske, R. (2010). *Entering Research: A Facilitator's Manual.* New York: W.H. Freeman & Company.

SCIENCE AND SOCIETY

Learning Objectives

Trainees will:

- Learn how science can be perceived differently by the scientific community and the general public
- Become aware of how science and society interact.
- Consider the social implications of research.
- Recognize the responsibility to communicate research to the general public.
- Develop strategies to translate research to the general public.

Consider the research that you are doing from the perspective of the general public. Based on the readings that you completed for today's session, consider the following questions:

1. What should people know about science?

2. Why is it important for the general public to understand your research?

3. How will you make your research presentation accessible to the general public?

Contributed by L. Adams and A. R. Butz with information from Branchaw, J. L., Pfund, C., and Rediske, R. (2010). *Entering Research: A Facilitator's Manual.* New York: W.H. Freeman & Company.

SCIENCE LITERACY TEST

Learning Objectives

Trainees will:

- ► Articulate what it means to be scientifically literate.
- ► Reflect on individual levels of scientific literacy.

Trainee Level

undergraduate or graduate trainees
novice trainees

Areas of Trainee Development

- ► Research Comprehension and Communication Skills
  - Develop research communication skills.
  - Develop disciplinary knowledge.

Activity Components and Estimated Time for Completion

- ► In Session Time: 1 hour

Total time: 1 hour

When to Use This Activity

This activity is best suited for novice trainees who are developing an understanding of what is considered "scientific research." Other activities that may be used with this activity include:

- ► Science and Society
- ► Discussion of the Nature of Science
- ► Science or Pseudoscience?
- ► What Happens to Research Results?

Inclusion Considerations

Discuss how science literacy not only allows people to understand and value the results of research, but can empower them to make informed decisions about issues that impact their lives (e.g., health, environment, technology). Invite trainees to consider how the research they are doing will positively impact people's lives in their own communities. What role can they play as a researcher in their community to ensure these positive impacts are realized?

Contributed by L. Adams and A. R. Butz with information from Branchaw, J. L., Pfund, C., and Rediske, R. (2010). *Entering Research: A Facilitator's Manual.* New York: W.H. Freeman & Co.

Implementation Guide

Workshop Session (1 hour)

► **Discussion** (10 minutes)

- What is scientific literacy?
- Does the public need to be scientifically literate?

► **Activity: Scientific Literacy Test** (25 minutes)

- Background: The Scientific Literacy Test items were taken and adapted from the National Science Board report, *Science and Engineering Indicators* (2006). These items were used as part of a survey that was administered to adults all over the world to assess and compare science literacy and to learn more about what people know about science and from where they obtain scientific information. Results from this survey appear in Chapter 7 (Science and Technology: Public Attitudes and Understanding) of this report and can be accessed at https://www.nsf.gov/statistics/seind/. "Science and Technology (S&T) questions asked in the biennial General Social Survey (GSS) are a major source of data for this chapter. The GSS is a high-quality, nationally representative data source on attitudes and behavior of the U.S. population. Questions about S&T information, knowledge, and attitudes have been included in the GSS since 2006 and have formed the basis of this chapter in *Science and Engineering Indicators* since 2008. The GSS collects data primarily through in-person interviews. Comparable survey data collected between 1982 and 2004 used telephone interviewing; prior to 1982, these data were collected via in-person interviews. Changes in data collection methods over these years, particularly prior to 2006, may affect comparisons over time." Survey description from Chapter 7 (https://www.nsf.gov/statistics/seind14/index.cfm/chapter-7/c7i.htm)
- Distribute and have trainees complete the science literacy test (a key for this test appears at the end of the facilitator guide). (10 minutes)
- In pairs, have trainees discuss any questions for which they have differing answers and try to come to an agreement on the "right" answers. (10 minutes)
- After the trainees discuss, show the answer key and talk through any items that are confusing. (5 minutes)

Alternative Implementation: The questions presented on the Science Literacy Test may be split up among small groups of trainees for discussion. Groups discuss their assigned question for 10 minutes, then present highlights of their discussion to the larger group before proceeding to the discussion questions below.

► **Discussion Questions** (20 minutes)

- Why might a person give the wrong answer?
- Does this test accurately assess scientific literacy?
- Is it more important for the general public to know scientific facts/content, or for them to understand the scientific process? Why?
- What questions would you ask on a scientific literacy test?

Optional activity: Short Film Discussion (20 minutes) Participants can be introduced to the value of science literacy by viewing and discussing a short film, such as "A Private Universe," that was made to highlight science literacy issues. This film can be viewed by participants prior to the session or can be incorporated into the session.

Contributed by L. Adams and A. R. Butz with information from Branchaw, J. L., Pfund, C., and Rediske, R. (2010). *Entering Research: A Facilitator's Manual.* New York: W.H. Freeman & Co.

Answer Key for Science Literacy Test

1. The center of the Earth is very hot.	**TRUE**
2. All radioactivity is man-made.	**FALSE**
3. It is the father's gene that decides whether the baby is a boy or a girl.	**TRUE**
4. Lasers work by focusing sound waves.	**FALSE**
5. Electrons are smaller than atoms.	**TRUE**
6. Antibiotics kill viruses as well as bacteria.	**FALSE**
7. The universe began with a huge explosion.	**TRUE**
8. The continents have been moving their locations for millions of years and will continue to move.	**TRUE**
9. Human beings have developed from earlier species of animals.	**TRUE**
10. The sun goes around the Earth.	**FALSE**

Contributed by L. Adams and A. R. Butz with information from Branchaw, J. L., Pfund, C., and Rediske, R. (2010). *Entering Research: A Facilitator's Manual.* New York: W.H. Freeman & Co.

SCIENCE LITERACY TEST

Learning Objectives

Trainees will:

- ► Articulate what it means to be scientifically literate.
- ► Reflect on individual levels of science literacy.

Scientific Literacy Test

True/False Questions

1. The center of the Earth is very hot.	TRUE	FALSE
2. All radioactivity is man-made.	TRUE	FALSE
3. It is the father's gene that decides whether the baby is a boy or a girl.	TRUE	FALSE
4. Lasers work by focusing sound waves.	TRUE	FALSE
5. Electrons are smaller than atoms.	TRUE	FALSE
6. Antibiotics kill viruses as well as bacteria.	TRUE	FALSE
7. The universe began with a huge explosion.	TRUE	FALSE
8. The continents have been moving their locations for millions of years and will continue to move.	TRUE	FALSE
9. Human beings have developed from earlier species of animals.	TRUE	FALSE
10. The sun goes around the Earth.	TRUE	FALSE

Note: This survey is part of the General Society Survey (http://gss.norc.org/) that is administered every two years. The current data regarding the performance of different groups in answering these questions are available on the National Science Foundation's Science and Engineering Indicators website: https://www.nsf.gov/statistics/seind/

Short Answer Questions

1. Some articles refer to the results of a scientific study. When you read or hear the term *scientific study*, do you have a clear understanding of what it means, a general sense of what it means, or little understanding of what it means? In your own words, describe what it means to study something scientifically.

Contributed by L. Adams and A. R. Butz with information from Branchaw, J. L., Pfund, C., and Rediske, R. (2010). *Entering Research: A Facilitator's Manual.* New York: W.H. Freeman & Co.

2. Two scientists want to know if a certain drug is effective in treating high blood pressure. The first scientist wants to give the drug to 1,000 people with high blood pressure and see how many experience lower blood pressure levels. The second scientist wants to give the drug to 500 people with high blood pressure and not give the drug to another 500 people with high blood pressure and see how many in both groups experience lower blood pressure levels. Which is the better way to test this drug? Why is it better to test the drug this way?

3. A doctor tells a couple that their "genetic makeup" means that they've got one in four chances of having a child with an inherited illness. Does this mean that if their first child has the illness, the next three will not? Does this mean that each of the couple's children will have the same risk of suffering from the illness?

Contributed by L. Adams and A. R. Butz with information from Branchaw, J. L., Pfund, C., and Rediske, R. (2010). *Entering Research: A Facilitator's Manual.* New York: W.H. Freeman & Co.

SCIENCE OR PSEUDOSCIENCE?

Learning Objectives

Trainees will:

- Evaluate scientific and pseudoscientific claims.

Trainee Level

undergraduate or graduate trainees
novice or intermediate trainees

Areas of Trainee Development

- Research Comprehension and Communication Skills
  - Develop disciplinary knowledge.
  - Develop logical/critical thinking skills.

Activity Components and Estimated Time for Completion

- In Session Time: 30 minutes

Total time: 30 minutes

When to Use This Activity

This activity can be implemented at all trainee stages, but is best suited for individuals with limited experience with scientific research. Other activities that may be implemented with this activity include:

- Discussion of the Nature of Science
- Science and Society
- Science Literacy Test
- What Happens to Research Results?

Inclusion Considerations

Encourage respect of various religious, cultural, or other backgrounds of trainees as they relate to understanding the history and thoughts of science including discoveries and their impact on society.

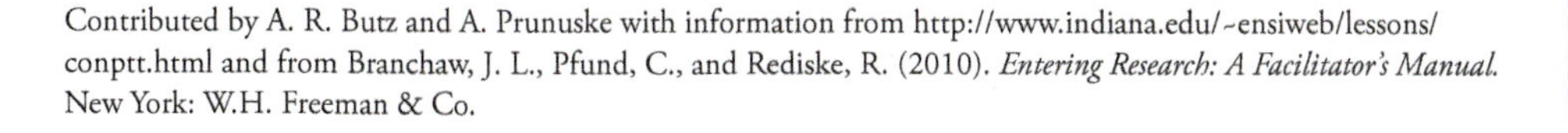

Contributed by A. R. Butz and A. Prunuske with information from http://www.indiana.edu/~ensiweb/lessons/conptt.html and from Branchaw, J. L., Pfund, C., and Rediske, R. (2010). *Entering Research: A Facilitator's Manual.* New York: W.H. Freeman & Co.

Implementation Guide

Implementation Guide

Workshop Session (30 minutes)

- Ask trainees to consider and discuss in small groups the question "What is science?"
- For novice trainees, start the activity by introducing suggested criteria for what constitutes science and ask them to discuss and evaluate these criteria. For example (http://www.indiana.edu/~ensiweb/lessons/conptt.pdf):
  - A natural cause can be used to explain why the observation is occurring.
  - Explanation can be used to predict and test future events.
  - Experimental test results are observable and reproducible.
  - Theories or explanations are tentative and can be modified based on future results.
  - Testability
  - Tentativeness
- **Activity** (10–15 minutes)
  - Distribute the list of statements included in the trainee materials and ask trainees to decide whether each is "scientific" or not, and to generate an argument to support their position. They should do this independently first, then gather in small groups to compare and discuss their conclusions. Alternatively, facilitators may choose statements that are currently relevant to trainees' area(s) of research or use headlines from news stories.

Alternative Implementation: To encourage interaction among trainees, write "science" and "pseudoscience" on opposite ends of the whiteboard (or post signs on opposite sides of the room) and have them move to the side that they believe is correct for each statement. Have each side of the room present arguments to try to convince individuals on the opposing side to switch. If your session room is not set up where all individuals can easily move around the room, you can also conduct this activity by giving individuals or small groups different colored sheets of paper (one labeled "science" and one labeled "pseudoscience") and have them hold up their answer for each statement.

- **Large-Group Discussion** (10–15 minutes)
  - What criteria did you use to determine whether each statement was scientific?
  - What is pseudoscience?
  - How can you tell the difference between science and pseudoscience?
  - Can you design experiments to test the statements you identify as scientific?
  - Must one be able to test a statement in order to consider it scientific?

Contributed by A. R. Butz and A. Prunuske with information from http://www.indiana.edu/~ensiweb/lessons/conptt.html and from Branchaw, J. L., Pfund, C., and Rediske, R. (2010). *Entering Research: A Facilitator's Manual.* New York: W.H. Freeman & Co.

SCIENCE OR PSEUDOSCIENCE?

Learning Objectives

► Students will evaluate scientific and pseudoscientific claims.

Choose whether each of the following statements is scientific or pseudoscientific and provide a rationale to explain your choice.

Statement	Classification (circle)	Rationale for Classification
Without sunlight (or comparable artificial light), green plants will die.	Science / Pseudoscience	
If you are a "Taurus," your horoscope for today is "You are getting along better than ever with coworkers and other folks you may be forced to deal with every day. Your lucky numbers are 16 and 24."	Science / Pseudoscience	
The Earth began about 6,000 years ago and nothing will change that.	Science / Pseudoscience	
Ships and planes passing through the Bermuda Triangle tend to sink and disappear.	Science / Pseudoscience	

Trainee Materials

Contributed by A. R. Butz and A. Prunuske with information from http://www.indiana.edu/~ensiweb/lessons/conptt.html and from Branchaw, J. L., Pfund, C., and Rediske, R. (2010). *Entering Research: A Facilitator's Manual.* New York: W.H. Freeman & Co.

SEARCHING ONLINE DATABASES

Learning Objectives

Trainees will:

- Learn how to use online resources to search for scholarly articles.

Trainee Level

undergraduate or graduate trainees
novice trainees

Areas of Trainee Development

- Practical Research Skills
  - Develop ability to design a research project.
- Research Comprehension and Communication Skills
  - Develop logical/critical thinking skills.

Activity Components and Estimated Time for Completion

- Trainee Pre-Assignment Time: 30 minutes
- In Session Time: 1 hour

Total time: 1 hour 30 minutes

When to Use This Activity

This activity is best suited for trainees who have joined a research group. In the case of graduate trainees in programs with rotations, this activity can be implemented during the first research rotations. Facilitators may elect to invite a campus librarian to co-lead this session.

Inclusion Considerations

Consider learning styles, differences, and disabilities when discussing best practices in searching online databases. Ask if trainees have concerns and empower them to talk about any of these barriers or concerns with their mentor or another advisor.

Contributed by A. Prunuske with information from Branchaw, J. L., Pfund, C., and Rediske, R. (2010). *Entering Research: A Facilitator's Manual.* New York: W.H. Freeman & Co.

Implementation Guide

Trainee Pre-Assignment (30 minutes)

- Before the session, trainees should talk with their research group members to learn how they identify and access relevant scientific articles. Trainees should also reflect on their research and generate a list of keywords.

Workshop Session (1 hour)

- Introduce the trainees to the databases most relevant to their research (e.g., PubMed for health research and Web of Science for general science research). A librarian can be invited as a guest facilitator to lead this session. Most libraries have webpages with the available databases and the subjects they cover. *If it is beneficial to the trainees, facilitators may wish to introduce reference manager software, how to save searches, or how to set up new record alerts as part of this activity.*
- Discuss how using databases to find peer-reviewed articles is more effective than using a generic search engine.
- Demonstrate how to access the relevant databases through the institution and make sure the trainees have the required usernames and passwords to access the articles.
- Hold the session in the library and/or use tutorials to facilitate learning.
  - PubMed tutorial: https://www.nlm.nih.gov/bsd/disted/pubmedtutorial/cover.html
  - Web of Science tutorial: http://wokinfo.com/training_support/training/web-of-knowledge/
- Discuss good keywords followed by some basic search strategies including how to use BOOLEAN or MeSH terms and how to limit the search to review articles.
- Trainees should have at least 5 minutes to practice searching for relevant articles using their keywords.
- **Wrap-up** (5 minutes)
  - At the end of the session, trainees can compare and contrast which search strategies worked well and how they refined their keyword search terms.

Contributed by A. Prunuske with information from Branchaw, J. L., Pfund, C., and Rediske, R. (2010). *Entering Research: A Facilitator's Manual.* New York: W.H. Freeman & Co.

SEARCHING ONLINE DATABASES

Learning Objectives

Trainees will:

- ► Learn how to use online resources to search for scholarly articles.

Searching the literature to find scholarly articles is something most researchers do daily. Identifying relevant articles to your research is necessary when planning research projects and interpreting the results of those projects. This activity introduces you to research databases and will help to develop the skills needed to identify and access relevant scholarly articles.

BEFORE the session:

1. Which online databases do members of your research group use to search the literature?

2. With your mentor or another member of your research group, generate a list of keywords relevant to your research project. Include important authors in the field, the organism or population of interest, techniques or interventions, molecules or drugs, and relevant diseases.

Contributed by A. Prunuske with information from Branchaw, J. L., Pfund, C., and Rediske, R. (2010). *Entering Research: A Facilitator's Manual.* New York: W.H. Freeman & Co.

DURING the session:

1. Write your own step-by-step guide on how to search the database most relevant to your research.

2. Test the keywords you generated with your mentor using the database. Identify at least one article you think would be helpful to your project. Write the article citation below.

3. Identify at least two strategies you found to be particularly effective in identifying relevant articles.

Contributed by A. Prunuske with information from Branchaw, J. L., Pfund, C., and Rediske, R. (2010). *Entering Research: A Facilitator's Manual.* New York: W.H. Freeman & Co.

CASE STUDY: SELECTION OF DATA

Learning Objectives

Trainees will:

- ► Discuss and practice ethical research decision making.

Trainee Level

undergraduate or graduate trainees
novice trainees

Areas of Trainee Development

- ► Research Ethics
  - Develop responsible and ethical research practices.
- ► Research Comprehension and Communication Skills
  - Develop logical/critical thinking skills.

Activity Components and Estimated Time for Completion

- ► In Session Time: 25 minutes

Total time: 25 minutes

When to Use This Activity

This activity is suitable for undergraduate and graduate level trainees and novice trainees who have chosen a research mentor and who have been conducting research for at least one month. Other activities that may be used with this activity include:

- ► Case Study: The Sharing of Research Materials
- ► Case Study: Credit Where Credit Is Due
- ► Ethics Case Discussion with Mentor
- ► Truth and Consequences Article

Inclusion Considerations

Discuss with trainees how understanding of ethical behavior may be different based on differences in cultural backgrounds or across generations. Facilitators can ask the trainees to consider the case study from different cultural or generational perspectives. Emphasize that it can sometimes be as difficult to determine whether behavior is unethical as it is to decide how to deal with that behavior in a sensitive and respectful manner. Encourage trainees to seek input from others, in particular those who can offer different perspectives, when dealing with potentially unethical situations in the research environment.

Contributed by A. R. Butz with information from Branchaw, J. L., Pfund, C., and Rediske, R. (2010). *Entering Research: A Facilitator's Manual.* New York: W.H. Freeman & Co.

Implementation Guide

Workshop Session (25 minutes)

- **Case Study: Selection of Data** (5 minutes): Distribute the case study and have one trainee read it aloud, or display the case on a projector screen and have trainees read and consider it silently.
- **Discussion Questions** (10 minutes)
  - How should the data from the two suspected runs be handled?

> - The data *should* be included in analyses, but the trainees should also mention the data anomalies in their discussion of the data and results.
> - The data *should* be included in analyses with no mention of the anomalies.
> - The data *should not* be included in analyses, but the trainees should mention the data anomalies in their discussion of the data and results.
> - The data *should not* be included in the analyses and should not be mentioned in the paper.

  - Should the data be included in tests of statistical significance? Why/why not?

> - Yes, it is appropriate to include all data points in this test.
> - No, anomalous data should be excluded from tests of significance because these data points are outliers that are not consistent with the expected ranges for data points.

  - What sources of information could Kathleen and Deborah use to help decide?

> - Previous research/papers.
> - Statistical/research methods publications.
> - Contact the national laboratory to see if others have reported similar anomalies in their own data.

- **Wrap-up** (5 minutes)
  - Summarize the main ideas generated from the discussion and generate an "action plan" for Kathleen and Deborah. Would you be comfortable carrying out all of these action items if Deborah was your direct mentor? Why or why not?
  - Invite trainees to ask their mentors how they determine what data to include and exclude in their own research.

Contributed by A. R. Butz with information from Branchaw, J. L., Pfund, C., and Rediske, R. (2010). *Entering Research: A Facilitator's Manual.* New York: W.H. Freeman & Co.

CASE STUDY: THE SELECTION OF DATA

Learning Objectives

Trainees will:

- Discuss and practice ethical research decision making.

Modified from: "On Being a Scientist: Responsible Conduct in Research," 2nd ed., National Academy Press, 1995

Trainees Deborah and Kathleen have made a series of measurements on a new experimental semiconductor material using an expensive neutron source at a national laboratory. When they get back to their own lab and examine the data, they get the following data points. A newly proposed theory predicts results indicated by the curve.

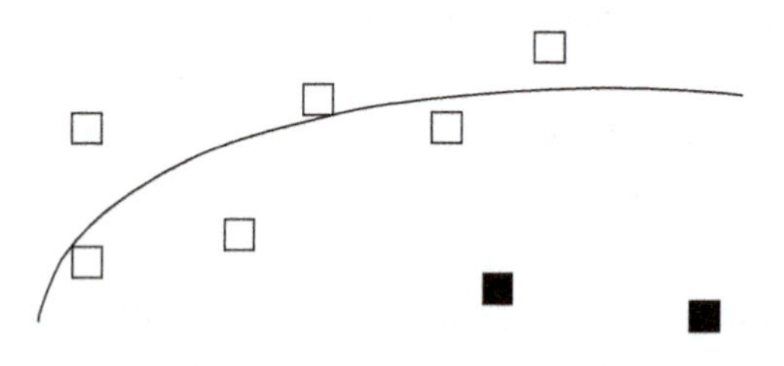

During the measurements at the national laboratory, Deborah and Kathleen observed that there were power fluctuations they could not control or predict. Furthermore, they discussed their work with another group doing similar experiments, and they knew that the group had gotten results confirming the theoretical prediction and were writing a manuscript describing their results.

In writing up their own results for their research project and hopefully for publication, Kathleen suggests dropping the two anomalous data points near the abscissa (the solid squares) from the published graph and from the statistical analysis. She proposes that the existence of the data points be mentioned in the paper as possibly due to power fluctuations and being outside the expected standard deviation calculated from the remaining data points. "These two runs," she argues to Deborah "were obviously wrong."

Discussion Questions

1. How should the data from the two suspected runs be handled?
2. Should the data be included in tests of statistical significance? Why/why not?
3. What sources of information could Kathleen and Deborah use to help decide?

Contributed by A. R. Butz with information from Branchaw, J. L., Pfund, C., and Rediske, R. (2010). *Entering Research: A Facilitator's Manual.* New York: W.H. Freeman & Co.

STEPS TO RESEARCHER INDEPENDENCE

Learning Objectives

Trainees will:

- Become familiar with the characteristics of researcher independence at various training stages.

Trainee Level

undergraduate or graduate trainees
novice or intermediate trainees

Areas of Trainee Development

- Researcher Confidence and Independence
  - Develop independence as a researcher.
- Professional and Career Development Skills
  - Explore and pursue a research career.

Activity Components and Estimated Time for Completion

- In Session Time: 15–20 minutes

Total time: 15–20 minutes

When to Use This Activity

Though this activity is appropriate for novice undergraduate or graduate trainees at any stage, it is most beneficial when implemented at the beginning of a new training stage, but after the trainee has some research experience for context about knowledge and skills that would indicate independence.

Inclusion Considerations

Confidence to make decisions and assume responsibility for a research project may vary widely among trainees, depending on their previous experiences, personalities, and cultural backgrounds. Encourage them to share concerns they have with the group and to discuss strategies to move toward independence at an appropriate rate with their research mentor.

Contributed by J. Branchaw, C. Pfund, and A. R. Butz with information from Pfund, C., Branchaw, J., and Handelsman, J. (2014). *Entering Mentoring.* New York: W.H. Freeman.

Implementation Guide

Workshop Session (15–20 minutes)

► **Activity/Discussion:** What does the term *independence* mean to you?

- Ask trainees: What knowledge and skills do you need to demonstrate to convey to your mentors that you are on the path to independence? (5–10 minutes)
  - **For undergraduate students:** Consider what independence should look like for a junior undergraduate student, first-year graduate student, and third-year graduate student.
  - **For graduate students:** Consider what independence should look like for a third-year graduate student, finishing graduate student, and postdoctoral researcher.
  - You can use one or all of these career stages, depending upon the needs of your group.
  - Facilitators should draw a line on a whiteboard or flipchart and mark on it the career stages they want trainees to consider. Facilitators can write ideas on the whiteboard/flipchart as participants describe characteristics at each career stage and draw lines across career stages when there is debate about where particular characteristic should be listed (see sample grids provided in this implementation guide).
- **Large-Group Discussion** (5–10 minutes)
  - Does the list they generated align with their mentors' expectations?
  - How would they know?

► **Optional Activity: Mapping Knowledge and Skills**

- Have trainees map a list of knowledge and skills they believe they need to demonstrate in order for their mentor to feel confident that they are on the path toward independence and discuss it with their mentors.

Independence Grid Examples Independence may look different at each stage depending on prior research experience, disciplinary differences in expectations, etc.

Contributed by J. Branchaw, C. Pfund, and A. R. Butz with information from Pfund, C., Branchaw, J., and Handelsman, J. (2014). *Entering Mentoring.* New York: W.H. Freeman.

Undergraduate Student Researchers

Novice (year 1–2)	**Intermediate** (year 2–3)	**Advanced** (year 3–4)
Eager to learn how to do experiments	Able to understand experimental design and individual experimental techniques	Able to assist in the design and coordination of a project and conduct experiments with minimal oversight
Aware of data analysis output	Understand data analysis output	Able to conduct data analysis
Has basic writing skills	Has basic research writing skills and able to draft an abstract	Able to write an abstract and draft a research report
Able to work when mentor is available to help	Able to do simple experimental procedures on own	Able to do more complex experimental procedures and trouble-shoot on own
Able to identify components of research articles that are unclear	Able to articulate and find answers to questions about research articles	Able to critically read the literature, find answers to questions, and summarize disciplinary articles
Able to describe the basic research goals or questions of the research group	Able to describe how own research project contributes to the group's research	Able to describe own research project and how the group's research contributes to the discipline
Able to attend and follow along at research team meetings	Able to ask relevant questions at research team meetings	Able to co-present at research team meetings
Able to ask questions about a poster or research talk	Able to present a poster on own research	Able to present a poster or research talk on own research

Contributed by J. Branchaw, C. Pfund, and A. R. Butz with information from Pfund, C., Branchaw, J., and Handelsman, J. (2014). *Entering Mentoring*. New York: W.H. Freeman.

Graduate Student Researchers

Novice (year 1–3)	**Intermediate** (year 2–4)	**Advanced** (year 3–5)
Able to assist in the design and coordination of a project and conduct experiments with minimal oversight	Able to take effective ownership of a project and to design a rigorous experiment	Able to articulate an independent research statement that could be used in applying for research positions
Able to conduct data analysis	Able to conduct data analysis and seek out new methods on their own	Able to design and conduct data analysis
Able to write an abstract and draft a research report	Able to write an abstract and draft a research paper for peer review	Able to write an abstract and be the lead author on a research paper for peer review
Able to do more complex experimental procedures and trouble-shoot on own	Able to plan and manage own experimental procedures	Able to plan and manage own experimental procedures, while also mentoring on other projects
Able to critically read the literature, find answers to questions, and summarize disciplinary articles	Able to critically review and synthesize disciplinary research articles and books relevant to project	Able to generate and integrate a broad literature review—to be "fluent" in the field
Able to describe own research project and how the group's research contributes to the discipline	Able to describe how own research project contributes to the group's research and how the group's research contributes to the discipline	Able to describe how multiple individual research projects contribute to the group's research, how the group's research contributes to the discipline and what the next steps in research should be
Able to present at research team meeting	Able to present at the research team meeting and at the department seminar	Able to present to professional audiences and articulate the skills and knowledge they have to contribute
Able to present a poster or research talk on own research	Able to present a poster or research talk on the group's research	Able to present a poster or research talk on the group's research

Contributed by J. Branchaw, C. Pfund, and A. R. Butz with information from Pfund, C., Branchaw, J., and Handelsman, J. (2014). *Entering Mentoring.* New York: W.H. Freeman.

STEPS TO RESEARCHER INDEPENDENCE

Learning Objectives

Trainees will:

- ► Become familiar with the characteristics of researcher independence at various training stages.

Instructions: What knowledge and skills do you need to demonstrate to convey to your mentors that you are on the path to independence? List the signs of an independent researcher at each stage of a trainee's education/training:

Novice undergraduate student researcher (year 1–2):

Intermediate undergraduate student researcher (year 2–3):

Advanced undergraduate student researcher (year 3–4):

Novice graduate student researcher (year 1–3):

Intermediate graduate student researcher (year 2–4):

Advanced graduate student research (year 3–5):

Contributed by J. Branchaw, C. Pfund, and A. R. Butz with information from Pfund, C., Branchaw, J., and Handelsman, J. (2014). *Entering Mentoring.* New York: W.H. Freeman.

STEREOTYPE THREAT

Learning Objectives

Trainees will:

- Define *stereotype threat* and identify the different types of stereotype threat that may be present in a research environment.
- Identify ways to mitigate stereotype threat in the research experience.

Trainee Level

undergraduate or graduate trainees
novice, intermediate, or advanced trainees

Areas of Trainee Development

- Equity and Inclusion Awareness and Skills
  - Develop skills to deal with personal differences in the research environment.
  - Advance equity and inclusion in the research environment.
- Researcher Confidence and Independence
  - Develop confidence as a researcher.
- Professional and Career Development Skills
  - Develop confidence in pursuing a research career.

Activity Components and Estimated Time for Completion

- Trainee Pre-Assignment Time: 15 minutes
- In Session Time: 45 minutes

Total time: 1 hour

When to Use This Activity

Depending on when this activity is implemented and the familiarity of trainees with the concept of stereotype threat, it may be useful to assign readings and/or background information to review in advance of this session.

It is best to implement this activity with an already established cohort of trainees or after a cohort has been meeting regularly for a few weeks. Ground rules for discussing challenging topics should have been already co-constructed and should be displayed on a whiteboard or flipchart throughout this session. In addition, because there are several references to self-efficacy, it may be beneficial to complete the "Fostering Your Own Research Self-Efficacy" activity before this session.

Inclusion Considerations

Encourage all trainees to share their experiences with the group, but do not require it. Trainees from groups historically underrepresented in research should not be asked or expected to speak for their respective identity group. All shared experiences should be presented as individual experiences, yet considered in light of the historical and social contexts the trainees learn about in the readings. Challenge all trainees to share in identifying challenges and barriers to creating more inclusive teams and diversifying the research workforce.

Contributed by A. Kaatz. (2018). *Stereotype threat.*

Implementation Guide

Trainee Pre-Assignment (10 minutes)

► Distribute the reading "Stereotype Threat," found in the Trainee Materials section, and instruct students to review the document prior to the session.

Workshop Session (45 minutes)

► **Small-Group Brainstorm Followed by Large-Group Sharing** (15 minutes)

- Based on what you learned from the reading, describe stereotype threat in your own words.
    - **Stereotype threat—definition:** *Stereotype threat* is the psychological experience of anxiety about performing in a way that reinforces a negative stereotype about your group (e.g., girls are bad at math).
- What kinds of stereotypes are there in academia?
    - In academia, there are strong negative stereotypes that women, racial/ethnic minorities, and first-generation college students lack the intrinsic ability to succeed in science, technology, engineering, and mathematics (STEM) fields, which make members of these groups highly vulnerable to stereotype threat. Stereotype threat can be activated by any means that makes stereotype-group membership salient. This includes emphasizing that a test is indicative of ability, stating that performance disparities between certain groups, providing demographic information before a test, being the only member of your social group in a classroom in a field where there are stereotypes about ability, or any other interaction or experience that makes a negatively stereotyped identity salient (e.g., pictures of prior faculty that only include White men).
- What are some ways to prevent activation of or reduce stereotype threat?
    - There are many ways to prevent the activation of stereotype threat and to protect students from underperforming. One of the most effective methods is to build strong self-efficacy beliefs. Self-efficacy is belief in your ability to perform the tasks necessary to succeed in a given domain, such as research, biology, medicine, mathematics, or leadership. Multiple studies have shown that students with high self-efficacy beliefs are less likely to underperform after their negatively stereotyped identity is made salient; strong self-efficacy beliefs can also lead to increased performance for students in negatively stereotyped groups under stereotype threat conditions—a state called *reactance*.
    - In the research experience, there are several strategies that can be used by trainees to mitigate the negative impact of stereotype threat on performance. These include:
        - learning about stereotype threat;
        - focusing on an alternate, unthreatened identity;
        - providing contextual reasons for research anxiety (e.g., low-grade noise);
        - reframing the research experience as not indicative of ability or talent (e.g., saying men and women perform similarly well in research); or emphasizing that they *have* the ability to perform well;
        - providing role models (e.g., inviting successful women in science to present in department colloquia);
        - adopting a *growth* vs. *fixed* view of intelligence (e.g., *anyone* can learn to do research vs. men are innately better at research than women);
        - completing a self-affirmation activity prior to tackling a challenging research task.

Contributed by A. Kaatz. (2018). *Stereotype threat.*

► **Case Studies** (15 minutes)

- Distribute the case study handout provided in the trainee materials. Ask trainees to review each case study alone or in pairs. Alternatively, you can display each case study to the large group and ask for a volunteer to read each case study aloud.
- Ask trainees to discuss each case study in pairs or small groups, identifying strategies that could be used in each case to reduce stereotype threat. Examples of strategies are provided on the first page of the trainee materials.
- As a large group, list the strategies that could be used in each case.

► **Reflection Exercise** (10 minutes)

- Ask trainees to use the space at the bottom of their handout to reflect on some negative stereotypes that apply to them.
- Invite trainees to share examples of how stereotype threat could have impacted their research experience.
  - How did they handle it?
  - What other coping strategies could have been used?

► **Wrap-up** (5 minutes)

- Summarize the key points raised in the discussion.
- Remind students to be mindful of situations in their research environment that could make them (or other trainees) vulnerable to stereotype threat.

STEREOTYPE THREAT

Undergraduate

Learning Objectives

Trainees will:

- Define *stereotype threat* and identify the different types of stereotype threat that may be present in a research environment.
- Identify ways to mitigate stereotype threat in the research experience.

Case Studies

1. You are enrolled in a large pre-med physics course at a competitive math and science school. You notice that the instructor has students share some demographic data about their gender identity, race, and ethnicity on the first page of the test. You also notice that there are very few women in the course, and only one Black student. You've recently read that physics is a domain where stereotypes describe White men as having strong ability and innate talent. You've also recently learned about stereotype threat and its impacts. What can be done to prevent the activation of stereotype threat?

   - How might being asked to share data about ***your*** gender identity, race, and ethnicity impact you?

   - What could ***you*** do to prevent the activation of stereotype threat for individuals in the course?

   - What could ***the instructor*** do to prevent the activation of stereotype threat for individuals in the course?

2. You are a female student enrolled in a large university in the western United States. On your way to the lab, you pause to notice the pictures of distinguished faculty in the hallway. You notice that all the pictures are of older White men. There is nobody who looks like you. You start to feel a little anxious, but you figure it is just nervousness about working with your new research mentor.

   - Why might this experience activate stereotype threat in you and what steps could you take to mitigate its impact for this female student?

Reflection exercise: On the back of this page, reflect on some negative stereotypes that apply to you. Can you think of a situation where any of these negative stereotypes could have impacted your performance? How did you handle it? What are some effective coping strategies you could have used? (*Examples:* American students are not as competitive as their foreign counterparts. Women are bad at math and science. Black students have low academic aptitude.)

Contributed by A. Kaatz. (2018). *Stereotype threat.*

STEREOTYPE THREAT

Graduate

Learning Objectives

Trainees will:

- Define *stereotype threat* and identify the different types of stereotype threat that may be present in a research environment.
- Identify ways to mitigate stereotype threat in the research experience.

Case Studies

1. You are a teaching assistant for a large pre-med physics course at a competitive math and science school. You notice that the instructor has students share some demographic data about their gender identity, race, and ethnicity on the first page of the test. You also notice that there are very few women in the course, and only one Black student. You've recently read that physics is a domain in which stereotypes describe White men display strong ability and innate talent. You've also recently learned about stereotype threat and its impacts.
   - What suggestions could you make to the instructor to help prevent the activation of stereotype threat?

2. You are a female student enrolled in a chemistry Ph.D. program at a large university in the western United States. On your way to your mentor's research lab, you pause to notice the pictures of distinguished faculty in the hallway. You notice that all the pictures are of older White men. There is nobody who looks like you. You start to feel a little anxious, but you figure it is just nerves about working with your research mentor.
   - How would this experience activate stereotype threat and what steps could be taken to mitigate its impact for this female student?

Reflection exercise: Use the space below to reflect on some negative stereotypes that apply to you. Can you think of a situation where any of these negative stereotypes could have impacted your performance? How did you handle it? What are some effective coping strategies you could have used? (*Examples:* American students are not as competitive as their foreign counterparts. Women are bad at math and science. Black students have low academic aptitude.)

Contributed by A. Kaatz. (2018). *Stereotype threat.*

STEREOTYPE THREAT

Stereotype threat is the psychological experience of anxiety about performing in a way that reinforces a negative stereotype about your group (e.g., girls are bad at math). It was first discovered by Dr. Claude Steele while he was at the University of Michigan. He noticed a troubling trend where White students would start to outperform Black students by their sophomore year, despite having similar ability and credentials when they started college. He attempted to re-create this performance gap in his lab, and after several failed attempts he identified the causal mechanism (Steele & Aronson, 1995). He gave two groups of similarly qualified White and Black students a portion of the verbal GRE. Prior to taking the test, he told the first group that it was a test of their ability; he told the second group that it was a problem-solving task (Steele & Aronson, 1995). Black students performed significantly worse than White students when they were told the exam was a test of ability. By comparison, Black and White students performed similarly when told the exam was a problem-solving task. These results remained even after controlling for students' prior standardized test scores.

Hundreds of studies and decades later, we now know that stereotype threat can be triggered by subtle cues that make membership of a negatively stereotyped-group salient, and that it undermines performance by causing anxiety. In Dr. Steele's study, simply saying the exam was a test of ability made salient the stereotype that Blacks have low academic competence. This led Black students to experience test anxiety, which took away some of their working memory, led them to underperform, and subsequently reinforced a negative stereotype about ability for their social group.

In academia, there are strong negative stereotypes that women, racial/ethnic minorities, and first-generation college students lack the intrinsic ability to succeed in science, technology, engineering, and mathematics (STEM) fields, which makes members of these groups highly vulnerable to stereotype threat. Stereotype threat can be activated by essentially any means that makes stereotype-group membership salient. This includes emphasizing that a test is indicative of ability (Steele & Aronson, 1995), stating that performance disparities between certain groups (e.g., men vs. women, Whites vs. Blacks; Spencer, Steele, & Quinn, 1999), providing demographic information before a test (e.g., gender, race/ethnicity; Steele & Aronson, 1995), being the only member of your social group in a classroom (e.g., the only woman, the only Black student) in a field where there are stereotypes about ability (Sekaquaptewa, Waldman, & Thompson, 2007), or any other interaction or experience that makes a negatively stereotyped identity salient (e.g., pictures of prior faculty that only include White men).

There are many ways to prevent the activation of stereotype threat and to protect students from underperforming. One of the most effective methods is to build strong self-efficacy beliefs.

Self-efficacy is belief in your ability to perform the tasks necessary to succeed in a given domain, such as biology, medicine, mathematics, or leadership. Multiple studies have shown that students with high self-efficacy beliefs are less likely to underperform after their negatively stereotyped identity is made salient (e.g., Spencer, Steele, & Quinn, 1999; Sekaquaptewa, Waldman, & Thompson, 2007; Hoyt & Blascovich, 2007); strong self-efficacy beliefs can also lead to increased performance for students in negatively stereotyped groups under stereotype threat conditions—a state called *reactance* (Hoyt & Blascovich, 2007).

Contributed by A. Kaatz. (2018). *Stereotype threat.*

Other ways to mitigate the negative impact of stereotype threat on student performance include educating students about stereotype threat (Johns, Inzlict, & Schmader, 2008); encouraging students to focus on an alternate, unthreatened identity (e.g., college student vs. *woman* in math; Ambaby et al., 2004); providing contextual reasons for test anxiety (e.g., low-grade noise); reframing a test as not indicative of ability or talent (e.g., saying men and women perform similarly well on this test; Steele & Aronson, 1995) or emphasizing that they *have* the ability to perform well on the test (Cohen, Steele, & Ross, 1999); providing role models (e.g., have the test proctor be a successful woman in science; Marx, Stable, & Muller, 2005); encouraging students to have a growth vs. *fixed* view of intelligence (e.g., *anyone* can learn math vs. men are innately better at math than women; Thoman et al., 2008); and having students do a self-affirmation activity prior to taking a test (Cohen et al., 2006).

References

Ambady, N., Paik, S. K., Steele, J., Owen-Smith, A., and Mitchell, J. P. (2004). Deflecting negative self-relevant stereotype activation: The effects of individuation. *Journal of Experimental Social Psychology, 40*: 401–408.

Cohen, G. L., Garcia, J., Apfel, N., and Master, A. (2006). Reducing the racial achievement gap: A social-psychological intervention. *Science, 313*: 1307–1310.

Cohen, G. L., Steele, C. M., and Ross, L. D. (1999). The mentor's dilemma: Providing critical feedback across the racial divide. *Personality and Social Psychology Bulletin, 25*: 1302–1318.

Good, C., Aronson, J., and Harder, J. A. (2008). Problems in the pipeline: Stereotype threat and women's achievement in high-level math courses. *Journal of Applied Developmental Psychology, 29*: 17–28.

Hoyt, C. L., and Blascovich, J. (2007). Leadership efficacy and women leaders' responses to stereotype activation. *Group Processes & Intergroup Relations, 10*(4): 595–616.

Johns, M., Inzlicht, M., and Schmader, T. (2008). Stereotype threat and executive resource depletion: Examining the influence of emotion regulation. *Journal of Experimental Psychology: General, 137*: 691–705.

Marx, D. M., Stapel, D. A., and Muller, D. (2005). We can do it: The interplay of construal orientation and social comparison under threat. *Journal of Personality and Social Psychology, 88*: 432–446.

Sekaquaptewa, D., Waldman, A., and Thompson, M. (2007). Solo status and self-construal: Being distinctive influences racial self-construal and performance apprehension in African American women. *Cultural Diversity and Ethnic Minority Psychology, 13*: 321–327.

Spencer, S. J., Steele, C. M., and Quinn, D. M. (1999). Stereotype threat and women's math performance. *Journal of Experimental Social Psychology, 35*: 4–28.

Steele, C. M., and Aronson, J. (1995). Stereotype threat and the intellectual test performance of African-Americans. *Journal of Personality and Social Psychology, 69*: 797–811.

Steele, C. M., Spencer, S. J., and Aronson, J. (2002). Contending with images of one's group: The psychology of stereotype and social identity threat. In M. Zanna (Ed.), *Advances in Experimental Social Psychology*. San Diego: Academic Press.

Thoman, D. B., White, P. H., Yamawaki, N., and Koishi, H. (2008). Variations of gender-math stereotype content affect women's vulnerability to stereotype threat. *Sex Roles, 58*: 702–712.

Contributed by A. Kaatz. (2018). *Stereotype threat.*

SUMMER UNDERGRADUATE RESEARCH PROGRAMS

Learning Objectives

Trainees will:

- Explore undergraduate research programs available to STEM students.
- Evaluate programs of interest.
- Compile information needed for applications.

Trainee Level

undergraduate trainees
novice or intermediate trainees

Activity Components and Estimated Time for Completion

- In Session Time: 55 minutes

Total time: 55 minutes

When to Use This Activity

This activity should be implemented one to two months before the undergraduate research program applications are due. For summer research programs, the due dates are typically early spring, so implementing this activity in late fall is ideal. Other activities that may be used with this activity include:

- Letter of Recommendation
- Developing a Curriculum Vitae
- Personal Statement

If these activities were implemented earlier in a course or workshop series, you may wish to revisit key points as you implement this activity.

Inclusion Considerations

Trainees from different cultural, socioeconomic, and other backgrounds might have family members who see research activities and experiences as unfamiliar and not in alignment with familial notions of success (i.e., working during the summer to make money vs. participating in a summer undergraduate research program). Invite trainees to discuss strategies to use when explaining the value of a summer research experience to their families and others in their support network.

Areas of Trainee Development

- Professional and Career Development Skills
  - Explore and pursue a research career.
  - Develop confidence in pursuing a research career.

Contributed by A. Bramson. (2018). *Summer Undergraduate Research Programs.*

Implementation Guide

Workshop Session (55 minutes)
The purpose of this discussion is to make students aware of summer undergraduate research programs. For example:

- National Science Foundation Research Experiences for Undergraduates (REU)
- NASA Summer Internships and Fellowships
- Medical Research Programs by the Association of American Medical Colleges
- Pathways to Science

Share websites of funded summer research programs that are relevant to the trainees who are attending this session. Demonstrate how to search for programs and where to find the type of research that each offers and the eligibility requirements.

- **Self-Exploration and Discussion:** Deciding whether to participate in a summer research program. (20 minutes)
  - Instruct trainees to consider whether applying to a summer research program is a good choice for them by answering the questions on page 1 of the trainee materials. (10 minutes)
    1. What would be the benefits of participating in a summer research program?

       - Get to try out a new project.
       - Learn new skills.
       - Live in a new location.
       - Experience a new mentoring style.
       - Get paid.
       - Narrow down research interests for graduate school.
       - Try a potential graduate school.

    2. What kind of research would you like to do? Would you continue with the same area of research that you have been doing, or try something new?
    3. Does a program with a narrow theme, in which everyone is working on similar problems, appeal to you, or would you rather attend a program with a broad range of topics?
    4. If you have the chance to identify a mentor in the program with whom you would like to do research, what factors will you consider when choosing?
    5. How would you convey your research and career interests in a personal statement?
  - Bring the students together to discuss their answers to the questions and to answer any additional questions they have. (10 minutes)

Contributed by A. Bramson. (2018). *Summer Undergraduate Research Programs.*

► **Program Exploration** (30 minutes)
- Instruct trainees to search the Internet for at least three programs of interest to them and use the second page of the trainee materials to document information about those programs.
  - Did all the programs you looked at offer the same benefits? Did they all require the same application materials?
  - Which programs did you list on your handout as your "Top 3"?
  - Why did you list these programs? Did all programs appeal to you for the same reasons or different?
- *Note:* Most applications require letters of recommendation. Trainees should contact their letter writers at least one month before the letters are due. An organized table with due dates and any other relevant information about each program should be prepared and given to each letter writer.

Optional Activity: Guest Speaker (20 minutes)

► Invite students who have participated in a summer research program to talk about their experience and answer questions.

Contributed by A. Bramson. (2018). *Summer Undergraduate Research Programs.*

SUMMER UNDERGRADUATE RESEARCH PROGRAMS

Learning Objectives

Trainees will:

- Explore undergraduate research programs available to STEM students.
- Evaluate programs of interest.
- Compile information needed for applications.

Deciding Whether to Participate in a Summer Research Program

Answer the following questions to explore whether applying to participate in a summer research program is a good choice for you.

1. What would be the benefits of participating in a summer research program?

2. What kind of research would you like to do? Would you continue with the same area of research that you have been doing, or try something new?

3. Does a program with a narrow theme, in which everyone is working on similar problems, appeal to you, or would you rather attend a program with a broad range of topics?

4. If you have the chance to identify a mentor in the program with whom you would like to do research, what factors will you consider when choosing?

5. How would you convey your research and career interests in a personal statement?

Contributed by A. Bramson. (2018). *Summer Undergraduate Research Programs.*

IDENTIFYING SUMMER RESEARCH PROGRAMS THAT INTEREST YOU

Program #1:

Application Due Date:
Website:
Start Date: End Date: Location:

What does the application require?

What are the benefits included with the program (stipend, housing/lodging, travel expenses, seminars, course credits, opportunity to present at a conference, etc.)?

Why does this program interest me?

Program #2:

Application Due Date:
Website:
Start Date: End Date: Location:

What does the application require?

What are the benefits included with the program (stipend, housing/lodging, travel expenses, seminars, course credits, opportunity to present at a conference, etc.)?

Why does this program interest me?

Program #3:

Application Due Date:
Website:
Start Date: End Date: Location:

What does the application require?

What are the benefits included with the program (stipend, housing/lodging, travel expenses, seminars, course credits, opportunity to present at a conference, etc.)?

Why does this program interest me?

Contributed by A. Bramson. (2018). *Summer Undergraduate Research Programs.*

Implementation Guide

THE NEXT STEP IN YOUR CAREER

Learning Objectives

Trainees will:

- Identify factors that are important when considering the next steps in professional development/career.

Trainee Level

undergraduate or graduate trainees
intermediate or advanced trainees

Areas of Trainee Development

- Professional and Career Development Skills
  - Explore and pursue a research career.
  - Develop confidence in pursuing a research career.
- Researcher Identity
  - Develop identity as a researcher.

Activity Components and Estimated Time for Completion

- In Session Time: 25–35 minutes

Total time: 25–35 minutes

When to Use This Activity

This activity is most appropriate for intermediate to advanced trainees who are nearing the end of their research experience and contemplating the next steps in their training or career. When this activity is used as part of a course using the *Entering Research* curriculum, it should be implemented towards the end of the course. Other activities that can be used with this activity include:

- Research Careers: The Informational Interview

Inclusion Considerations

Trainees from different cultural, socioeconomic, and other backgrounds might have family members who see research activities and experiences as unfamiliar and not in alignment with familial notions of success (i.e., pursuing a graduate degree or doing postdoctoral research vs. getting a permanent job). Invite trainees to discuss strategies to use when explaining their plans to pursue a research career to their families and others in their support network.

Contributed by A. Bramson with information from Branchaw, J. L., Pfund, C., and Rediske, R. (2010). *Entering Research: A Facilitator's Manual.* New York: W.H. Freeman & Co.

Implementation Guide

Workshop Session (25–35 minutes)

► **Activity: The Next Step in Your Career** (10–15 minutes)
- Allow trainees 5–10 minutes to cut out and individually rank the factors.
- Have the trainees pair up and explain their rankings to their partners.

► **Discussion** (10–15 minutes)
- Survey the group to identify and discuss the most common top priorities.
  - If the group is small, ask each student to share their top two priorities with the whole group.
  - If the group is large, consider writing the priorities on the board or large Post-it Note and asking students to place a tally or dot sticker by their top 2 choices.
- Why did you select ______ as a priority?
  - Alternatively, if you have data about the class priorities (see above), you can ask students about why they chose the most common factor and what made other factors less of a priority.
- How can you learn about or investigate the factors that are most important to you?

> - Find a mentor in the field to guide you.
> - Do research online.
> - Talk to current students/employees in the programs/businesses of interest.
> - Ask the graduate training or professional program administrator.

- Complementary discussion topic:
  - What skills have you learned that will transfer to the next phase of your training or career?

Contributed by A. Bramson with information from Branchaw, J. L., Pfund, C., and Rediske, R. (2010). *Entering Research: A Facilitator's Manual.* New York: W.H. Freeman & Co.

THE NEXT STEP IN YOUR CAREER

Undergraduate

Learning Objectives

Trainees will:

- Identify factors that are important when considering the next steps in professional development/career.

Each box lists a factor that you may want to consider when selecting a graduate or professional school program or a job. There are also two blank boxes in which you may add factors that are important to you and are missing from the list. Cut out the boxes and rank the factors in order of importance to you.

Opportunity to work with a specific advisor, mentor, physician, or teacher	Coursework requirements and/or professional development opportunities
Climate of the training environment	Location
Relative value of teaching and research training	Alignment of personal goals with the offerings/ opportunities
Reputation of a specific advisor, mentor, or coworker	Funding
Reputation of the department, office, program, or institution	Happiness of other graduate/medical students/ coworkers in the program
Range of academic opportunities to engage beyond graduate study	Feeling of inclusivity—seeing that there are others like you there
Type of preliminary/qualifying exam	Type of curriculum (e.g., case-based, traditional lecture, clinical)
Other:	Other:

Contributed by A. Bramson with information from Branchaw, J. L., Pfund, C., and Rediske, R. (2010). *Entering Research: A Facilitator's Manual.* New York: W.H. Freeman & Co.

THE NEXT STEP IN YOUR CAREER

Graduate

Learning Objectives

Trainees will:

► Identify factors that are important when considering the next steps in professional development/career.

Each box lists a factor that you may want to consider when selecting your next training opportunity or job. There are also two blank boxes in which you may add factors that are important to you and are missing from the list. Cut out the boxes and rank the factors in order of importance to you. *Note:* Some factors may not apply to your specific career goals.

Opportunity to work at a specific institution or company	Additional schooling or training required
Fits with my specific interests and skills	Location
Opportunities for teaching and/or research training	Alignment of personal goals with the offerings/ opportunities
Reputation of an institution or company	Available funding/salary; benefits; potential for promotion
Job security	Happiness of students/trainees/coworkers in the program/lab/company
Job and/or school opportunities available for spouse, partner, or child	Feeling of inclusivity—seeing that there are others like you there
Expectations for success in training, publications, and/or funding	Type of work (primarily research, teaching, working alone vs. part of a team, etc.)
Other:	Other:

Contributed by A. Bramson with information from Branchaw, J. L., Pfund, C., and Rediske, R. (2010). *Entering Research: A Facilitator's Manual.* New York: W.H. Freeman & Co.

THE POWER OF SOCIAL PERSUASION

Learning Objectives

Trainees will:

- Assess the influence that mentors can have on confidence in abilities.
- Devise strategies to cope with and respond to feedback that negatively influences trainee confidence.

Trainee Level

undergraduate or graduate trainees
novice, intermediate, or advanced trainees

Areas of Trainee Development

- Researcher Confidence and Independence
  - Develop confidence as a researcher.
- Research Comprehension and Communication Skills
  - Develop effective interpersonal communication skills.

Activity Components and Estimated Time for Completion

- In Session Time: 50 minutes

Total time: 50 minutes

When to Use This Activity

This activity is designed for use with trainees who are in a position to receive feedback from their mentors on their work and/or writing. Other activities that may be used with this activity include:

- Fostering Your Own Research Self-Efficacy
- Case Study: Responding to Feedback
- Messages Sent and Received

Inclusion Considerations

Discuss the potential implications or perceptions of cross-cultural/cross-gender feedback. Trainees from underrepresented and stereotyped groups in research may carry some unconscious "threats" with them that will influence how they perceive critiques of their research. Discuss how the similarities or differences between the backgrounds of the mentor and the trainee might influence how criticism is intended and how it is perceived. Share information about stereotype threat and imposter syndrome to help trainees learn to overcome and push past these barriers with confidence in one's ability to complete tasks with success. Encourage trainees to ask questions of their mentors, including "What else should I know or be doing to be successful on this project?"

Contributed by A. R. Butz. (2015). *The Power of Social Persuasion.*

Implementation Guide

Workshop Session (50 minutes)

Self-efficacy refers to the confidence that individuals have in their ability to perform a given task. Individuals evaluate their self-efficacy based on their past accomplishments and experiences, the successes and failures of others, their emotional and physiological state, and the messages that they receive from others (i.e., social persuasion). This activity focuses on the messages that trainees receive from their mentor. Feedback and criticism are an inevitable part the mentoring relationship, especially when it comes to writing up and presenting research. In this activity, we will consider the influence that feedback from mentors may have on a trainee's self-efficacy, and devise strategies to help them maintain confidence in the face of criticism.

► **Activity: The Power of Social Persuasion** (40 minutes)

- The goal of this activity is for trainees to assess the influence that mentors have on their self-efficacy and to devise strategies to cope with and respond to feedback that can negatively influence self-efficacy.
- Have trainees read the first version of the mentor's feedback on the handout quietly to themselves or out loud in small groups (pairs or triads). (3 minutes)
- Have the participants discuss the following questions in their small groups, or have the following discussion with the large group. (10 minutes)
  - How do you feel right now? Write down some of the emotions and/or physical responses you are feeling.
  - How would this feedback influence your confidence in your ability to continue to prepare this manuscript for publication?
  - How would it influence your confidence in your ability to write future successful manuscripts?
  - How might you go about looking to other sources (i.e., individuals, messages, or experiences) that could increase your self-efficacy to revise this manuscript?
- If using small groups, have participants share highlights of their discussions with the large group. (5 minutes)
- Large-group discussion questions. (20 minutes)
  - How might this feedback influence your mentoring relationship? How might it influence your desire to collaborate with this individual in the future?
  - How might this feedback be perceived differently if you were at an earlier stage in your career/training (e.g., as a first year undergraduate or graduate student) or at a later stage (e.g., as a full professor or senior scientist)?
  - What are the assumptions that you find yourself making about the person giving you this feedback?
  - How do you maintain your confidence in the face of criticism?
  - Would you address how this feedback made you feel with your mentor? Why or why not?
- Hand out and invite trainees to read the second version of the mentor's feedback.
  - How does your reaction to this feedback differ from the first example?
  - What are the assumptions that you find yourself making about the person
  - giving you this feedback?
  - What is the intent behind each of these feedback examples? Is it the same?

Contributed by A. R. Butz. (2015). *The Power of Social Persuasion.*

► **Wrap-up** (5 minutes)

- Teachers and mentors are in a position to have a profound influence on how trainees perceive their capabilities, for better or for worse. One way to deal with critical feedback that may lower self-efficacy is to consider the intent of what the teacher or mentor said.
- Social persuasions are just one source of self-efficacy; at any given time, one or all four of the sources of self-efficacy may influence trainee confidence. Tell trainees that when they receive feedback that lowers their self-efficacy, they should look to other sources to verify their capabilities (e.g., I have successfully navigated the ups and downs of preparing a manuscript before).

Contributed by A. R. Butz. (2015). *The Power of Social Persuasion.*

THE POWER OF SOCIAL PERSUASION

Learning Objectives

Trainees will:

- Assess the influence that mentors have on confidence in abilities.
- Devise strategies to cope with and respond to feedback that negatively influences trainee confidence.

Self-efficacy refers to the confidence that you have in your ability to perform a given task. Individuals evaluate their self-efficacy based on their past accomplishments and experiences, the successes and failures of others, their emotional and physiological state, and the messages that they receive from others (i.e., social persuasions). In this activity, we are going to focus on the messages that you receive from your mentor. Feedback and criticism are an inevitable part of the mentoring relationship, especially when it comes to writing and presenting research. In this activity, consider the influence that feedback from mentors may have on your self-efficacy, and devise strategies to maintain confidence in the face of criticism.

Directions: You have written the first draft of a manuscript for which your mentor is a coauthor. You spent a lot of time working on the manuscript and are really pleased with the progress that you have made on this paper. You send the manuscript to your mentor for feedback. Imagine that you have received an email from your mentor with this feedback:

> *I have included some edits for grammar and clarity in the document. The manuscript needs substantial work before I see it again. You have cited a lot of prior research in the introduction and literature review, but it is disorganized and difficult to follow. The methods and results are okay, but the manuscript will not be ready to submit to the editor until the discussion is further developed. Once you have made these changes, let me know and I will take another look. I do not want to waste any more of my time reviewing this until the manuscript has been drastically improved.*

Questions for Discussion

- How do you feel right now? Write down some of the emotions and/or physical responses you are feeling.
- What are the assumptions that you find yourself making about the person giving you this feedback?
- How would this feedback influence your confidence in your ability to continue to prepare this manuscript for publication?
- How would it influence your confidence in your ability to write successful manuscripts in the future?
- How might you go about looking to other sources (i.e., individuals, messages, or experiences) that could increase your self-efficacy to revise this manuscript?

Contributed by A. R. Butz. (2015). *The Power of Social Persuasion.*

Consider the same feedback framed in a different way:

This is a good first draft of the manuscript. I have included some edits for grammar and clarity in the document. I can tell that you have put in a lot of time and effort into reviewing the literature. The methods and results are clearly articulated and are explained in a way that should be accessible to a broad audience, which should please the journal editor when we submit it. The discussion section needs some work, particularly where you are trying to make the case for how our study extends what is currently known. I think you could also spend a little more time in the introduction setting up the study and doing a little foreshadowing for the reader. I would like to review the manuscript again once you have addressed these comments, but I have every confidence that you can get this manuscript to where it needs to be.

► How do you feel after receiving this feedback?
► What are the assumptions that you find yourself making about the person giving you this feedback?

Contributed by A. R. Butz. (2015). *The Power of Social Persuasion.*

CASE STUDY: THE SHARING OF RESEARCH MATERIALS

Learning Objectives

Trainees will:

- Discuss and practice ethical research decision making.

Trainee Level

undergraduate or graduate trainees
novice trainees

Activity Components and Estimated Time for Completion

- In Session Time: 25 minutes

Total time: 25 minutes

When to Use This Activity

This activity is suitable for undergraduate and graduate level-novice trainees who have chosen a research mentor and who have been conducting research for at least one month. Other activities that may be used with this activity include:

- Case Study: Credit Where Credit Is Due
- Case Study: Selection of Data
- Ethics Case Discussion with Mentor
- Truth and Consequences Article

Inclusion Considerations

Discuss with trainees how understanding of ethical behavior may be different based on differences in cultural backgrounds or across generations. Facilitators can ask trainees to consider the case study from different cultural or generational perspectives. Emphasize that it can sometimes be as difficult to determine whether behavior is unethical as it is to decide how to deal with that behavior in a sensitive and respectful manner. Encourage trainees to seek input from others, in particular those who can offer different perspectives, when dealing with potentially unethical situations in the research environment.

Areas of Trainee Development

- Research Ethics
  - Develop responsible and ethical research practices.
- Research Comprehension and Communication Skills
  - Develop an understanding of the research environment.

Contributed by A. R. Butz with information from Branchaw, J. L., Pfund, C., and Rediske, R. (2010). *Entering Research: A Facilitator's Manual.* New York: W.H. Freeman & Co.

Implementation Guide

Workshop Session (25 minutes)

- **Role-Playing Exercise** (5 minutes)
  - Distribute and act out the role-playing exercise included in the trainee materials. Ask for volunteers to play the roles of Ed, Maya, and Abdul, and someone to read the narrated sections of the role-playing exercise in italics.
- **Large-Group Discussion Questions** (10 minutes)
  - What kinds of information are appropriate for researchers to share with their colleagues? What about when they change research groups?

    - Specific techniques or materials that are developed by a researcher should only be shared with the permission of that person.
    - Norms about sharing information between colleagues or research groups could be different at different institutions or within different fields of research. It is important to understand the "culture" around sharing resources before disclosing any information.
    - Information should only be shared when it is publicly available (e.g., in published form or presented at a conference).
    - Information can be shared with any colleague in the same area of research.

  - Where can Ed go for help in obtaining materials?

    - Ed can ask Abdul to intervene on his behalf with the former professor.
    - Ed can ask Maya for additional guidance or ask if she knows of any other individuals who could help.

  - Should Ed have contacted the professor at the other university directly himself?

    - The professor may have been more open to sharing information if they knew that Abdul was involved in the research group. An alternative approach could have been to ask Abdul if he would mind introducing Ed to his professor via email first to help make the connection.

  - Are there risks involving other people in this situation?

    - The denied request could cause a rift in Abdul's relationship with his former research group.

- Alternatively, trainees can discuss these questions in small groups and report highlights from their discussion to the larger group.
- **Wrap-up** (5 minutes)
  - Summarize the main ideas generated from the large-group discussion.
  - Encourage trainees to talk with their mentors about their policies and philosophies regarding sharing work with others outside of the research group.

Contributed by A. R. Butz with information from Branchaw, J. L., Pfund, C., and Rediske, R. (2010). *Entering Research: A Facilitator's Manual.* New York: W.H. Freeman & Co.

CASE STUDY: THE SHARING OF RESEARCH MATERIALS

Learning Objectives

Trainees will:

- Discuss and practice ethical research decision making.

(Modified from: *On Being a Scientist: Responsible Conduct in Research*, 2nd ed., National Academy Press, 1995)

The Players

- ***Ed,*** an undergraduate student majoring in biochemistry doing a senior honors thesis.
- ***Maya,*** a 4th year graduate student, is his direct mentor in the lab.
- ***Abdul,*** a postdoctoral fellow, who just joined the lab.

The Situation

Ed is frustrated because he is having trouble getting an assay to work correctly. He has tried everything he can think of and now turns to his mentor for help.

Ed: "Maya, I still can't get that assay to work reliably. Sometimes it does, but usually I get nothing. Even when I do get data, I don't know if it is reliable. Can you think of anything else to try?"

Maya: "You've tried everything I can think of. When we did the experiment together, your technique was flawless, so it must be that catalyzing reagent. Have you talked to Abdul about it?"

Ed: "Abdul?"

Maya: "Oh that's right, you haven't met him yet. He's a new post-doc in the lab, who just started this week. The lab he did his Ph.D. in used similar assays. I bet they used that catalyzing reagent, or something like it. You should ask him."

Ed: "Thanks, I'll go introduce myself."

Ed finds Abdul in Professor Lowrey's office. When he approaches, they invite him in and Professor Lowrey introduces them to one another.

Ed: "Nice to meet you, Abdul, and welcome to the lab. I'd love to talk to you about my project sometime and learn about yours."

Abdul: "Me, too. I'm anxious to learn about all the different projects going on in the lab. Wanna' have lunch tomorrow?"

Ed: "Sounds good. I have class until 11:45. I'll come find you after that."

Abdul: "Great."

The next day at lunch, Ed describes his project, and the problems he's having with the catalyzing reagent to Abdul.

Abdul: "You should contact my old lab. They use that reagent and have developed a form of it that yields good results. I'm sure they would send you some if you asked."

Ed: "Thanks for the tip! I'll send them an email today."

Contributed by A. R. Butz with information from Branchaw, J. L., Pfund, C., and Rediske, R. (2010). *Entering Research: A Facilitator's Manual.* New York: W.H. Freeman & Co.

Ed writes the professor in charge of Abdul's former lab, which is at another university. She writes back and says they are still struggling to develop and characterize the reagent and are not ready to share it yet. Ed is frustrated.

Ed: "Your old mentor claims that their catalyzing reagent isn't ready for prime time, so she doesn't want to share it yet."

Abdul: "That's ridiculous. They just don't want to give you a break."

Questions for Discussion

1. What kinds of information are appropriate for researchers to share with their colleagues? What about with their research groups?

2. Where can Ed go for help in obtaining materials?

3. Should Ed have contacted the professor at the other university directly himself?

4. Are there risks involving other people in this situation?

Contributed by A. R. Butz with information from Branchaw, J. L., Pfund, C., and Rediske, R. (2010). *Entering Research: A Facilitator's Manual.* New York: W.H. Freeman & Co.

THREE MENTORS

Learning Objectives

Trainees will:
- ▶ Identify different mentoring styles.
- ▶ Identify preferred mentoring styles.

Trainee Level

undergraduate or graduate trainees
novice trainees

Areas of Trainee Development

- ▶ Professional and Career Development Skills
  - Develop confidence in pursuing a research career.
- ▶ Research Comprehension and Communication Skills
  - Develop effective interpersonal communication skills.
- ▶ Researcher Identity
  - Develop identity as a researcher.

Activity Components and Estimated Time for Completion

- ▶ Trainee Pre-Assignment Time: 20 minutes
- ▶ In Session Time: 35 minutes

Total time: 55 minutes

When to Use This Activity

This activity is appropriate for novice researchers in the process of selecting a research mentor. Other activities that may be used with this activity include:

- ▶ Finding a Research Mentor
- ▶ Finding Potential Research Rotation Groups and Mentors

A modified description of the three mentors for graduate trainees is provided. The facilitation instructions described below are appropriate for both career stages.

Inclusion Considerations

Trainees from backgrounds historically underrepresented in STEM might perceive typical lab or mentoring dynamics, such as limited access to a mentor or a hands-off style of mentoring, as a negative reflection of their mentor's interest in them. Likewise, power and authority dynamics may be factors that cause distance between a trainee and their mentor. Regardless of mentoring style, encourage all trainees to establish clear expectations to which both they and their mentor can agree. Help them understand that, even with different communication styles and backgrounds from their mentor, they can develop a positive and trusting relationship.

Contributed by E. Frazier, C. Pfund, and A. R. Butz with information from Branchaw, J. L., Pfund, C., and Rediske, R. (2010). *Entering Research: A Facilitator's Manual.* New York: W.H. Freeman & Co.

Implementation Guide

Trainee Pre-Assignment (20 minutes)

► **Trainee Reflection Worksheets**

- Trainees should answer the questions on the "Trainee Reflection Worksheet," located in the trainee materials, in preparation for the in-class discussion.

Workshop Session (35 minutes)

► **Three Mentors Activity** (5 minutes)

- Ask students to individually complete the Three Mentors Activity

► **Small-Group Discussion** (10 minutes)

- In small groups, have trainees discuss the advantages and disadvantages of working with each type of mentor and share the qualities of their ideal mentors from the Trainee Reflection assignment, located in the trainee materials. Alternatively, facilitators may wish to bring the large group together for this discussion.

Alternative Implementation: Trainees can be divided into three small groups with each group discussing one of the professors (1, 2, or 3) for 5–7 minutes. The facilitator can then open the large-group discussion by asking each small group to present their discussion.

► **Large-Group Discussion** (20 minutes)

- Using the questions below, discuss trainees' answers and reflect on which mentoring strategy would work best for each trainee given his/her learning style and level of independence.
  - What are some strategies you discussed that would be helpful when working with each of the professors? Below are a few examples:

> - Professor 1—If the trainee feels that weekly meetings are too frequent, they could ask the professor to meet once every two weeks or once a month that after they have demonstrated signs of independence.
> - Professor 2—The trainee may want to schedule a once a month meeting to meet with the professor to give her/him an update or to ask questions about other aspects of academia.
> - Professor 3—If a trainee needs more guidance, they should take the initiative to arrange additional monthly meetings with the mentor or whenever the trainee feels the need for more guidance.

  - What are the most important elements or characteristics that you would like in your mentor–trainee relationship to have now?
  - What does your selection of mentor tell you about how you learn best?
  - Do you anticipate that this will change in the future? If so, how?
  - What can you do to ensure that you get what you need out of this relationship with your mentor?

Contributed by E. Frazier, C. Pfund, and A. R. Butz with information from Branchaw, J. L., Pfund, C., and Rediske, R. (2010). *Entering Research: A Facilitator's Manual.* New York: W.H. Freeman & Co.

THREE MENTORS

Undergraduate

Learning Objectives

Trainees will:

- ► Identify different mentoring styles.
- ► Identify preferred mentoring styles.

Trainee Reflection Worksheet

1. What kind of learner are you? Do you learn best by looking at diagrams, graphs, and figures (visual learner), by listening (auditory learner), and/or by handling materials (kinesthetic learner)? Do you learn best by using a combination of these methods?

2. How can your mentor support your learning style in your research experience?

3. How much guidance do you feel that you need when joining a new laboratory?
   a. Consistent individual guidance throughout my entire time in the lab
   b. A great deal of guidance until I learn the procedures, but after that I can be left alone and would just like to report back my findings and ask for help if I need it
   c. Just enough to learn the procedures and then I like to be left alone to do my work and will report back or ask for help if I need it

4. Based on your reflections above, describe YOUR ideal mentor.

5. Why did you pick these characteristics in your mentor?

Contributed by E. Frazier, C. Pfund, and A. R. Butz with information from Branchaw, J. L., Pfund, C., and Rediske, R. (2010). *Entering Research: A Facilitator's Manual.* New York: W.H. Freeman & Co.

"THREE MENTORS"

Read the descriptions of three different types of research mentors. From your perspective, describe the advantages and disadvantages of working with each mentor. Outline strategies you would use to overcome the disadvantages. After completing this exercise, revise the characteristics of your ideal mentor on the trainee reflection worksheet.

Mentor 1

This mentor is very hands-on and likes to be the primary mentor for undergraduate researchers in the lab. The mentor works directly with trainees much of the time and wants to know everything that goes on all the time. The mentor sets up weekly individual meetings and engages in frequent dialogue about the research.

Advantages:

Disadvantages:

What qualities/attributes or skills are needed for a trainee to succeed in this environment?

Mentor 2

This mentor is very famous and travels a great deal. Because of this, they have a formal system in which the senior researchers in the group act as mentors for the newer trainees. The mentor keeps up-to-date on the progress of each trainee via frequent emails and meetings when they are in town.

Advantages:

Disadvantages:

What qualities/attributes or skills are needed for a trainee to succeed in this environment?

Contributed by E. Frazier, C. Pfund, and A. R. Butz with information from Branchaw, J. L., Pfund, C., and Rediske, R. (2010). *Entering Research: A Facilitator's Manual.* New York: W.H. Freeman & Co.

Mentor 3

This mentor is hands-off. The mentor is around, but likes to give trainees space to see how they handle independence. The mentor typically has senior trainees informally mentor the newer ones. This mentor meets with each trainee once a month and holds regular structured lab meetings.

Advantages:

Disadvantages:

What qualities/attributes or skills are needed for a trainee to succeed in this environment?

Contributed by E. Frazier, C. Pfund, and A. R. Butz with information from Branchaw, J. L., Pfund, C., and Rediske, R. (2010). *Entering Research: A Facilitator's Manual.* New York: W.H. Freeman & Co.

THREE MENTORS

Graduate

Learning Objectives

Trainees will:

- ► Identify different mentoring styles.
- ► Identify preferred mentoring styles.

Trainee Reflection Worksheet

1. What kind of learner are you? Do you learn best by looking at diagrams, graphs, and figures (visual learner), by listening (auditory learner), and/or by handling materials (kinesthetic learner)? Do you learn best by using a combination of these methods?

2. How can your mentor support your learning style in your research experience?

3. How much guidance do you feel that you need when joining a new laboratory?
   a. Consistent individual guidance throughout my entire time in the lab
   b. A great deal of guidance until I learn the procedures, but after that I can be left alone and would just like to report back my findings and ask for help if I need it
   c. Just enough to learn the procedures and then I like to left alone to do my work and will report back or ask for help if I need it

4. Based on your reflections above, describe YOUR ideal mentor:

5. Why did you pick these characteristics in your mentor?

Trainee Materials

Contributed by E. Frazier, C. Pfund, and A. R. Butz with information from Branchaw, J. L., Pfund, C., and Rediske, R. (2010). *Entering Research: A Facilitator's Manual.* New York: W.H. Freeman & Co.

"THREE MENTORS"

Read the discussions of three different types of mentors and describe the advantages and disadvantages of working with each from your perspective. Outline strategies you would use to be successful in this research environment.

Mentor 1

This mentor is very hands-on and likes to be the primary mentor for the graduate trainees in the lab. Trainees have to work on questions relating to the funded research grants and cannot deviate from such research questions. The mentor works directly with trainees much of the time and wants to know everything that goes on all the time. The mentor keeps tabs of the graduate students and questions them if they are not in the laboratory working on their research projects. The mentor sets up weekly individual meetings and engages in frequent dialogue about the research.

Advantages:

Disadvantages:

What qualities/attributes or skills are needed for a trainee to succeed in this environment?

Mentor 2

This mentor is very famous and travels a great deal. Because of this, they have a formal system in which the senior researchers in the group act as mentors for the newer trainees. Although this research mentor hands the graduate students their research questions related to the funded grants, the mentor is open to the discussion of other research questions that relate to the mentor's topic of research. The mentor keeps up-to-date on the progress of each graduate student via frequent emails and meetings when in town.

Advantages:

Disadvantages:

What qualities/attributes or skills are needed for a trainee to succeed in this environment?

Contributed by E. Frazier, C. Pfund, and A. R. Butz with information from Branchaw, J. L., Pfund, C., and Rediske, R. (2010). *Entering Research: A Facilitator's Manual.* New York: W.H. Freeman & Co.

Mentor 3

This mentor is hands-off. The mentor is around, but likes to give students space to see how they handle independence. The mentor typically has senior trainees informally mentor the newer ones. The mentor typically hands graduate students research questions that are related to funded grants, but is open to discuss other research questions as long as they relate to the mentor's topic of research. This mentor meets with each graduate student once a month and holds regular structured lab meetings.

Advantages:

Disadvantages:

What qualities/attributes or skills are needed for a trainee to succeed in this environment?

Contributed by E. Frazier, C. Pfund, and A. R. Butz with information from Branchaw, J. L., Pfund, C., and Rediske, R. (2010). *Entering Research: A Facilitator's Manual.* New York: W.H. Freeman & Co.

THREE-MINUTE RESEARCH STORY

Learning Objectives

Trainees will:

- ► Practice communicating research to a general audience in three minutes or less.

Trainee Level

undergraduate or graduate trainees
novice, intermediate, or advanced trainees

Areas of Trainee Development

- ► Research Comprehension and Communication Skills
  - Develop research communication skills.
- ► Researcher Identity
  - Develop identity as a researcher.

Activity Components and Estimated Time for Completion

- ► Trainee Pre-Assignment Time: 0–1 hour (if abstract was not developed prior to session)
- ► In Session Time: 25 minutes

Total time: 25 minutes–1 hour, 25 minutes

When to Use This Activity

This activity can be implemented at any time in the research experience, but it should be introduced following a discussion on the importance of science communication. The aim of this activity is to help trainees develop skills in how to effectively communicate research to a public audience in an oral (elevator pitch) format through an iterative process of practice and revision.

This activity works best when trainees come to the session with a polished draft of an abstract or summary of their research for the general public. Other activities that may be used with this activity include:

- ► General Public Abstract
- ► Communicating Research to the General Public

An important follow-up is an assignment in which the trainees come back to class with a polished short form of science communication that is designed for a specific audience.

Inclusion Considerations

Some trainees may come from backgrounds where humility is valued, and they have been acculturated not to speak of their achievements freely. Talk about why it is important to speak about one's research positively and with confidence, and reassure them that they can do this while remaining true to their ideals around humility.

To build trainees' confidence in presenting research, especially in the presence of those in authority, provide plenty of positive feedback on their presentations. Consider using a "compliment sandwich" approach: Start with what was done well, insert what can be improved, and close with a positive comment.

Contributed by L. Adams (2018). *Three-Minute Research Story.*

Implementation Guide

Trainee Pre-Assignment (1 hour if not already completed)

- Approximately one week before the session, distribute the "Three-Minute Research Story" assignment (see trainee materials). Trainees should write and bring to class a general public abstract or brief paragraph that describes their research, which may be completed as part of the "General Public Abstract" activity.
- Write possible audiences on slips of paper and place them in a vessel. Trainees choose an audience to present to by drawing a slip of paper from the vessel and adapt their "Three-Minute Research Story" to this audience for presentation during the next session.
  - Possible Audiences:
    - 4th–8th grade science fair
    - High school science fair
    - Incoming first-year undergraduate science majors
    - Sierra Club/Nonprofit group
    - Reporter for TV news station
    - YouTube audience
    - Citizen scientists
    - Group of scientists (outside of your discipline)
    - Donors
    - State legislature/policymakers
    - Your family
    - During a job interview for a general science position
    - During an interview for graduate school

Workshop Session (25 minutes)

- **Introduction** (5 minutes)
  - Trainees can present and provide feedback during a typical 50-minute session. Alternatively, trainees may post their presentation or general public paragraph online before the session and time in session can be used to provide more extensive feedback.
  - Briefly discuss what an elevator pitch is and why it is important. Explain that although many elevator pitches are 30–60 seconds long, we will be practicing a longer 2- to 3-minute version.
  - *Different approaches to crafting an **elevator pitch***:
    - Outline four key topics: the problem, why it matters, potential solutions, and the benefits of addressing it. List all the points that you might want to make under these topics, and then narrow them to the most important.
    - Follow the basic structure of a story (e.g., set-up, complicating action, development, and climax): set up your subject, give it a source of tension, reveal a possible solution, and combine all of the content to release the tension.
- **Three-Minute Research Story Practice and Feedback** (15 minutes)
  - Have the trainees pair up and take turns practicing their 2- to 3-minute research story, and give one another feedback. Alternatively, if the group is small, each trainee may present to the entire class.
  - When giving feedback, trainees should be instructed that as a "listener," they should be prepared to paraphrase back to the speaker the main take-away message they heard. The "listener" can then follow up with constructive feedback on how to improve the message. Having the listener paraphrase back the message they heard will ensure that two-way communication is taking place.

- As time allows, trainees can rotate partners, which provides multiple opportunities to practice their pitch in one sitting.

► **Assessment Rubric**

- The assessment rubric can be used in multiple ways:
  - The facilitator can use it to assess the content and delivery of the 3-minute research stories.
  - The facilitator can distribute it to trainees with the assignment as a guide.
  - Trainees can use it to provide feedback to their peers (peer review).

THREE-MINUTE RESEARCH STORY

Assessment Rubric

Name of presenter: Title/Topic:

	0	1	2	3
Introduction	Absent	The presenter: • did not clearly and concisely state their name, intended audience, and style of pitch • did not introduce audience to the "big picture" of the research	The presenter: • clearly stated their name, and somewhat clearly and concisely stated the intended audience, and style of pitch • may have introduced audience to the "big picture" of the research	The presenter: • clearly and concisely stated their name, intended audience, and style of the pitch • quickly engaged the audience by introducing the "big picture" of the research
Content	Absent	• somewhat identified the problem and importance of their work • did not propose a hypothesis or possible solution to the research question • may have illustrated the methodology, but not the potential impact of the research • did not provide a summary of the project	• somewhat identified the problem and importance of their work • may have proposed a hypothesis or possible solution to the research question • may have illustrated the methodology and potential impact of the research • provided an unclear summary of the project	• identified the problem and importance of their work • proposed a possible solution to the research question • illustrated the methodology and potential impact of the research • provided a clear summary of the project
Delivery	Absent	• articulated key points in a lengthy or unprofessional manner • conveyed minimal confidence and enthusiasm about their research project • did not utilize nonverbal communication to enhance the pitch • took longer than three minutes to present	• articulated key points in a lengthy but professional manner • somewhat conveyed confidence and enthusiasm about their research project • utilized nonverbal communication to enhance the pitch • concluded presentation within three minutes	• articulated key points in a concise professional manner • conveyed confidence and enthusiasm about their research project • utilized nonverbal communication to enhance the pitch • concluded presentation within three minutes

Notes:

Contributed by L. Adams (2018). *Three-Minute Research Story.*

THREE-MINUTE RESEARCH STORY

Learning Objectives

Trainees will:

► Practice communicating research to a general audience in three minutes or less.

The purpose of this assignment is to practice oral research communication skills by crafting and delivering an elevator pitch; a three-minute research story that communicates some sort of research in a short, concise, but engaging manner to a specific audience.

The intent of an elevator pitch is to get someone interested in what you have to say and to engage in further discussions, not to tell them everything there is to know.

The assignment: Prepare to communicate a piece of research (most likely your own) to a specific audience in three minutes or less.

At the start of your talk, please state your name, intended audience, and brief intent of the message (e.g., informative or persuasive).

Strategies you may use to craft your elevator pitch:

1. Outline four key topics: the problem or research question, why it matters, potential solutions or answers, and the benefits of fixing it or answering it. List all the points that you might want to make under these topics, and then winnow them down to the most important ones.
2. Follow the basic structure of a story (e.g., set-up, complicating action, development, and climax): set up your subject, give it a source of tension, reveal a possible solution, and combine all of the content to release the tension.

Additional Resources to Help You Get Started

Kwok, R. (2013, February 6). Communication: Two minutes to impress. *Nature, 494*. Retrieved from http://www.nature.com/naturejobs/science/articles/10.1038/nj7435-137a

Shaik-Lesko, R. (2014, August 1). Scientific elevator pitches. *The Scientist.* Retrieved from http://www.the-scientist.com/?articles.view/articleNo/40702/title/Scientific-Elevator-Pitches/

Uyen. (2013, August 18). Elevator pitches for scientists: What, when, where and how [Blog post]. Retrieved from http://thepostdocway.com/content/elevator-pitches-scientists-what-when-where-and-how

Contributed by L. Adams (2018). *Three-Minute Research Story.*

TIPS FOR TECHNICAL WRITERS

Learning Objectives

Trainees will:

- Learn strategies for improving technical writing.
- Learn strategies and criterion for giving feedback on technical writing.

Trainee Level

undergraduate or graduate trainees
novice, intermediate, or advanced trainees

Areas of Trainee Development

- Research Comprehension and Communication Skills
  - Develop research communication skills.

Activity Components and Estimated Time for Completion:

- Trainee Pre-Assignment Time: 30 minutes
- In Session Time: 40 minutes

Total time: 1 hour, 10 minutes

When to Use This Activity

This activity may be used with undergraduate and graduate students at any level of research experience. It is recommended that this activity is implemented before trainees perform any kind of technical writing (e.g., prior to writing a research proposal). The resource included in the trainee materials for this activity may also be used by for trainees when providing feedback on one another's technical writing.

Inclusion Considerations

Novice trainees, especially those who have traveled nontraditional academic pathways, may have had limited opportunity to practice research writing. Reassure trainees that their research writing skills will improve with practice and even the most experienced writers use an iterative process of review and revision when writing. If trainees have challenges with writing (e.g., disabilities), talk about these challenges and encourage them to seek help from professionals and to share this information with their research mentor, so that he or she is aware and can help.

Contributed by A. Sokac. (2018). *Tips for Technical Writers.*

Implementation Guide

Trainee Pre-Assignment (30 minutes)

- Assign trainees to write one paragraph summarizing one task or experiment that they have completed in the lab (an organic synthesis reaction, a cell fixation and staining protocol, an equipment calibration, etc.).
- The paragraph should include why they did the task or experiment, what they did, and what they got as a final product (i.e., rationale, brief methods, and result). Tell trainees to bring two printed copies of their paragraph to class.

Workshop Session (40 minutes)
Tell trainees that communicating information by writing clear, concise, engaging text (papers, abstracts, proposals, etc.) is critical to research success.

- Distribute the "Technical Writing Tips" resource found in the Trainee Materials and tell trainees that being aware of the strategies in this handout can help writers at any ability level to avoid common mistakes that undermine the effectiveness of their technical writing. In addition, the "Tips" document also provides good criteria for giving feedback on the technical writing of others.
- **Activity: Tips for Technical Writers** (30 minutes)
  - Ask trainees to swap paragraphs with a partner.
  - Tell trainees to carefully read the paragraph that they received and complete the rubric. They can use the "Technical Writing Tips" resource, found in the Trainee Materials, to evaluate the paragraph. (10 minutes)
  - Ask trainees next to evaluate and complete the rubric for their own paragraph. (10 minutes)
  - Pair with their partner and discuss the evaluation of their paragraph. (8–10 minutes)
  - **Discussion Questions**
    - What were the strengths of each paragraph? What were the weaknesses?
    - What was the biggest challenge for you in writing this paragraph?
    - What did you learn from evaluating your partner's paragraph?
    - How can your paragraph be improved?
- **Wrap-up** (1–2 minutes)
  - Ask the trainees how they would improve their paragraph based on the feedback they received.
  - Encourage trainees to run their paragraph through a free editing software, such as Grammarly; and/or calculate the readability statistics using the tools in Microsoft Word.
  - Encourage trainees to reference the "Technical Writing Tips" handout when completing other writing assignments or reviewing the work of peers.
- **Optional Assignment: Revise and Resubmit Paragraph**
  - Ask the trainees to edit their paragraph using (1) the feedback that they received in class; (2) the "Technical Writing Tips" resource; and (3) the feedback gained from Grammarly and the readability statistics from Word. They should turn in the original paragraph, rubrics, and the edited paragraph and be graded or given feedback on how effectively they incorporated the feedback and how much improvement was made. The rubric can also be used by facilitators to evaluate trainee paragraphs.

Contributed by A. Sokac. (2018). *Tips for Technical Writers.*

TIPS FOR TECHNICAL WRITERS

Learning Objectives

Trainees will:

- ► Learn strategies for improving technical writing.
- ► Learn strategies and criterion for giving feedback on technical writing.

Technical Writing Tips

Use the first person.

Use of "I" or "We" is common and encouraged in the biomedical sciences, but may be discouraged in other STEM disciplines. Make sure to ask a researcher in your area about the use of "I" and "We" in scientific communications.

Pay attention to publications in your field of study and notice if and how often "I" and "We" are used. Do not use the first person in the methods section of a paper in any discipline.

Avoid ambiguous references.

We used an ATP hydrolysis assay to test the function of the enzyme, and it did not work. (What did not work? The assay or the enzyme?)

Use active voice.

Early embryos expressed zygotic genes by cell cycle 14. (Active)

Zygotic genes were expressed in early embryos by cell cycle 14. (Passive)

Be concise, and use simple language.

Our work showed the network motifs present in complex networks by determining the patterns of interconnections that occur within complex networks as opposed to randomized networks.

We identified network motifs specific to complex networks by comparing connectivity patterns found in complex networks with those in randomized networks.

Be specific, not vague.

Small changes in the dosage of Gene X impact the outcome of cell fate decisions.

A 30% reduction in Gene X dosage increases the chance that cells make the wrong cell fate decision.

Clearly define what "this" and "these" refer to

This suggests that Myosin-2 motor activity is not required for contraction.

This result suggests that Myosin-2 motor activity is not required for contraction.

Contributed by A. Sokac. (2018). *Tips for Technical Writers.*

Use consistent formatting for items in a series.

The research project will provide Viola ample opportunity for generating data and publication of research articles.

The research project will provide Viola ample opportunity for generating data and publishing research articles.

Use the "Readability Statistics" function of Word.

Aim for no more than 15–20 words per sentence.

Aim for a reading level appropriate to your audience and to the purpose of your text. Aim for sentences that vary in structure (some short, some long).

Make sure not all sentences start with the same word such as "The."

Make sure that every paragraph has a topic sentence and all other sentences in that paragraph support the topic sentence.

Use transition words or transitions between sentences and between paragraphs

Use of transitions establishes relationships between sentences and paragraphs that improve clarity.

Recognize that writing is an iterative process.

After writing your text, read it as if you are the audience (not the author). Often, it even helps to read your text out loud to catch awkward sentences and typos. Do not be afraid to edit your own drafts ruthlessly—try not to become too attached to what you have written as it may not be the best way to communicate an idea, a result, etc. Always be on the lookout for ways to improve. If it is unclear in any way, fix it. If there are typos or grammatical errors, fix them. At least as much time may be spent on editing as on writing.

Practice, practice, practice.

Look for opportunities to improve your technical writing skills. Examples of writing opportunities include weekly research summaries of your work, travel award grants, utilizing writing centers on campus, fellowship applications, reading scientific literature, and mimicking the style of papers that you find particularly well written. Also, look for opportunities to volunteer to give other people feedback on their abstracts, applications, and proposals.

Get feedback.

Ask for feedback from advisors, classmates, and lab mates. In some cases, even a family member can give valuable feedback regarding readability and clarity.

*We acknowledge the valuable input of Susan Marriott, Ph.D., founder and President of Bioscience Writers, Inc. (www.biosciencewriters.com).

Contributed by A. Sokac. (2018). *Tips for Technical Writers.*

TECHNICAL WRITING TIPS

Assessment Rubric

Instructions: Use the rubric below to evaluate your own and your partner's draft paragraph.

	0	1	2	3
Professional appearance of the text	Unacceptable; no evidence of any proofreading	Many grammatical and spelling errors; no title	Few grammatical and spelling errors; no title	No grammatical or spelling errors; an appropriate title is included
Clarity of the text	The writing does not communicate any information clearly	The writing communicates some information clearly, but it is challenging to understand many sentences	The writing communicates most information clearly	The writing communicates all information clearly
Readability of the text	Text is long, complex, and difficult to read	Text is wordy with many run-on sentences; word choices are overly complex and sometimes miss the intended meaning	Text is somewhat wordy, but with no run-on sentences; word choices are reasonable, but could be simplified; punctuation aids readability	Text is concise; there are no run-on sentences; word choices are simple and appropriate; punctuation aids readability
Precision of the language	Information provided is completely superficial with no details; vague words and ambiguous language are used throughout; not clear what "this" and "that" refer to	Little detail is provided; some vague words and ambiguous language are scattered throughout; not clear what "this" and "that" refer to	Some or too much detail is provided; there are few examples of vague words or ambiguous language; it is usually clear what "this" and "that" refer to	An appropriate level of detail is provided; words and language are specific; it is always clear what "this" and "that" refer to
Organization and use of one topic sentence per paragraph	Paragraphs have no obvious structure, and there is no topic sentence; poor organization of sentences and ideas overall	A topic sentence per paragraph is often difficult to find; organization of sentences and ideas is somewhat logical	Paragraphs contain one topic sentence and the other sentences mostly support the topic sentence; organization of sentences and ideas are logical with only occasional gaps	Paragraphs contain one topic sentence and all other sentences support the topic sentence; organization of sentences and ideas are logical with no gaps

Contributed by A. Sokac. (2018). *Tips for Technical Writers.*

	0	1	2	3
Level of reader engagement achieved by the text	No engagement: Text is written in the passive voice; text is difficult to read, and the message is so unclear that the audience "tunes-out" immediately	Little engagement: Text is written in the passive voice, and is somewhat understandable; audience "tunes-out" by the end	Moderate engagement: Text is written in the first-person and mostly active voice; text is understandable; audience may want to ask a follow-up question after reading	High engagement: Text is written in the first-person and active voice, and is compelling; meaning is completely clear; audience wants to read or discuss more

Discussion Questions

- What were the strengths of each paragraph? What were the weaknesses?

- What was the biggest challenge for you in writing this paragraph?

- What did you learn from evaluating your partner's paragraph?

- How can your paragraph be improved?

Contributed by A. Sokac. (2018). *Tips for Technical Writers.*

TRUTH AND CONSEQUENCES ARTICLE

Learning Objectives

Trainees will:

- Explore academic misconduct and the impact it can have on mentors and their trainees.

Trainee Level

undergraduate or graduate trainees
novice, intermediate, or advanced trainees

Areas of Trainee Development

- Research Ethics
    - Develop responsible and ethical research practices.
- Research Comprehension and Communication Skills
    - Develop an understanding of the research environment.
- Researcher Identity
    - Develop an identity as a researcher.

Activity Components and Estimated Time for Completion

- Trainee Pre-Assignment Time: 15 minutes
- In Session Time: 20 minutes

Total time: 35 minutes

When to Use This Activity

This activity can be used with undergraduate and graduate trainees at any career stage, though it may be most relevant for trainees who have had some experience working in a lab or prior experience with a research mentor or PI.

Other activities that may be used with this activity include:

- Case Study: Selection of Data
- Case Study: Credit Where Credit Is Due
- Case Study: The Sharing of Research Materials
- Ethics Case: Discussion with Mentor

Inclusion Considerations

Discuss with trainees how understanding of ethical behavior may be different based on differences in cultural backgrounds or across generations. Facilitators can ask trainees to consider the "Truth and Consequences" situation from different cultural or generational perspectives. Emphasize that it can sometimes be as difficult to determine whether behavior is unethical as it is to decide how to deal with that behavior in a sensitive and respectful manner. Encourage trainees to seek input from others, in particular those who can offer different perspectives, when dealing with potentially unethical situations in the research environment.

Contributed by A. R. Butz with information from Branchaw, J. L., Pfund, C., and Rediske, R. (2010). *Entering Research: A Facilitator's Manual.* New York: W.H. Freeman & Co.

Implementation Guide

Trainee Pre-Assignment (15 minutes)

► Read article "Truth and Consequences" http://science.sciencemag.org/content/313/5791/1222.full Couzin, J. (September 2006). **"Truth and Consequences,"** *Science*, *313*: 1222–1226. doi:10.1126/science.313.5791.1222

Workshop Session

- Trainees should read the article prior to participating in the discussion.

► **Discussion Questions** (15 minutes)

1. What would you do if you suspected your PI or mentor of falsifying data?

> - Talk to your supervisor.
> - Talk to graduate students or other senior students in the research group.
> - Talk to your PI.
> - Approach the department chair or other trusted faculty members.
> - Visit your institution's research office on responsible conduct of research and research ethics. Make an appointment to have a confidential conversation with someone in this office.

2. Do you think the students handled the situation correctly? Why or why not?
3. Would the stage of your academic career (e.g., sophomore undergraduate vs. fourth-year graduate student) impact your decision about what you would do? How?

> - Impact of the power-differential between mentor and trainee (i.e., undergraduate student approaching a PI versus a graduate student or post-doc).
> - Different nature of the relationships (more advanced trainees often have more contact with the PI than novice trainees).
> - A more senior graduate student or post-doc may have greater fear of the consequences on their career.

4. What might motivate someone to falsify data?

> - Pressure to publish, get grant funding.
> - To impress a mentor or supervisor in order to get a good performance review or a letter of recommendation.

5. Does your school have procedures for dealing with academic misconduct?

> - Facilitators may wish to direct trainees to where they can locate policies and procedures on their institution's website.

Contributed by A. R. Butz with information from Branchaw, J. L., Pfund, C., and Rediske, R. (2010). *Entering Research: A Facilitator's Manual.* New York: W.H. Freeman & Co.

6. Additional questions for discussion:
   - What other ethical issues in addition to falsification of data can arise in scientific research?
   - What ethical issues could you experience in your own research?
   - Where can you find the code of ethics in your discipline?

► **Optional Activity: Retraction Watch**

- Encourage trainees to visit retractionwatch.com (or pull up the site during the session) to emphasize the point that falsifying data and retracting published studies is pretty common.

► **Wrap-up** (3 minutes)

- Trainees often ask what happened to the professor from this article. Dr. Goodwin pled guilty to falsifying data and was sentenced to two years of probation, directed to pay a $500 fine and $50,000 restitution to both the U.S. Department of Health and Human Services and the University of Wisconsin. Dr. Goodwin also agreed to not be involved in any federal government research for three years. After leaving the University of Wisconsin, Dr. Goodwin held a job with a research-based pharmaceutical company. See this press release from the Milwaukee division of the FBI for more information: https://archives.fbi.gov/archives/milwaukee/press-releases/2010/mw090310.htm)

Contributed by A. R. Butz with information from Branchaw, J. L., Pfund, C., and Rediske, R. (2010). *Entering Research: A Facilitator's Manual.* New York: W.H. Freeman & Co.

TRUTH AND CONSEQUENCES ARTICLE

Learning Objectives

Trainees will:

- ► Explore academic misconduct and the impact it can have on mentors and their trainees.

Read article "Truth and Consequences." Couzin, J. (September 2006). **"Truth and Consequences,"** *Science*, *313*: 1222–1226. doi:10.1126/science.313.5791.1222, http://science.sciencemag.org/content/313/5791/1222.full

The goal of this discussion is to explore a "real-life" incident of academic misconduct and the impact it had on the people involved.

Discussion Questions

1. What would you do if you suspected your PI or mentor of falsifying data?

2. Do you think that the students handled the situation correctly? Why or why not?

3. Would the stage of your academic career (e.g., sophomore undergraduate vs. fourth-year graduate student) impact your decision about what you would do? How?

4. Does your school have procedures for dealing with academic misconduct?

Contributed by A. R. Butz with information from Branchaw, J. L., Pfund, C., and Rediske, R. (2010). *Entering Research: A Facilitator's Manual.* New York: W.H. Freeman & Co.

UNDERGRADUATE THESIS 1: COMPONENTS OF AN UNDERGRADUATE RESEARCH THESIS

Learning Objectives

Trainees will:

- ► Learn about the components of an undergraduate research thesis.
- ► Analyze research theses of previous or graduating students.

Trainee Level

undergraduate trainees
intermediate or advanced trainees

Areas of Trainee Development

- ► Research Comprehension and Communication Skills
  - • Develop research communication skills.

Activity Components and Estimated Time for Completion

- ► In Session Time: 1 hour

Total time: 1 hour

When to Use This Activity

This activity can be used with undergraduate trainees who are considering or have committed to writing a research thesis. Undergraduate students who have a working understanding (intermediate level) of their research project will gain the most from this workshop. Other activities that may be used with this activity include:

- ► Undergraduate Thesis 2: Thesis Writing Discussion Panel
- ► Undergraduate Thesis 3: Developing a Thesis Writing Plan

Inclusion Considerations

Novice trainees, especially those who have traveled nontraditional academic pathways, may have had limited opportunity to practice research writing and may be overwhelmed at the prospect of writing a thesis. Reassure trainees that their research writing skills will improve with practice and even the most experienced writers use an iterative process of review and revision when writing.

Contributed by J. M. Bhatt (2018). *Undergraduate Thesis 1: Components of an Undergraduate Research Thesis.*

Implementation Guide

Facilitator Preparation

- ► Prior to implementing this activity, obtain a copy of the thesis guidelines recommended by the university or the programs of the participating trainees. These will help students understand the thesis format requirements.
- ► If possible, share "model" theses submitted by previous students who have graduated from the trainees' programs. Alternatively, the theses could be from different fields such that students can easily identify the similarities and differences in the documents.

Workshop Session (1 hour)

- ► Introduce students to the value of student research theses in developing their science communication skills. If you have a group of trainees from diverse disciplines, introduce the idea that research theses look different across fields. For example, a psychology thesis could look very different than one in neuroscience.
- ► **Brainstorm: What is in a thesis?** (5 minutes)
  - Ask trainees to brainstorm the different components of a thesis. Some possible components include:

    - Title
    - Abstract
    - Introduction
    - Theoretical framework
    - Methods
    - Results
    - Discussion
    - Conclusion
    - Figures and/or tables
    - Acknowledgments
    - Citations

- ► **Example Thesis Documents** (20 minutes)
  - Distribute example theses.
  - Ask trainees to get in small groups of two or three, review an example thesis, and identify the components of the document. (15 minutes)
  - Invite trainees to share the different components of the thesis that they reviewed. Record their responses on a whiteboard or flipchart, noting components that are mentioned more than once. The facilitator should be prepared to add in any components that were not identified by trainees. (5 minutes)
- ► **Thesis Guidelines** (25 minutes)
  - Review the university or program thesis guidelines.
    - What surprised you about the format and/or components of the thesis you reviewed? Did it follow the university or program guidelines?
    - What field-to-field differences and/or similarities did you notice between theses?

- Facilitators should convey the importance of adhering to the standards that are practiced in their respective fields and academic programs.

► **Wrap-up** (10 minutes)
- Invite trainees to ask any remaining questions they have about the thesis format?
- Summarize the main ideas generated from the large-group discussion.

Note: If the facilitators' university or program does not have recommended thesis guidelines, alternative resources include:

► *https://biochem.wisc.edu/undergraduate_program/thesis-undergraduate-program*
► *https://www.jou.ufl.edu/grad/forms/Guidelines-for-writing-thesis-or-dissertation.pdf*
► *http://www.hdfs.chhs.colostate.edu/students/undergraduate/documents/HonorsUndergraduat eThesisGuidelines.pdf*

Note: There are no trainee materials for this activity.

Contributed by J. M. Bhatt (2018). *Undergraduate Thesis 1: Components of an Undergraduate Research Thesis.*

UNDERGRADUATE THESIS 2: THESIS WRITING DISCUSSION PANEL

Learning Objectives

Trainees will:

- ▶ Learn about the thesis writing process with research mentors and graduated or soon to graduate trainees who have successfully written an undergraduate research thesis.
- ▶ Identify writing strategies to guide the thesis writing process.

Trainee Level

undergraduate trainees
intermediate or advanced trainees

Areas of Trainee Development

- ▶ Research Comprehension and Communication Skills
  - Develop research communication skills.

Activity Components and Estimated Time for Completion

- ▶ Trainee Pre-Assignment Time: 30 minutes
- ▶ In Session Time: 1 hour-1 hour, 30 minutes

Total time: 1 hour, 30 minutes–2 hours

When to Use This Activity

This activity can be used with undergraduate trainees who are considering or have already committed to writing a research thesis. Undergraduate trainees who have chosen a research mentor and have a working understanding of their own project will benefit the most from this activity. Other activities that may be used with this activity include:

- ▶ Undergraduate Thesis 1: Components of an Undergraduate Research Thesis
- ▶ Undergraduate Thesis 3: Developing a Thesis Writing Plan

Prior knowledge of the format and requirements of a research thesis in the field and/or academic program prior to engaging in this activity is beneficial.

Inclusion Considerations

Include mentors or recently graduated trainees from diverse backgrounds and with diverse abilities to discuss writing their theses on the panel. Encourage panelists to share the challenges and fears they faced when writing their theses and how they overcame them. Ask trainees who have concerns about writing an undergraduate thesis to ask the panelists for advice about the challenges they anticipate they will face.

Contributed by J. M. Bhatt (2018). *Undergraduate Thesis 2: Thesis Writing Discussion Panel.*

Implementation Guide

Facilitator Preparation:

- Invite two or three research mentors and two or three trainees who have recently completed a thesis in the areas of research in which your trainees are involved to serve as panelists. You may wish to ask mentors to identify recent thesis completers from their program to participate on the panel. It is not necessary to have mentor–trainee pairs on the panel, but pairs will provide the opportunity for trainees to learn about the mentoring relationship in the context of writing a thesis.
- Prior to the panel discussion, it may be useful to work with trainees to generate a set of questions for the panel. Invite trainees to write questions down on index cards, or collectively come up with a list of questions that will benefit the group.

Trainee Pre-Assignment (30 minutes)

- Trainees should familiarize themselves with the format and requirements of a thesis in their program prior to engaging in this activity. Alternatively, facilitators can implement the activity "Components of an Undergraduate Research Thesis" prior to this session to familiarize trainees with thesis requirements.

Workshop Session (1 hour–1 hour, 30 minutes)

- **Panelists Describe Their Research** (15 minutes)
  - Ask each panelist to introduce themselves and describe their research. (2–3 minutes per panelist)
- **Panel Discussion** (30–50 minutes)
  - If the trainees are shy at the beginning or they run out of questions, the facilitator can utilize the questions or points below to keep the discussion going:
    - **Trainee Questions**
      - Why did you decide to write an undergraduate research thesis? What motivated you?
      - What conversations did you have with your mentor about the thesis writing process before you began writing?
      - What was your first step in writing your thesis (e.g., outlining)?
      - In what order were the individual sections written?
      - What was your timeline/writing plan? How closely did you stick to this timeline?
      - Did you submit drafts to your mentor? If so, how many?
      - What comments arose from your mentor and the committee? How long did it take to fix them?
      - How did you avoid plagiarism?
      - Did you have to obtain copyright clearance to reproduce figures (or tables) for the introduction section of your thesis? If so, what steps did you have to take?
      - What was the final length of your thesis and of each chapter?
    - **Mentor Questions**
      - When should a trainee begin planning to write an undergraduate thesis?
      - How can trainees best prepare to write a thesis?
      - What writing strategies do you recommend for trainees new to this type of writing?
      - What does the feedback process look like in your research group? Do you give written feedback, verbal feedback, etc.?
      - Do you have any specific advice, or can you suggest any writing strategies for trainees for whom English is not their first language?

Contributed by J. M. Bhatt (2018). *Undergraduate Thesis 2: Thesis Writing Discussion Panel.*

- **Mentor and Trainee Questions**
  - What writing strategies work best for you (writing in small chunks every day, dedicating several hours to writing, etc.)? What strategies do not work well?
  - Did you ever hit a writer's block? How did you overcome it?
  - How did you manage your references (EndNote, RefWorks, Mendeley, Zotero, etc.)? Did you cite as you were writing or at the end?

► **Wrap-up** (10 minutes)

- Summarize the thesis writing tips generated from the discussion with the panel.
- Share contact information of panelists (with their permission).
- Discuss availability of the writing center to help trainees.

Note: There are no trainee materials for this activity.

Contributed by J. M. Bhatt (2018). *Undergraduate Thesis 2: Thesis Writing Discussion Panel.*

UNDERGRADUATE THESIS 3: DEVELOPING A THESIS WRITING PLAN

Learning Objectives

Trainees will:

- Develop a thesis writing plan.

Trainee Level

undergraduate trainees
intermediate or advanced trainees

Areas of Trainee Development

- Research Comprehension and Communication Skills
  - Develop research communication skills.

Activity Components and Estimated Time for Completion

- In Session Time: 1 hour

Total time: 1 hour

When to Use This Activity

This activity should be implemented with undergraduate trainees who have collected and analyzed enough research data to begin the thesis writing process. Trainees must possess an in-depth understanding of their own project and the literature within which the study is embedded. Other activities that may be used with this activity include:

- Undergraduate Thesis 1: Components of an Undergraduate Thesis
- Undergraduate Thesis 2: Thesis Writing Discussion Panel

Inclusion Considerations

Novice trainees, especially those who have traveled nontraditional academic pathways, may have had limited opportunity to practice research writing and may be overwhelmed at the prospect of writing an undergraduate thesis. Beyond this, some trainees may have learning disabilities that could affect their writing. Encourage them to talk about any challenges or concerns they have with their mentor and to ask for the support they need to be successful. Reassure trainees that their scientific writing skills will improve with practice.

Consider using a "compliment sandwich" approach when providing feedback on thesis drafts: Start with what was done well, insert what can be improved, and close with a positive comment. Reassure trainees that their writing skills will improve with practice.

Contributed by J. M. Bhatt. (2018). *Undergraduate Thesis 3: Developing a Thesis Writing Plan.*

Implementation Guide

Workshop Session (1 hour)

- Briefly review the thesis format requirements and/or ask students if they have any questions about the expectations before beginning.
- **SMART Goals** (20 minutes)
  - In small groups, ask trainees to generate a list of strategies and considerations to facilitate the writing process.
  - Invite trainees to share their ideas with the large group, writing them down on a whiteboard or flip-chart. Some examples of considerations that trainees may have include:

> - What sections need to be written first?
>   - *Note:* When working with a group of trainees from different disciplines, facilitators should be aware that the preferred ordering of sections can vary between disciplines.
> - How much time should each section take?
> - What are their daily/weekly writing goals?
> - Which days and times work best for writing?
> - Who will hold you accountable for your daily/weekly writing progress? Mentor? Friend? Study-buddy?

  - Introduce the concept of SMART goals: A SMART goal is defined as one that is specific, measurable, achievable, realistic, and time-bound.
- Trainees develop their writing plan. (15 minutes)
  - Trainees work individually on their timelines to set SMART goals using the template provided in the trainee materials.
- **Small-Group Discussion** (10 minutes)
  - In small groups, trainees share their timelines as a way of encouraging accountability and setting realistic goals.
- **Wrap-up** (5 minutes)
  - Summarize the main ideas generated from the large-group discussion.
  - Invite trainees to share any challenges that they encountered in crafting their timeline or SMART goals.

Note: If trainees will be going through the thesis process as a cohort, it may be useful to revisit SMART goals mid-way through the process to see if the timeline and/or goals need revision. Facilitators can invite trainees to consider whether their estimated timeline was accurate and to identify the factors that kept them on track or made them fall behind on their self-imposed deadlines.

Contributed by J. M. Bhatt. (2018). *Undergraduate Thesis 3: Developing a Thesis Writing Plan.*

UNDERGRADUATE THESIS 3: DEVELOPING A THESIS WRITING PLAN

Learning Objectives

Trainees will:

- Develop a thesis writing plan.

Instructions: Use this worksheet as a starting point to develop a backward timeline and SMART goals for writing your thesis. Remember SMART goals are:

- **Specific**—There is a specific objective outlined; e.g., "*Write three pages of my methods section.*" versus "*Write my thesis.*"
- **Measurable**—You are able to clearly determine whether you met the goal or not. For example, if you write three pages of your methods section, you know you will have met your goal.
- **Achievable**—You can achieve the goal. Writing three pages in a week is likely an achievable goal. Writing an entire thesis in that amount of time may not be attainable.
- **Realistic**—The goal is attainable given the difficulty of the task and the timeframe in which you have to complete it.
- **Time bound**—There is a timeline for the successful completion of your goal. To make the goal of writing three pages time bound, we need to include a deadline, such as "Write three pages *a week.*"

Last Month: ______________

SMART Goal(s): Complete all requirements for thesis paper.

Tasks to achieve goal:

Month: ______________

SMART Goal(s):

Tasks to achieve goal:

Month: ______________

SMART Goal(s):

Tasks to achieve goal:

Month: ______________

SMART Goal(s):

Tasks Relevant to SMART Goals:

Contributed by J. M. Bhatt. (2018). *Undergraduate Thesis 3: Developing a Thesis Writing Plan.*

Month: ______________

SMART Goal(s):
Tasks Relevant to SMART Goals:

Month: ______________

SMART Goal(s):
Tasks Relevant to SMART Goals:

First Month: ______________

SMART Goal(s):
Tasks Relevant to SMART Goals:

Contributed by J. M. Bhatt. (2018). *Undergraduate Thesis 3: Developing a Thesis Writing Plan.*

EXAMPLE THESIS WRITING PLAN

Goal: Complete thesis and present at Undergraduate Symposium by Spring Semester

April

SMART Goal: Complete all requirements for thesis paper.

Tasks Relevant to SMART Goals:

- ► Present thesis research at Undergraduate Symposium.
- ► Address all the feedback provided by mentor.
- ► Submit the final approved product to the program/department and the university library.

March

SMART Goal: Submit final thesis to mentor and prepare for presentation.

Tasks Relevant to SMART Goals:

- ► Do final proofread of thesis for spelling, grammar, and formatting.
- ► Rehearse presentation for symposium in front of peers and incorporate feedback.

February

SMART Goal #1: Prepare complete draft of thesis.

Tasks Relevant to SMART Goals:

- ► Submit a complete thesis draft to research mentor to obtain preliminary feedback.
- ► Address all the mentor comments.

January

SMART Goal #1: Address all feedback obtained from mentors and peers.
SMART Goal #2: Compile all the sections together into one document with the correct formatting as required by the program or department.

December

SMART Goal #1: Complete the methods and introduction sections.
SMART Goal #2: Write the discussion, conclusion, and abstract sections.
SMART Goal #3: Cite all the relevant references that were used during the study.
SMART Goal #4: Ask for feedback from mentor and peers.

November

SMART Goal #1: Revise results section of thesis based on feedback.

- ► Ask research mentor and peers in the program to give critical feedback.

SMART Goal #2: Draft the methods and introduction sections.
SMART Goal #3: Outline the discussion and conclusion sections.

Contributed by J. M. Bhatt. (2018). *Undergraduate Thesis 3: Developing a Thesis Writing Plan.*

October

SMART Goal #1: Complete results section of thesis.

- ► Convert data collected into results tables and figures for the thesis.
- ► Write the first draft of the results section of the thesis, incorporating tables, figures, and figure legends.

September

SMART Goal #1: Complete all data collection required for the thesis project.

Tasks Relevant to SMART Goals:

- ► Identify what data is needed to complete the research project.
- ► Collect all data.

SMART Goal #2: Collect and organize all the relevant references that will be needed for the thesis in reference management software.

Contributed by J. M. Bhatt. (2018). *Undergraduate Thesis 3: Developing a Thesis Writing Plan.*

UNIVERSALISM IN STEM: CASE STUDY AND ANALYSIS

Learning Objectives

Trainees will:

- Consider the social, political, and historical influences behind actions.
- Examine how historical narratives can help us understand contemporary issues in STEM.
- Explore how history can help the STEM community create a more inclusive environment.

Trainee Level

undergraduate or graduate trainees
advanced undergraduate trainees or graduate trainees at any level

Areas of Trainee Development

- Equity and Inclusion Awareness and Skills
  - Advance equity and inclusion in the research environment.
- Research Comprehension and Communication Skills
  - Develop understanding of the research environment.

Activity Components and Estimated Time for Completion

- In Session Time: 1 hour, 20 minutes
- Post-Session Assignment: 1 hour

Total time: 2 hours, 20 minutes

When to Use This Activity

This activity works best when implemented after establishing an inclusive, safe or "brave" environment where trainees feel comfortable talking about social issues in the context of STEM (see activity "Setting the Stage for Inclusive Discussions"). Trainees should be familiar with the peer-review process or have completed "Research Writing 5: The Peer-Review Process" activity prior to participating in this activity.

Inclusion Considerations

Encourage all trainees to share their experiences with the group, but do not require it. Trainees from groups historically underrepresented in research should not be asked or expected to speak for their respective identity group. All shared experiences should be presented as individual experiences, yet considered in light of the historical and social contexts discussed. Challenge all trainees to share in identifying challenges and barriers to creating more inclusive teams and diversifying the research workforce.

Contributed by C. R. C. Long. (2018). *Universalism in STEM: Case Study & Analysis.*

Implementation Guide

Workshop Session (1 hour 20 minutes)

- Many STEM students do not have the opportunity to take a history class because of curriculum restrictions and/or time. However, all fields of study were created by humans, so understanding human history provides insight into the context of the discipline (e.g., influences of who, what, when, where, and how).
- Following the session, trainees should be encouraged to explore the references provided in the activity reading to learn more about the history of their field.

- **Case Study: "Inviting Speakers to Campus"** (20 minutes)
  - Distribute the case study included in the trainee materials. Have trainees read the case silently to themselves or ask for one or two volunteers to read the case aloud to the group. If working with a larger group of trainees, break out into smaller groups.
  - In pairs or small groups, have trainees discuss the questions included with the case study:
    - How would you describe the interactions between the trainees in this case study?
    - Do you think that Andy and Erin's comments about Charnell's suggested speakers were fair? Why or why not?
    - What criteria other than those mentioned in the case might you use to select a guest speaker for your research group or program?
  - Bring the whole group together and invite trainees to share one or two key highlights of their discussion.

- **In-Class Reading: "Analysis: Reading between the Lines"** (10 minutes)
  - Have trainees silently read the handout "Reading between the Lines," included in the trainee materials, and write down their initial reactions.

- **Role-play: "Inviting Speakers to Campus"** (10–15 minutes)
  - Organize trainees into small groups. Ask each group to revisit the case study "Inviting Speakers to Campus" and to consider the evidence provided in the "Reading between the Lines" analysis. Each group can be encouraged to revisit the case study from the perspective of one section of the analysis (i.e., 1, 2, 3, or 4).
  - Invite groups to act out a modified case study using evidence from their assigned section of the analysis to develop a response to Andy's dismissal of Charnell and Muhammad's speaker choices. Have group members assign themselves to play the part of one individual in the case study (Andy, Erin, Charnell, Muhammad, or Heather).
  - In larger groups, additional group members can be assigned the role of observer/spokesperson and can report out on the interactions that they witnessed. In smaller groups, encourage trainees to appoint a spokesperson at the start of the role-play exercise who can help summarize the key points raised in the role-play exercise.

Note: If facilitators notice that groups are reluctant to come up with responses on their own, they may want to encourage trainees to invite others in their group to "tag in" with a response as ideas develop.

Contributed by C. R. C. Long. (2018). *Universalism in STEM: Case Study & Analysis.*

► **Small-Group Presentations** (15 minutes)

- Have the spokesperson for each group summarize the responses that were portrayed in their modified case study role-play. Invite other groups to ask questions (2–3 minutes/group). Facilitators or spokespersons can record example responses on a whiteboard, flipchart, or online document for future reference.

► **Large-Group Discussion** (10 minutes)

- Push the group to explore strategies for dealing with situations like the one presented in the case study, drawing upon the ideas shared in the session thus far.
  - How are decisions made about who should be invited to the institution to give a talk?

> ▪ How well individuals and/or their work are known in the field.
> ▪ Individuals from prestigious universities and/or research labs.
> ▪ Research that is considered cutting edge and/or has a large amount of financial support.
> ▪ Researchers who have won national awards, merits, and/or recognitions.

  - What types of researchers are often adversely affected by the perceived neutrality of universalism?

> ▪ Early career researchers.
> ▪ Women.
> ▪ Researchers at Historically Black Colleges & Universities (HBCU), Hispanic Serving Institutions (HSI), Minority Serving Institutions (MSI).
> ▪ Researchers from minoritized racial/ethnic groups.
> ▪ Researchers from countries not established as major contributors to research.

  - How does the race and gender-based educational tracking system impact how scientific work is accessed?

> ▪ Race or gender can affect the types of resources at an institution.
> ▪ Race or gender can harm or improve access to scientific research opportunities.
> ▪ Race or gender can determine how one accesses scientific research (i.e., school, after-school activities, summer programs, parents, financial resources, etc.).

  - How are race and gender hidden within objective practice in peer review?

> ▪ The type of research conducted by people of a specific race or gender may not be viewed as rigorous.
> ▪ People from a specific race or gender have traditionally attended and worked at specific universities, which may not be considered prestigious.
> ▪ As a result of K–12 tracking systems, people from a specific races or genders may have had limited access to top-ranked universities.

***Note:** Montague Cobb's work was published in the Journal of Negro Education even though it was a scientific study. Challenge trainees to think about why Cobb's work may have been published in this journal and why the Journal of Negro Education was founded.*

Contributed by C. R. C. Long. (2018). *Universalism in STEM: Case Study & Analysis.*

- What can be done to reduce bias in the STEM community?

  - Double-blind reviews (i.e., when the identities of the authors and the reviewers are not known to one another).
  - Intentionally elevating marginalized voices (e.g., inviting scholars from historically underrepresented or marginalized groups in your discipline to give research seminars).

- What can be done to promote inclusion in the STEM community?

  - Invite guest speakers from HBCU, HSI, MSI, TCU.
  - Intentionally cite people of color, women, early career researchers in publications.
  - Intentionally establish collaborations with early career researchers, people of color, and women.
  - Read research across different tier journals.

► **Wrap-up** (5 minutes)

- Emphasize that trainees are moving into an environment where things are assumed to be objective. Yet the structures in place to evaluate the merit of work are not objective, often for historical reasons.

► **Trainee Post-Session Assignment: Reflection on Discussion of Prestige** (1 hour)

- Have trainees investigate how quality scientific research, rigor, prestige, and peer review are defined and conducted in their research group and/or discipline more broadly. They should talk to the individual responsible for the seminar/brown bag in their program/department and ask how they determine who will be invited to speak. What are the application/selection criteria?
- Assign trainees to write a short reflection essay including:
  - the definition, features or characteristics they identified in their department/program,
  - whether they agree or are comfortable with this definition, and
  - how they would ideally modify the definition.

► **(Optional) Follow-up session:** Trainees share their reflections with one another.

Contributed by C. R. C. Long. (2018). *Universalism in STEM: Case Study & Analysis.*

UNIVERSALISM IN STEM: CASE STUDY AND ANALYSIS

Learning Objectives

Trainees will:

- Consider the social, political, and historical influences behind actions.
- Examine how historical narratives can help us understand contemporary issues in STEM.
- Explore how history can help the STEM community create a more inclusive environment.

Case Study: "Selecting Speakers to Invite to Campus"

Background: A committee of five second-year graduate trainees must select one guest speaker for this year's seminar series. Each trainee on the committee gives a presentation about two researchers that he or she would like to invite to campus. After all presentations, the committee members cast a secret ballot.

Case: Charnell was very excited at the prospect of inviting two amazing scientists to campus this year to give a talk. She spent all night preparing slides, making sure to illustrate how important their work is to the field. However, following her presentation, her graduate trainee colleagues did not share her enthusiasm. One of the trainees, Andy, made a comment about the scientists not working at highly ranked universities. Another, Erin, mentioned that neither of the scientists were Nobel laureates or members of the National Academy of Sciences. Heather and Muhammad did not have anything negative to say and felt that the selections should be added to their ballot for voting. Overall, Charnell was very hurt by the committee's comments and felt as though the committee members did not value her opinion.

Another graduate trainee, Muhammad presented after Charnell. Like Charnell, Muhammad spent all night preparing his slides. He was very nervous about the presentation because graduate trainees have not always treated his ideas with respect. He noticed that most of the presenters last year were from the same universities and Muhammad wanted to seize this opportunity to bring someone from a different university to campus. After his presentation, the committee was silent for a while. Then Heather spoke up and endorsed Muhammad's selection. Immediately after Heather's endorsement, Andy dismissed the scientists by saying he had never heard of them, so their work must not that be that important. Erin agreed with Andy's comment. Charnell was still upset about the response to her own presentation and remained silent.

Andy presented two very well-known American scientists as his choices to nominate. Erin and Heather immediately endorsed the selections. Although Charnell and Muhammad were familiar with the work of these two scientists and would love to hear either of them speak, they were hurt that their selections were not well received by the group. Heather was not enthusiastic about the scientists selected by Andy, but endorsed them anyway because Andy is in her research group. Erin endorsed Andy's selection because both of the scientist work at Erin's alma mater.

Questions for Discussion

1. How would you describe the interactions between the trainees in this case study?

2. Do you think that Andy and Erin's comments about Charnell's suggested speakers were fair? Why or why not?

3. What criteria might you use to select a guest speaker for your research group or program?

Contributed by C. R. C. Long. (2018). *Universalism in STEM: Case Study & Analysis.*

UNIVERSALISM IN STEM: CASE STUDY AND ANALYSIS

Analysis: Reading Between the Lines

Universalism

Universalism is the idea that researchers' achievements in the field are assessed solely on their merits. For universalism to work, the community must assess a researcher's contributions objectively, without influence of socially constructed attributes such as class, gender, race, religion, nationality, or personality because these might interfere with the objective evaluation of their work (Merton, 1973). However, universalism often does not impact all researchers equally. Below are a few examples of how structural and systemic bias led to unequal opportunities in higher education for researchers.

1. **A researcher's position in the social structure of the scientific community can impact the evaluation of their work.**
   - Publication productivity is linked to the amount of resources at an institution. Several studies have shown that researchers at more "prestigious," well-funded universities are more productive than researchers at other institutions. A researcher's productivity greatly impacts how they are recognized within a field through citations and awards (Johnson & Hermanowicz, 2017).
   - Impact factor is often the most important factor in how often an article is cited (Callaham, Wears, & Weber, 2002). However, studies have found that journals attempt to maintain or increase their impact factor through strategic self-citation. This has made evaluation of scientific quality using impact factor unreliable and problematic (Johnson & Hermanowicz, 2017).
   - Professional age (i.e., how long an individual has been in the field) may affect editors' evaluations by influencing the types of material (preferred subject matter, theory, etc.) to which they respond favorably and their personal ties with other researchers (Johnson & Hermanowicz, 2017).
   - Doctoral training and academic affiliations influence personal ties between researchers. This directly impacts their evaluation of scientific work (Johnson & Hermanowicz, 2017).

2. **Universities have not provided equal opportunity or access to STEM fields.**
   - Land grant institutions were established as the places where the laboring class was to learn how to do science, specifically to support the agriculture-based economy. These schools were identified as A&M (Agricultural & Mechanical), which differentiated them from the perceivably more "prestigious" liberal arts colleges.
   - Many southern states created separate land grant institutions for African Americans, but this was not required by law until the Morrill Act of 1890 (Geiger, Finlay, Sorber, & Fairbanks, 2015).
   - The Smith-Lever Act of 1914 provided additional funding for land grant institutions but left distribution of funds up to the states. This led to the underfunding of Black land grant institutions (Slaton, 2010).
   - In the 1930s, educational offerings for White people at land grant institutions included architecture, dentistry, engineering, forestry, journalism, law, library science, medicine, pharmacy, veterinary medicine, nursing, commerce, and business. During this time, there was a scientization of agricultural education that was not seen at Black land grant institutions (Slaton, 2010).
   - According to Chapman's dissertation, Howard University, North Carolina Agricultural & Technical State University, and Hampton University were the only HBCUS that offered engineering courses during the 1930s (Slaton, 2010).
   - In addition, university administrators promoting separate but equal education were not encouraging the inclusion of training programs in engineering or medical professions at Black land grant institutions (Slaton, 2010).

Contributed by C. R. C. Long. (2018). *Universalism in STEM: Case Study & Analysis.*

3. Research has been used to perpetuate racial and gender inequality.

- Research studies were used to justify acts of violence against communities of color and to justify their lowered social position (Kendi, 2017; Saini, 2017).
  - For example, during the 1980s and 1990s there was growing understanding within the academic community that tools to accurately measure intelligence do not exist. This new revelation threatened policies that kept people of color out of the best-funded colleges and jobs. This motivated Charles Murray and Richard Herrnstein to publish the *Bell Curve: Intelligence and Class Structure in American Life* in 1994. In this book, Murray and Herrnstein argued that there were inherent differences in intelligence based on race and class. This book has been used to perpetuate the idea that persons of color may not be well suited for degree programs and jobs often associated with a high level of intelligence (i.e., STEM careers; Kendi, 2017).
- Scientists challenged racist and sexist studies by pointing out the flaws in the research. However, their work was often overlooked or disregarded because they did not have the same social capital as other researchers in the field.
  - Montague Cobb, a scientist at Howard University, conducted studies and found no evidence to show a measurable difference between Black and White body types. He concluded that Black people were not inherently inferior to White people. Cobb's findings were published in the *Journal of Negro Education* in 1934, which did not reach as broad of an audience as the scientific journals that had published the studies he was refuting (Baker, 1998).
  - William Alexander Hammond claimed that women's brains were lighter than men's brains at birth. Helen Hamilton Gardener refuted these claims by pointing out there was no way to measure differences between male and female brains at birth. Gardener did not have the same social capital as Hammond. Consequently, her work was not widely accepted in the community (Saini, 2017).

4. Although Black and White women both face gender discrimination, access to scientific research differed between these two racial groups.

- White women studying science have come largely from the middle and upper middle class families that could afford to send their daughters to college. Traditionally, White women took math and science courses to become educators. However, there was a decline in enrollment in science courses by White women during the early 1900s as a result of the social mobility afforded to them through clerical and professional occupations (Tolley, 2014).
- The majority of Black women gained access to science and math in their education courses at HBCUs. Teaching was seen as a good profession for Black women during the public high school movement; however, when clerical positions opened up to women during the 1900s, racist ideas inhibited Black women from clerical occupations, so they were encouraged to pursue domestic science (Shaw, 2010).

Contributed by C. R. C. Long. (2018). *Universalism in STEM: Case Study & Analysis.*

References

Baker, L. D. (1998). *From savage to Negro: Anthropology and the construction of race, 1896–1954.* University of California Press.

Callaham, M., Wears, R. L., and Weber, E. (2002). Journal prestige, publication bias, and other characteristics associated with citation of published studies in peer-reviewed journals. *JAMA, 287*(21): 2847–2850.

Geiger, R. L., Finlay, M. R., Sorber, N. M., and Fairbanks, R. B. (2015). *Science as service: Establishing and reformulating American land-grant universities, 1865–1930* (Vol. 1). University of Alabama Press.

Johnson, D. R., and Hermanowicz, J. C. (2017). Peer review: From "Sacred ideals" to "Profane realities." In *Higher education: Handbook of theory and research* (pp. 485–527). Springer, Cham.

Kendi, I. (2017). *Stamped from the beginning: The definitive history of racist ideas in America.* Random House.

Merton, R. K. (1973). *The sociology of science: Theoretical and empirical investigations.* University of Chicago Press.

Saini, A. (2017). *Inferior: How science got women wrong and the new research that's rewriting the story.* Beacon Press.

Shaw, S. J. (2010). *What a woman ought to be and to do: Black professional women workers during the Jim Crow era.* University of Chicago Press.

Slaton, A. E. (2010). *Race, rigor, and selectivity in U.S. engineering: The history of an occupational color line.* Harvard University Press.

Tolley, K. (2014). *The science education of American girls: A historical perspective.* Routledge.

Contributed by C. R. C. Long. (2018). *Universalism in STEM: Case Study & Analysis.*

Implementation Guide

VISITING PEER RESEARCH GROUPS

Learning Objectives

Trainees will:

- Develop an appreciation for differences in culture among research groups.
- Identify valuable research group attributes.

Trainee Level

undergraduate trainees
novice or intermediate trainees

Areas of Trainee Development

- Research Comprehension and Communication Skills
  - Develop effective interpersonal communication skills.
  - Develop an understanding of the research environment.
- Equity and Inclusion Awareness and Skills
  - Develop skills to deal with personal differences in the research environment.

Activity Components and Estimated Time for Completion

- Trainee Pre-Assignment Time: 30 minutes–1 hour
- In Session Time: 20 minutes

Total time: 50 minutes–1 hour, 20 minutes

When to Use This Activity

This activity is appropriate for novice to intermediate undergraduate trainees and can be used at any point after a trainee has joined a research group.

Inclusion Considerations

Encourage all trainees to consider whether they think they would feel comfortable as members of the research groups they visit. What do they notice that makes them think they would "fit in" and what do they notice that makes them think they might not? Trainees from diverse backgrounds may notice markers of attention to diversity and inclusion when they visit other research groups.

Contributed by J. Gleason with information from Branchaw, J. L., Pfund, C., and Rediske, R. (2010) *Entering Research: A Facilitator's Manual.* New York: W.H. Freeman & Co.

Implementation Guide

Trainee Pre-Assignment (30 minutes–1 hour)

- Distribute the handout provided in the trainee materials. Ask trainees to arrange with two of their peers in the class to visit their research groups. Encourage trainees to find groups that study very different questions or use different approaches than their group.
- Ask trainees to complete step 1 of the handout and share their paragraph describing their research group with the trainees who will be visiting.

Workshop Session (20 minutes)

- Ask trainees to share one observation about their visit to a peer's research group, particularly to highlight one similarity and one difference between their group and their peer's group (1–2 minutes per student). Observations may include differences in the following areas:
  - research group size
  - diversity of the research group members
  - tidiness of research space
  - collaborative vs. individual work
  - amount of talking within the group on a daily basis
  - repetitive tasks vs. new tasks
  - applied vs. basic research
  - amount of funding (e.g., make own stock solutions vs. purchasing solutions)
  - lab hierarchies vs. equal responsibilities
  - type of research (e.g., field-based, lab-based, computer-based)
- Discuss the benefits and disadvantages of different kinds of groups and what types of groups appeal to trainees. This may be particularly important for trainees who are considering joining a new research group or applying for a job or graduate school position. (10 minutes)

Contributed by J. Gleason with information from Branchaw, J. L., Pfund, C., and Rediske, R. (2010) *Entering Research: A Facilitator's Manual.* New York: W.H. Freeman & Co.

VISITING PEER RESEARCH GROUPS

Learning Objectives

Trainees will:

- Develop an appreciation for differences in culture among research groups.
- Identify valuable research group attributes.

Arrange to visit the research groups of two of your peers. Try to find groups that study very different questions or use different approaches than your group.

1. Before you visit the other groups, write a short paragraph describing the research of your own group to share with the peers who will visit your group. What are the major questions that your research group addresses? What is the nature of the research that you do (e.g., lab, field, computational)? How do members of your research team interact with one another?
2. Before visiting the other research groups, read the paragraphs of your peers to anticipate what you will experience. After your visits, write a short essay comparing the three research groups. Address the following points in your essay:
   a. How is the research space set up? Is it a laboratory, or some other kind of workspace? Does most of the research occur in the workspace or elsewhere?
   b. Who are the research group members and what kinds of interactions do you observe between them? Are they undergraduates, graduate students, or post-docs? How many people are in the research group? Is the PI in the lab or elsewhere?
   c. What different types of research activities occur in each research group? What types of approaches or methodologies do the groups use?
   d. Would you enjoy working in these other groups? Why or why not?

Contributed by J. Gleason with information from Branchaw, J. L., Pfund, C., and Rediske, R. (2010) *Entering Research: A Facilitator's Manual.* New York: W.H. Freeman & Co.

WHAT HAPPENS TO RESEARCH RESULTS?

Learning Objectives

Trainees will:

- ► Identify ways research findings are communicated to the public.
- ► Reflect on how research findings affect society.

Trainee Level

undergraduate or graduate trainees
novice trainees

Areas of Trainee Development

- ► Research Comprehension and Communication Skills
  - • Develop disciplinary knowledge.
- ► Research Ethics
  - • Develop responsible and ethical research practices.

Activity Components and Estimated Time for Completion

- ► Trainee Pre-Assignment Time: 30 minutes
- ► In Session Time: 30 minutes

Total time: 1 hour

When to Use This Activity

This activity is best suited for novice undergraduate or graduate researchers who are at the beginning of their research experience. It can be used regardless of whether trainees are currently working with a mentor. Other activities that may be implemented with this activity include:

- ► Discussion of the Nature of Science
  - • What Happens to Research Results?
- ► Science and Society
- ► Science Literacy Test
- ► Science or Pseudoscience

Inclusion Considerations

Trainees from nonscience backgrounds may not be familiar with the many ways that research findings can affect their lives. Discuss how research literacy not only allows people to understand and value the results of research, but can empower them to make informed decisions about issues that impact their lives (e.g., health, environment, technology). Ask the trainees to consider what role scientists should play in communicating their research findings to nonscientists.

Contributed by E. Frazier with information from Branchaw, J. L., Pfund, C., and Rediske, R. (2010). *Entering Research: A Facilitator's Manual.* New York: W.H. Freeman & Co.

Implementation Guide

Trainee Pre-Assignment (30 minutes)

- Assign trainees to complete the worksheet included in the trainee materials.

Workshop Session (30 minutes)

- **Small-Group Discussion** (10 minutes)
  - Prompt trainees to discuss questions from the assignment in small groups.
- **Large-Group Report-Out** (15 minutes)
  - Each group should select an individual who will summarize key points from their discussion for the larger group. Some possible answers to each question are provided below and can be used to help guide discussion.
    - What are three different ways in which research findings can be reported to the public?

> - Peer-reviewed publications.
> - Articles in nonscientific media outlets such as newspapers, magazines, Internet sites, etc.
> - Reports to federal, state, and local government.
> - Reports to private organizations such as chemical and pharmaceutical industries, for example.
> - Social media such as Twitter.

- What kind of information is found in your textbooks? Where does that information come from?

> - Textbooks are written by scientists who summarize current knowledge in the field and is based on research publications.

- How does research affect society?

> - Research is used in establishing federal and state guidelines in many areas (e.g., health, conservation, education, military, and science in general).
> - Research is used in the private sector for marketing and in developing new drugs by pharmaceutical companies.

- Identify at least five different fields or professions that are affected by research.

> - Pharmaceutical industry
> - Medicine
> - Education
> - Defense (military)
> - Cyber defense

- **Large-Group Discussion** (5 minutes)
  - In what ways, if any, did this discussion change your view about researchers?

Contributed by E. Frazier with information from Branchaw, J. L., Pfund, C., and Rediske, R. (2010). *Entering Research: A Facilitator's Manual.* New York: W.H. Freeman & Co.

WHAT HAPPENS TO RESEARCH RESULTS?

Learning Objectives

Trainees will:

- ▶ Identify ways that research findings are communicated to the public.
- ▶ Reflect on how research findings affect society.

Instructions: Respond to each of the prompts below.

1. Identify three ways in which research findings can be reported to the public.

2. What kind of information is found in textbooks? Where does that information come from?

3. In what ways do research findings affect society?

4. Identify at least five different fields or professions that are affected by research.

Contributed by E. Frazier with information from Branchaw, J. L., Pfund, C., and Rediske, R. (2010). *Entering Research: A Facilitator's Manual.* New York: W.H. Freeman & Co.

CASE STUDY: "WHATEVER YOU DO, DON'T JOIN OUR LAB."

Learning Objectives

Trainees will:

- Reflect on the factors to consider when selecting a thesis mentor and research group.
- Discuss strategies to use when faced with conflicting advice.

Trainee Level

graduate trainees
novice trainees

Areas of Trainee Development

- Professional and Career Development Skills
  - Explore and pursue a research career.
- Research Comprehension and Communication Skills
  - Develop effective interpersonal communication skills.

Activity Components and Estimated Time for Completion

- In Session Time: 25 minutes

Total time: 25 minutes

When to Use This Activity

This activity is suitable for graduate-level trainees who are deciding which group to join to conduct their thesis research. Other activities that may be used with this activity include:

- Three Mentors
- Research Group Diagram
- Finding Potential Research Rotation Groups and Mentors
- Research Rotation Evaluation

Inclusion Considerations

Encourage trainees to consider all factors that will impact their success and well-being when selecting a thesis mentor and research group. Suggest that trainees may want to ask prospective mentors about their management style and their thoughts about equity, inclusion, and diversity in research. They could ask about the diversity of the research group and the outcomes of trainees who have trained with this mentor.

Contributed by J. Branchaw. (2018). *Case Study: "Whatever you do, don't join our lab."*

Implementation Guide

Workshop Session (25 minutes)

► **Case Study:** "Whatever you do, don't join our lab" (5 minutes)

- Distribute the case study included in the trainee materials. Have one trainee read the case study aloud, display the case on a projector screen, or have trainees read and consider the case study silently.

► **Discussion** (10 minutes)

- As a group, discuss the following questions. Alternatively, if there are more than 10 trainees, they can discuss the questions in small groups of three or four and report highlights from their discussion to the larger group. Example responses appear below each question.
  - What should José do with the information he has learned from Sarah? Should he incorporate it into his decision-making process?

    - Yes, he should incorporate it. Everything must be considered.
    - No, he should not incorporate it. Obviously, Sarah is just mad about something at the moment and made an off-the-cuff remark.
    - Maybe, but José needs more information to determine how Sarah's experience should influence his decision.

  - Who could José talk to about the doubts that Sarah's comment has raised?

    - José could ask Sarah more specifically about her experience to better understand why she is miserable.
    - Other students in the Baldwin research group: Are they miserable?
    - Professor Baldwin
    - Training program director
    - Other mentors in José's network

  - How can graduate students know for sure whether they've chosen the right group for their thesis research?

    - It is impossible to know for sure, but systematically gathering and considering information about the research group can help one make a good decision.
    - No research group is perfect. There will always be challenges, so selecting a mentor and group with which one feels comfortable working through the challenges is very important.

► Facilitators may extend the case study discussion to consider issues that may be important to trainees from diverse backgrounds.

- Is there evidence of attention to inclusion in the research environment?
- Does the research group empower trainees of color, women, those with disabilities, and others under-represented in research to be successful?
- What should José do if he learns that "everyone is miserable" because they feel excluded, isolated, or unsupported on this research team?

► **Wrap-up** (5 minutes)

- Summarize the main ideas generated from the large-group discussion by generating an "action plan" for José. What should he do next?

Contributed by J. Branchaw. (2018). *Case Study: "Whatever you do, don't join our lab."*

CASE STUDY: "WHATEVER YOU DO, DON'T JOIN OUR LAB."

Learning Objectives

Trainees will:

- Reflect on the factors to consider when selecting a thesis mentor and research group.
- Discuss strategies to use when faced with conflicting advice.

As a first-year graduate student, José just finished his research rotations and now must select the lab he will join to do his thesis research. Overall, things went well and all of the professors he rotated with have invited him to join their groups. It was a difficult decision, but after carefully considering possible research projects, how he got along with each professor and their lab members, as well as the research funding available to support him, he has decided to join Professor Baldwin's group. He has made an appointment for next week to tell her and feels relieved that he has made his decision. Most of his peers are still agonizing over whose research group they will join.

Later that afternoon, José sees Sarah in the copy room, a senior graduate student from the Baldwin lab. She asks him whether he has selected a research group yet. He says, "Not yet," because he wants to tell Professor Baldwin first. "Well whatever you do, don't join our lab," says Sarah. "It looks great on the outside, but everyone is miserable." As she walks out of the copy room, José feels like he got punched in the stomach. Should he reconsider his decision? Should he keep his appointment with Professor Baldwin next week? He was feeling great just a few minutes ago, but now is questioning himself.

1. What should José do with the information he has learned from Sarah? Should he incorporate it into his decision-making process?

2. Who could José talk to about the doubts that Sarah's comment has raised?

3. How can graduate students know for sure whether they've chosen the right group for their thesis research?

Contributed by J. Branchaw. (2018). *Case Study: "Whatever you do, don't join our lab."*

WHY DIVERSITY MATTERS IN STEM RESEARCH

Learning Objectives

Trainees will:

- Learn why including individuals from diverse backgrounds in the STEM community is important.
- Be able to communicate to others the importance of diversity in STEM research.

Trainee Level

undergraduate or graduate trainees
novice, intermediate or advanced trainees

Areas of Trainee Development

- Equity and Inclusion Awareness and Skills
  - Advance equity and inclusion in the research environment.

Activity Components and Estimated Time for Completion

- In Session Time: 1 hour

Total time: 1 hour

When to Use This Activity

This activity can be introduced at any time during the research experience. While students may state that diversity is important in STEM research, they may not be able to articulate reasons for this statement. The aim of the activity is to help students understand and communicate how diversity of backgrounds and perspectives contribute to advancing our understanding through research in STEM fields.

Facilitator Preparation

Read:

- American Physical Society (APS) News regarding Supreme Court oral arguments about the importance of diversity in physics (https://www.aps.org/publications/apsnews/201602/supreme.cfm)
- The Open letter to SCOTUS from nearly 2500 Physicists (http://eblur.github.io/scotus/)

Preview:

- Dr. Michael Gavin from Colorado State University gives a Tedx examining the importance that history, language, and tradition have in the preservation of culture. In doing so, he highlights the scientific discoveries that may never have happened if we only used one lens—one cultural perspective—to try to solve scientific problems. (https://www.youtube.com/watch?v=48RoRi0ddRU)

Contributed by G. Chavira. (2018). *Why Diversity Matters in STEM Research.* Information from materials developed for NIH BUILD PODER, California State University, Northridge.

Inclusion Considerations

Encourage all trainees to share their experiences with the group, but do not require it. Trainees from groups historically underrepresented in research should not be asked or expected to speak for their respective identity group. All shared experiences should be presented as individual experiences, yet considered in light of the historical and social contexts the trainees learn about in the readings. Challenge all trainees to share in identifying challenges and barriers to creating more inclusive teams and diversifying the research workforce.

Contributed by G. Chavira. (2018). *Why Diversity Matters in STEM Research.* Information from materials developed for NIH BUILD PODER, California State University, Northridge.

Implementation Guide

Workshop Session (55 minutes–1 hour)

► **Introduction** (5–10 minutes)

- Ask the trainees to consider the question "What unique perspective does a minority student bring to a physics class?" Have them reflect silently on the question and write their response to the question on a piece of paper. Tell the group that you will revisit this question at the end of the session.

► **Video:** "Why Cultural Diversity Matters" (30 minutes)

- View the Tedx Talk, "Why cultural diversity matters," by Dr. Michael Gavin at Colorado State University. (17:52; available on YouTube, https://www.youtube.com/watch?v=48RoRi0ddRU)
- Ask trainees to discuss in pairs their response to the question.
  - "What unique perspective does a minority student bring to a physics class?"
  - Did your response to the question change after watching the Tedx Talk?
  - Why we should care about the loss of cultural diversity in STEM?

► **Large Group Discussion** (20 minutes)

- Share that the question "What unique perspective does a minority student bring to a physics class?" was brought up by Supreme Court Chief Justice John Roberts during oral arguments on December 9, 2015, during the affirmative action case, *Fisher v. University of Texas.*
- Discuss how the physics community reacted to Chief Justice Roberts' comment. You can even bring up the open letter to SCOTUS for students to read (http://eblur.github.io/scotus/).
- Talk briefly with students about why the physicist community was outraged.
- **Large-Group Discussion Questions**
  - Ask trainees to share their responses to the question "What unique perspective does a minority student bring to a physics class?"
  - Why should we care about the loss of cultural diversity in STEM?
  - Dr. Gavin talks about cultural diversity at a global level. Give an example of how cultural diversity matters in the United States.
  - Consider how the structures and systems present at your institution support diversity in STEM. What values and beliefs are implicitly or explicitly conveyed to students in your field at your institution?
  - Consider multiple forms of diversity (e.g., visible and invisible disabilities, gender identity, sexual orientation, religion, age). Are all forms of diversity supported at your institution?

Note: The term "minority" is used here to refer to individuals from historically underrepresented racial and ethnic groups. It is important to note that individuals from these groups may not always be in the minority in every context, but may still be impacted by implicit and/or explicit bias and prejudice.

► **Wrap-up** (5 minutes)

- This video and activity focus on cultural diversity, but other forms of diversity can also have important implications for STEM. These include:
  - Visible and invisible disabilities
  - Gender identity
  - Sexual orientation
  - Religion
  - Age
- The points raised in this discussion make the case for why many different types of diversity are crucial to continued innovation in STEM fields.

Note: There are no trainee materials for this activity.

Contributed by G. Chavira. (2018). *Why Diversity Matters in STEM Research.* Information from materials developed for NIH BUILD PODER, California State University, Northridge.

YOUR RESEARCH GROUP'S FOCUS

Learning Objectives

Trainees will:

- ► Learn about the research and methods used by the research group.
- ► Become familiar with the researchers in the group.
- ► Create a graphical abstract that represents the research of the group.

Trainee Level

undergraduate or graduate trainees
novice or intermediate trainees

Areas of Trainee Development

- ► Research Comprehension and Communication Skills
  - Develop disciplinary knowledge.
  - Develop effective interpersonal communication skills.
  - Develop research communication skills.
- ► Practical Research Skills
  - Develop ability to design a research project.

Activity Components and Estimated Time for Completion

- ► Pre-Assignment Time: 1–2 hours
- ► In Session Time: 1 hour, 20 minutes

Total time: 2–3 hours

When to Use This Activity

This activity is useful for undergraduate or graduate trainees who are early in their research experience. It can be implemented either before a trainee has selected a mentor (in the case of trainees who are rotating through research groups) or after a trainee has chosen a research group and mentor. Facilitators who are emphasizing science communication in their courses or workshops can use this activity as an opportunity to introduce trainees to the importance of "elevator speeches," or the ability to briefly summarize the focus and scope of what they do in the research group. Other activities that may be used with this activity include:

- ► Three-Minute Research Story
- ► Communicating Research to the General Public

Inclusion Considerations

Creation of graphical abstracts provides an opportunity to discuss why it is important for trainees to develop the ability to communicate research in diverse ways and to diverse audiences.

Encourage them to engage their own unique backgrounds in developing their graphical abstracts, consider how their graphical representations may be interpreted or perceived by those from different backgrounds, and how they will most effectively convey the meaning of their graphical representations.

Contributed by E. Frazier and D. Wassarman with information from Branchaw, J. L, Pfund, C., and Rediske, R. (2010). *Entering Research: A Facilitator's Manual.* New York: W.H. Freeman & Co.

Implementation Guide

Trainee Pre-Assignment (1–2 hours)

- Prior to the session, distribute the trainee materials, which include a two-step pre-assignment. Trainees should complete this pre-assignment before the session so they can invest time in understanding the research conducted in the laboratory/research group.
    - Have trainees write one paragraph that provides an overview of the research in their mentor's research group.
    - Have trainees draw a graphical abstract that provides an overview of the research in their mentor's research group. A graphical abstract is a single, concise pictorial, and visual summary.
    - Emphasize to trainees that the intent of the activity is to facilitate discussion about research, not to spend a lot of time and energy learning to use software to create sophisticated images. Simple drawings on whiteboards/chalkboards, large poster boards, or flowcharts created using simple software programs are sufficient.
    - Facilitators are encouraged to speak with the lab director or graduate students in the research group prior to this session to better understand the research that trainees will be presenting.

Note: Facilitators can revise Question 1 on the pre-assignment to guide trainees to write their paragraph more like an abstract of their work to set trainees up for expanding on their abstract in a research paper/presentation later in a course.

Workshop Session (1 hour, 30 minutes)

- The objective of this activity is for trainees to learn about the major research questions in the research group. This includes smaller projects and how they relate to the larger aims of the research group, the researchers conducting the projects, and the research methods commonly utilized in the group.
- The description of the activity below can be used for in-class discussions with undergraduate trainees or as part of research group rotations with graduate trainees.

- **Activity: Chalk Talks** (1 hour)
    - Each trainee presents her graphical abstract that addresses the research questions and methods utilized in her research group and explains any technical language. Example graphical abstracts and guidelines can be found here: http://www.cell.com/pb/assets/raw/shared/figureguidelines/GA_guide.pdf
    - Depending upon the setup of the room, a projector or whiteboard/chalkboard may be used for presentations.
    - Trainees who are not presenting can fill out a note card or sheet of paper answering the following questions:
        - What did the presenter do well?
        - What should the presenter improve upon?
        - What two questions do you still have about their research?

Note: Facilitators wishing to assess chalk talks can use the rubric included at the end of the facilitator notes.

Contributed by E. Frazier and D. Wassarman with information from Branchaw, J. L, Pfund, C., and Rediske, R. (2010). *Entering Research: A Facilitator's Manual.* New York: W.H. Freeman & Co.

► **Discussion** (20 minutes)

- How does your project fit within the bigger research question in the group?
- Are you comfortable with the methods you will be using in your research project?
  - For example, novice students may not be aware that they have to conduct surgery on animals as part of their research. Are they comfortable with the idea of doing animal surgery? Other students in the medical field may expect to be working directly with patients when in reality they will spend most of their time in front of a computer. Field researchers may have to take helicopters and be left in a remote area and picked up later in the day. Are they comfortable with that idea? If not, they may want to consider another research group.

► **Wrap-up** (5 minutes)

- Summarize the key points of the discussion. Encourage trainees to discuss any concerns that they have about research with their mentor.

Contributed by E. Frazier and D. Wassarman with information from Branchaw, J. L, Pfund, C., and Rediske, R. (2010). *Entering Research: A Facilitator's Manual.* New York: W.H. Freeman & Co.

YOUR RESEARCH GROUP'S FOCUS

Assessment Rubric

This rubric can be used for peer review and to assess chalk talks given by trainees.

Name of presenter: **Title/Topic:**

	0	1	2	3
Research Questions	Absent	The presenter did not clearly state their research group's research questions or hypotheses.	The presenter somewhat clearly stated their research group's research questions or hypotheses.	The presenter clearly stated their research group's research questions or hypotheses.
Methods	Absent	The presenter explained the methods used by their research group in a way that was unclear.	The presenter explained the methods used by their research group, but some components were unclear.	The presenter clearly explained the methods used by their research group.
Jargon	Absent	The presenter used jargon frequently and often did not explain technical language.	The presenter used some jargon and explained most of the technical language used.	The presenter did not use jargon and translated technical language so that the audience could understand.

Overall Score (circle one):

Poor Fair Good Very Good Excellent

Suggestions for Improvement:

Contributed by E. Frazier and D. Wassarman with information from Branchaw, J. L, Pfund, C., and Rediske, R. (2010). *Entering Research: A Facilitator's Manual.* New York: W.H. Freeman & Co.

YOUR RESEARCH GROUP'S FOCUS

Learning Objectives

Trainees will:

- Learn about the research and methods used by the research group.
- Become familiar with the researchers in the group.
- Create a graphical abstract that represents the research of the group.

The purpose of this activity is to help you understand your research group's focus and to practice using multiple modes of communication to convey this understanding to a diversity of audiences. It will also help you get to know other researchers on your research team and how their projects are related.

Pre-Session Assignment:

1. Write one paragraph, in your own words, describing the focus of your group's research. Be sure to include the group's major research questions, smaller projects within the major research question, as well as the names of the researchers responsible for each project, the general methods involved and what areas of this research are interesting to you. You may want to interview your research mentor and members of the research team to get a better understanding of the research.
2. Take the paragraph you've written and create a graphical abstract, which should convey the same points described in the paragraph, but in visual form. A link to an example of a graphical abstract appears on the next page.

In-class activity: In the next session, you will give a "chalk talk" (informal oral presentation) of your research group's work in which you will present your graphical abstract and explain any technical language.

To prepare for your chalk talk consider:

- How does your research project fit into the larger research question(s) of the group?
- How does your research project relate to other projects in the group?
- Have you given thought to the methods used by your team to address the research questions? Are you comfortable conducting these procedures?

Contributed by E. Frazier and D. Wassarman with information from Branchaw, J. L, Pfund, C., and Rediske, R. (2010). *Entering Research: A Facilitator's Manual.* New York: W.H. Freeman & Co.

PRE-SESSION ASSIGNMENT: GRAPHICAL ABSTRACT

Draw a graphical abstract that provides an overview of the research in your group. A graphical abstract is a single, concise, pictorial, and visual summary. Often, graphical abstracts are used to summarize the main findings of an article. They can also be used to summarize methods or hypotheses. For more information and examples, visit https://www.elsevier.com/authors/journal-authors/graphical-abstract

Contributed by E. Frazier and D. Wassarman with information from Branchaw, J. L, Pfund, C., and Rediske, R. (2010). *Entering Research: A Facilitator's Manual.* New York: W.H. Freeman & Co.

About the Authors

Janet Branchaw

Janet L. Branchaw is an Assistant Professor of Kinesiology in the School of Education and the Faculty Director of the Wisconsin Institute for Science Education and Community Engagement (WISCIENCE) at the University of Wisconsin–Madison (UW–Madison). She earned her B.S. in Zoology from Iowa State University and her Ph.D. in Physiology with a focus on cellular neurophysiology from the University of Wisconsin–Madison. After completing postdoctoral training and a lectureship in undergraduate and medical physiology at the UW–Madison's School of Medicine, she joined the University's then Center for Biology Education, which she now directs as WISCIENCE. Her research as a faculty member in the Department of Kinesiology and her programming work at the Institute focus on the development, implementation, and evaluation of innovative approaches to undergraduate science education, with a special emphasis on undergraduate research, assessment of student learning, and broadening participation in science.

In addition to developing the *Entering Research* curriculum, she has developed a curriculum to train research mentors, *Entering Mentoring*, and led the development and validation of the *Entering Research* Learning Assessment (ERLA). She has developed and directed Research Experience for Undergraduate (REU) and Undergraduate Research and Mentoring (URM) programs funded by the National Science Foundation and served as the Chairperson of the Biology REU Leadership Council as well as a member of the 2017 National Academies of Science, Engineering and Medicine's consensus committee on Undergraduate Research in STEM. She served as the Associate Director of the National Institutes of Health's National Research Mentoring Network's (NRMN) Mentorship Training Core and currently oversees Mentee Training Initiatives at the UW–Madison's Center for the Improvement of Mentored Experiences in Research (CIMER). Most recently she is leading UW–Madison's Howard Hughes Medical Institute Inclusive Excellence project to catalyze institutional change to support 2- to 4-year STEM transfer students.

Amanda Butz

Amanda R. Butz is the Director of Evaluation and Research for the Wisconsin Institute for Science Education and Community Engagement (WISCIENCE) at the University of Wisconsin–Madison. She received her Ph.D. in Educational Psychology from the University of Kentucky, where her research focused on academic motivation, self-efficacy, and the beliefs and aspirations of rural potential first-generation college students. She also holds an M.A. in Adult and Higher Education from Morehead State University and a B.A. from Indiana University.

As a member of the NRMN Mentorship Training Core, Butz has worked on the development, adaptation, and evaluation of mentor and mentee training curricula and assessment tools. As a postdoctoral research associate, she led the evaluation efforts for mentor training modules on promoting mentee research self-efficacy and culturally aware mentoring. She also worked with the What Matters in Mentoring project on the development and validation of an instrument to measure cultural diversity awareness in mentors and to study the motivation of mentors to address race/ethnicity in their research mentoring relationships. Butz coordinated the development, pilot testing, evaluation, and research on the second edition of *Entering Research* and was the lead researcher in the development and validation of the *Entering Research* Learning

Assessment (ERLA). She coordinates the evaluation efforts of WISCIENCE programs and initiatives and is a member of the team conducting evaluation on the UW–Madison's Howard Hughes Medical Institute Inclusive Excellence project.

Amber Smith

Amber R. Smith is the Associate Director of the Wisconsin Institute for Science Education and Community Engagement (WISCIENCE) at the University of Wisconsin–Madison (UW–Madison). She earned her B.S. in Biology from Carroll College and her Ph.D. in Plant Breeding Plant Genetics from the University of Wisconsin–Madison.

As a postdoctoral fellow at the Institute for Biology Education (now WISCIENCE), Smith developed first-year transition programs for Biology students at UW–Madison before continuing her educational development work as an instructional consultant in the Center for Research on Learning and Teaching (CRLT) at the University of Michigan. At CRLT she led inclusive teaching professional development trainings for graduate student instructors, postdoctoral fellows, and faculty. Smith returned to UW–Madison to direct campus-wide programming for research mentor and mentee training through WISCIENCE. In this role, she supports undergraduates to find and succeed in research opportunities through a series of professional development seminars and workshops using the *Entering Research* curriculum. Additionally, she directs two summer undergraduate research programs at UW–Madison that broaden access to underrepresented students. Smith offers mentor training opportunities for graduate students, postdoctoral fellows, and faculty mentors through the *Entering Mentoring* seminar, Culturally Aware Mentor training, and tailored mentor/mentee training workshops for graduate training programs. She is helping to lead the transfer student success programming for UW–Madison's Howard Hughes Medical Institute Inclusive Excellence project to support 2- to 4-year STEM transfer students. Smith is certified as a Master Facilitator and Master Consultant through the National Research Mentoring Network.

The Navigating Research and Mentoring Book Series

sirup/Getty Images

The Navigating Research and Mentoring Book Series is a collection of curricula aimed at improving research trainee success and research mentoring relationships. The Series includes training curricula for research mentors and trainees at the undergraduate, graduate student, post-doctoral and faculty levels across science, technology, engineering, mathematics and medical disciplines. Authored by well-known science educators, the Series provides readings, resources and active learning activities that support trainees as they develop research skills and learn to navigate the research environment, as well as activities that encourage mentors and trainees to reflect upon and improve their relationship. Each training curriculum includes an easy-to-follow, detailed facilitation guide for educators to use and adapt for their own settings.

Ongoing information and resources to support research mentors and trainees are located at the Center for the Improvement of Mentored Experiences in Research (CIMER, www.cimerproject.org/#/).

Co-editors for the Navigating Research and Mentoring Book Series:

- Christine Pfund, Wisconsin Center for Education Research, Center for the Improvement of Mentored Experiences in Research, and Institute for Clinical and Translational Research, University of Wisconsin–Madison
- Jo Handelsman, Wisconsin Institute for Discovery, Department of Plant Pathology, University of Wisconsin-Madison

Lisa Lockwood,
Executive Program Manager,
Life Sciences,
Macmillan Learning

Index

M

P

V